PALEOPATHOLOGICAL DIAGNOSIS
AND INTERPRETATION

Paleopathological Diagnosis and Interpretation

BONE DISEASES IN
ANCIENT HUMAN POPULATIONS

By

R. TED STEINBOCK
Harvard Medical School
Boston, Massachusetts

With a Foreword by

T. Dale Stewart, M.D.
Anthropologist Emeritus
Smithsonian Institution

CHARLES C THOMAS • PUBLISHER
Springfield • Illinois • U.S.A.

Published and Distributed Throughout the World by
CHARLES C THOMAS • PUBLISHER
Bannerstone House
301-327 East Lawrence Avenue, Springfield, Illinois, U.S.A.

This book is protected by copyright. No part of it may be reproduced in any manner without written permission from the publisher.

© 1976, by CHARLES C THOMAS • PUBLISHER
ISBN 0-398-03512-1
Library of Congress Catalog Card Number: 75-34701

With THOMAS BOOKS *careful attention is given to all details of manufacturing and design. It is the Publisher's desire to present books that are satisfactory as to their physical qualities and artistic possibilities and appropriate for their particular use.* THOMAS BOOKS *will be true to those laws of quality that assure a good name and good will.*

Library of Congress Cataloging in Publication Data

Steinbock, R. Ted.
    Paleopathological diagnosis and interpretation.

    Includes bibliographies and index.
    1. Paleopathology. 2. Bones—Diseases—Diagnosis.
I. Title. [DNLM: 1. Bone diseases—History.
2. Paleopathology. QZ11.5 S819p]
R134.8.S73   616.07   75-34701
ISBN 0-398-03512-1

Printed in the United States of America
W-2

TO MY PARENTS

FOREWORD

I FIRST MET Ted Steinbock in 1972 while he was attending the Paleopathology Seminar at the National Museum of Natural History. Some of the circumstances of our association throw light on the way in which his book developed and on what makes his book such an important contribution to paleopathology.

Ted and I were drawn together by a shared interest in a small but excellently preserved skeletal collection from Indian Knoll, Kentucky. Clarence B. Moore, the famous gentleman-archaeologist from Philadelphia, had donated this collection to the Museum in 1915 at the conclusion of one of his annual steamboat expeditions exploring Indian sites along the Green River.

Ted and I knew, of course, that during the WPA days (1939-40) the Indian Knoll site had been further explored by field parties under the direction of Major Webb of the University of Kentucky, and the number of recovered skeletons extended to well over 1000. Moreover, the site had been determined to be some 5000 years old. Thus, the combined Indian Knoll collections constitute the largest assemblage of carefully excavated and well-preserved skeletons from any Indian site of such great age. Obviously, the correct identification of the diseases afflicting this, and any other, early population is a matter of great historical importance.

Ted was interested in the Indian Knoll population through his archaeological work in Kentucky and anthropological training at Harvard. My interest, on the other hand, was mainly in seeing that the research potential of the skeletons from this site was fully realized. In fact, only three years before Ted's visit to Washington I had called attention to the little known circumstance that someone at the Army Medical Museum (now the Armed Forces Medical Museum) had arranged with Moore to save pathological bones for that institution. This could mean, I felt, that from the pathological standpoint the Indian Knoll collection in the National Museum is not a random sample.

Ted's concern about this matter was one of the first subjects we discussed. He wanted to know whether I had looked at the bones from Indian Knoll in the Armed Forces Medical Museum and, if not, whether I had any objection to his doing so. My answers to both questions being in the negative, in due course I received from him a detailed account of his findings. Later still he kindly reported on the condition of the larger part of the skeletal collection housed in Lexington, Kentucky, and on the examples of particular pathological conditions he had examined there. I could not help but be impressed by the systematic and thorough way in which he was pursuing the subject.

All this led in 1973 to an honors thesis which received a unanimous *summa* when submitted to the Department of Biology at Harvard. Ted then went on to Harvard Medical School and continued his interest in paleopathology. While still in medical school he has converted the thesis into book form. Although the book bears the same title and nearly the same table of contents as the thesis, the text and illustrations have been extensively revised and expanded. The earlier emphasis on the bone lesions present at Indian Knoll has given way to an extensive review of the paleopathological literature. Also, since the book is designed to provide a systematic approach to diagnosing bone lesions in excavated skeletal series and interpreting their significance in prehistoric human populations, the new emphasis is on documented clinical specimens.

A book of this sort has been needed, because few clinicians can claim competence in diagnosing diseases from gross bones alone, whereas the anthropologists who encounter such bones in great quantities rarely have sufficient knowledge of pathology to properly interpret osseous lesions. I congratulate Ted on the successful outcome of his endeavors and predict that as a result of his book the relatively new science of paleopathology will receive contributions of more lasting value than heretofore.

T. Dale Stewart, M.D.

INTRODUCTION

PALEPATHOLOGY AS DEFINED here is the study of diseases in ancient human populations as revealed by their skeletal remains. In certain instances, mummified tissues, ancient art forms, and literature can also be utilized. Paleopathology is important in providing information on the health status of ancient human populations. Furthermore, it provides new knowledge of the antiquity of specific diseases affecting bone and the end results of such diseases in human populations without medical technology.

The term paleopathology was popularized if not invented by Sir Marc Armand Ruffer at the turn of this century while working with the extensive collections of Egyptian mummies and skeletons. With few exceptions, the literature since then has consisted of short reports describing isolated specimens of bone pathology. The specimens are often regarded as interesting curios without attempting to interpret the significance of such osseous lesions in the populations represented by the individuals. More recently an epidemiological or population approach has been utilized to study the skeletal pathology of ancient human populations, and this provides much more meaningful information.

The purpose of this book is to provide a basic framework for those interested in diagnosing and interpreting bone lesions found in excavated skeletal series. The literature concerning the archaeological evidence of disease is reviewed here as a background for the presentation of the relevant clinical data on the major diseases of bone. This is primarily written for physical anthropologists who are the major contributors to the field of paleopathology. Pathologists and radiologists are more concerned with the microscopic and radiographic picture of bone disease than with the gross morphology of the dried bone specimen.

However, workers in both areas have made valuable contributions to paleopathology, and this book owes much to their clinical studies of bone disease.

This book is of necessity limited in scope and breadth. The gross morphology and radiographic appearance of the macerated bone specimen is emphasized here. Other techniques such as microscopic examination of sectioned bone are of questionable value in archaeological specimens, and further advances in this area are urgently needed. Mummified tissues offer an exciting opportunity for studying diseases which do not affect bone or bone diseases with soft tissue manifestations. Rehydration and staining techniques for dessicated tissues and fecal specimens are not described here as they could easily be the subject of an entire book. Only the major infectious, nutritional, metabolic, degenerative, and neoplastic diseases of bone are discussed here. These diseases account for the great majority of pathological lesions in excavated specimens.

Finally, although diagnosis is emphasized in this book, this does not imply that one can always specify the disease illustrated by the dried bone specimen. Such an attitude can only produce more confusion than enlightenment in the field of paleopathology. A more rational approach is to state the most likely diagnosis followed by a list of possible alternatives in order of decreasing likelihood. Again, I cannot overemphasize the importance of carefully considering the archaeological and epidemiological context of the pathological specimen in establishing the antiquity of disease and its prevalence in ancient human populations.

R. Ted Steinbock
Harvard Medical School

ACKNOWLEDGMENTS

Many people have helped me either directly or indirectly in the preparation of this book during the past four years. Professor Jonathan Friedlaender at the Harvard Peabody Museum initiated my interest in paleopathology and physical anthropology. Professor Farish Jenkins at the Harvard Museum of Comparative Zoology deserves special thanks and appreciation for his readiness to help in all aspects of my education and for his constant encouragement during the hectic years of medical school. Dr. Alan Schiller, orthopedic pathologist at the Massachusetts General Hospital, kindly read the manuscript and offered many helpful suggestions.

My research in gross bone pathology includes skeletal material from several prominent institutions, and I owe much to the equally prominent people associated with these institutions. At the Smithsonian Institution I wish to thank Dr. Donald Ortner, Associate Curator of the Division of Physical Anthropology, for allowing me to attend the Smithsonian's seminar in paleopathology and inviting me to return and undertake my own research project. Special thanks and appreciation are extended to Dr. T. Dale Stewart, anthropologist emeritus at the Smithsonian. His pioneering studies in paleopathology and skeletal age changes have been sources of great inspiration and information for me. At the Armed Forces Institute of Pathology, I wish to thank Dr. Lent C. Johnson, chief of orthopedic pathology, for allowing me to examine the pathological specimens in the affiliated Army Medical Museum. Many interesting cases of bone pathology were examined at the Warren Anatomical Museum of Harvard Medical School. Mr. David Gunner, Assistant Curator of the collection, was extremely helpful in making the entire collection available for my research.

Certain chapters in this book greatly benefited from my contact with workers in these areas. Dr. Don Fawcett, Professor of Anatomy at Harvard Medical School, has generously allowed me to use several of his illustrations in the first chapter on bone biology. Dr. Stanley Garn at the University of Michigan has provided excellent illustrations of transverse lines in living subjects. The difficult chapter on syphilis and other treponemal infections owes much to the advice of Dr. Cecil J. Hackett and Dr. E. Herndon Hudson. Dr. Hackett has also generously allowed me to use several illustrations from his two valuable studies on the bone lesions of yaws. Dr. Dan Morse kindly reviewed my chapter on tuberculosis and sent several photographs of paleopathological specimens for other chapters. Dr. Vilhelm Møller-Christensen of Denmark and Dr. Michel Lechat of Belgium provided helpful information and illustrations on archaeological and clinical examples of leprosy. Dr. Charles Short, Consultant to the Arthritis Unit at Massachusetts General Hospital, kindly reviewed the chapter on arthritis and provided several clinical radiographs. Dr. Jack Dreyfuss, Associate Professor of Radiology at Harvard Medical School, generously assisted in the final stages of manuscript preparation.

Several illustrations were graciously supplied by Mr. Don R. Brothwell, Dr. John E. Moseley, and Dr. Calvin Wells. Dr. E. B. D. Neuhauser, chief of radiology emeritus at Children's Hospital of Boston, and Mr. James Conway both provided excellent clinical roentgenograms of certain bone disorders. Many other people have helped with this book in various ways. They include Mrs. Sterling Ament, Mr. Lazlo Meszoly, Mr. Richard Parker, and Miss Priscilla White. To all these people I am grateful.

R. Ted Steinbock

CONTENTS

	Page
Foreword	vii
Introduction	ix
Acknowledgments	xi

Chapter
- I BONE AS A LIVING TISSUE 3
  - Gross Bone Anatomy 3
  - Basic Bone Reactions 8
  - Bone Formation and Resorption in Pathological Conditions.. 10
  - Summary 14
  - Bibliography 15
- II TRAUMA 17
  - Fractures 17
  - Crushing Injuries 24
  - Bone Wounds Caused by Sharp Instruments 24
    - Arrow and Spear Wounds 24
    - Scalping 24
    - Trephination (Trepanation) 29
    - Sincipital T 35
    - Amputation 36
  - Dislocations 39
  - Transverse Lines (Growth Arrest Lines) 43
  - Bibliography 55
- III NONSPECIFIC INFECTION: PYOGENIC OSTEOMYELITIS. 60
  - Acute Osteomyelitis 61
    - Acute Hematogenous Osteomyelitis 61
    - Acute Pyogenic Osteomyelitis by Direct Infection 72
  - Chronic Osteomyelitis 74
    - Chronic Pyogenic Osteomyelitis 74
    - Brodie's Abscess 74
    - Chronic Nonsuppurative Osteomyelitis 76

Chapter	Page
Pyogenic Arthritis	78
Differential Diagnosis	79
Paleopathology	81
Bibliography	83
IV TREPONEMA INFECTION: SYPHILIS, BEJEL, AND YAWS.	86
Infectious Disease: Specific Infections	86
Theories Relating to the Spread of Syphilis	87
Columbian Theory	87
Pre-Columbian Theory	87
Unitarian Theory—The Biological Evidence	90
The Biological and Epidemiological Evidence	91
The Paleopathological Evidence	94
Venereal Syphilis	98
Congenital Syphilis	98
Early Congenital Syphilis	98
Late Congenital Syphilis	101
Acquired Syphilis	108
Incidence of Bone Lesions in Acquired Syphilis	109
Prevalence of Syphilis in Early Human Populations	110
Gross Pathology of Osseous Syphilis	113
A Differential Diagnosis of Syphilis	136
Nonvenereal Syphilis (Endemic Syphilis, Bejel)	138
Yaws (Framboesia, Pian, Bubas)	142
Bibliography	160
V TUBERCULOSIS	170
Paleopathological Evidence	171
Diagnosis of Tuberculosis	175
Tuberculosis of the Vertebral Column (Pott's Disease)	176
Tuberculosis of the Joints and Metaphyses	182
Differential Diagnosis	186
Bibliography	188
VI LEPROSY	192
History	192
Paleopathological Evidence	193
Biology and Epidemiology	195

Chapter	Page
Frequency of Bone Changes	198
Gross Bone Pathology	201
Differential Diagnosis	208
Bibliography	209

VII HEMATOLOGIC DISORDERS—THE ANEMIAS 213
   The Problem of Osteoporosis Symmetrica
      (Spongy Hyperostosis) 213
   Thalassemia Major and Thalassemia Minor 220
   Sickle-cell Anemia 225
   Iron Deficiency Anemia 230
   Malaria and the Hereditary Anemias 234
   Iron Deficiency Anemia and American Indians 236
   Cribra Orbitalia (Usura Orbitae) 239
   Bibliography 248

VIII METABOLIC BONE DISEASE 253
   Diseases Due to Inadequate Osteoid Synthesis-Osteopenia...253
      Dietary Osteopenia—Scurvy 253
      Endocrine Osteopenia—Senile and Postmenopausal
         Osteopenia 256
      Stress Deficiency Osteopenia—Atrophy 260
      Congenital Osteopenia—Osteogenesis Imperfecta 261
   Inadequate Osteoid Mineralization 262
      Rickets 262
      Osteomalacia 272
   Bibliography 273

IX ARTHRITIS 277
   Osteoarthritis (Degenerative Joint Disease) 278
   Vertebral Osteophytosis 287
   Traumatic Arthritis 289
   Rheumatoid Arthritis 289
   Ankylosing Spondylitis 294
   Infectious Arthritis 299
   Gout 300
   Paleopathology of Arthritic Conditions 300
   Bibliography 312

Chapter	Page
X TUMORS AND TUMOR-LIKE PROCESSES OF BONE 316	
Bone Tumor Prevalence and Incidence 318	
Benign Tumors of Bone 319	
Osteochondroma (Solitary Osteocartilaginous Exostosis)...319	
Chondroma (Enchondroma) 325	
Ear Exostoses 329	
Osteoid-Osteoma 333	
Fibrous Dysplasia 336	
Nonossifying Fibroma and Fibrous Cortical Defect 340	
Solitary Bone Cyst 344	
Giant-cell Tumor (Osteoclastoma) 346	
Hemangioma 350	
Meningioma 353	
Histiocytosis X (Reticuloendotheliosis) 355	
Malignant Tumors of Bone 362	
Osteosarcoma (Osteogenic Sarcoma) 362	
Ewing's Sarcoma 371	
Multiple Myeloma (Plasma Cell Myeloma, Myelomatosis).374	
Metastatic Carcinoma 385	
Bibliography 397	
Author Index 403	
Subject Index 411	

PALEOPATHOLOGICAL DIAGNOSIS
AND INTERPRETATION

Chapter I

BONE AS A LIVING TISSUE

THIS CHAPTER PROVIDES a brief introduction to bone as a living tissue and to the main features of gross bone anatomy. A thorough knowledge of normal bone growth and remodeling and normal bone structure is necessary before one can appreciate the response of bone to disease and the consequent changes in its gross appearance. For further information on normal bone physiology and anatomy, the reader should consult the bibliography at the end of this chapter. General references in paleopathology are also provided.

GROSS BONE ANATOMY

Bone is a connective tissue specially modified to provide a rigid framework. It is composed of living cells imbedded in an extracellular matrix of collagenous fibrils made rigid by calcium salts. Bone provides for the internal support of the body, protection of vital organs, and attachment for muscles necessary in locomotion. Bone encloses the hematopoietic tissue or bone marrow which produces the blood elements. In addition, the skeleton acts as the main source of mobilizable calcium for the maintenance of calcium levels in the blood.

The bones of the skeleton fall into four morphologic groups: long, short, flat, and irregular. The long bones are the main components of the limbs, supporting the weight of the body and providing attachment for the muscles of locomotion. The humerus, radius, ulna, femur, tibia, and fibula are in this category.

The short bones consist of all the bones in the hands and feet. Some prefer to describe the metatarsals and metacarpals as short tubular bones because of their similarities to the long tubular bones of the limbs. The flat bones include the bones of the cranial vault, ribs, sternum, scapulae, and pelvic bones. These bones provide protection and wide areas for muscle attachment. They are also the major sites of red bone marrow or active hematopoietic tissue in the adult. The irregular bones consist of the vertebrae and many bones of the skull. The vertebrae are also a major source of red bone marrow.

A typical long bone illustrates many of the main features of gross anatomy (see Fig. 1). At the end of the bone is an *epiphysis* which bears articular cartilage in the living state. The epiphysis contains the secondary center of ossification and

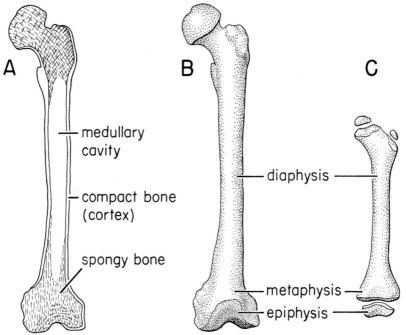

Figure 1. Gross anatomy of the femur. (A.) Longitudinal section revealing the medullary cavity and spongy or cancellous bone. (B.) External view of adult femur. (C.) External view of infant femur showing the epiphyses separate from the metaphyses and diaphysis.

epiphyseal disk and fuses to the *metaphysis* at some point during maturation. The metaphysis is immediately beneath the epiphyseal disk in the growing bone and extends into the medullary cavity as a loosely packed tissue which melds into the cortex of the *diaphysis* or shaft of the bone. The diaphysis is the central portion of a tubular bone and extends from the metaphysis of one end to the metaphysis of the opposite end.

Except where covered by cartilage for articulation with other bones, the entire bone is enveloped by a thin connective tissue membrane known as the *periosteum*. The periosteum does not persist in the dried bone specimen, but its osteogenic potential is very important in certain pathologic processes and in fracture repair. The periosteum adheres firmly to the underlying bone via bundles of collagen fibers and a rich network of capillaries and arterioles which permeate the bone wall or *cortex* through its system of vessel canals. During the growing years, the inner portion of the periosteal membrane contains numerous osteoblasts which lay down new bone. The number of periosteal osteoblasts is considerably decreased in adults but is still capable of producing new bone in response to inflammation or other stimulant.

Marrow Spaces and Endosteum

As mentioned earlier, the bones contain marrow for the synthesis of blood elements. The marrow is found mainly in the *medullary cavity* or canal of a long bone, but it is also located in all the cancellous spaces of the bone including their extensions along *Haversian canals* and other small vascular channels. The medullary cavity and cancellous spaces are theoretically lined by an osteogenic membrane called the *endosteum*. Although the endosteum is not a clearly defined and detachable membrane like the periosteum, the osteoblasts and potential osteoblasts of the marrow reticulum are important sources of new bone formation.

Bone Architecture

Bone can be divided into two main histological types: woven and lamellar bone. *Woven bone,* also known as fibrous or non-lamellar bone, is more primitive phylogenetically and consists of collagen fibrils forming an irregular matrix. Woven bone forms the embryonic skeleton and is gradually replaced by lamellar bone by the age of one year. Woven bone may also be formed in many pathological states where there is a stimulus for rapid bone formation. For example, woven bone is the first bone produced in fracture repair and is subsequently replaced by the lamellar bone. Bone produced by the periosteum in response to infection is primarily woven until replaced by lamellar bone. Bone-forming tumors such as osteosarcoma may produce large amounts of the coarse-fibered woven bone.

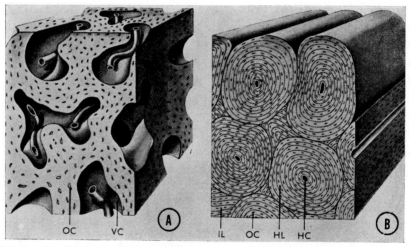

Figure 2. Diagram showing differences in structure of woven (A.) and lamellar bone (B.) HC, Haversian canal; HL, Haversian lamella; IL, interstitial lamella; OC, osteocyte; VC, vessel canal. (Courtesy of Dr. N. M. Hancox, 1972.)

In *lamellar bone*, the collagen fibrils are oriented parallel to each other within the plane of each lamella. The direction of the fibers changes in successive lamella to form a "laminated" sheet of tremendous strength. Virtually the entire skeleton is composed of lamellar bone which forms two main types of gross bone architecture: compact and cancellous or spongy bone.

Compact bone is dense lamellar bone making up the full thickness of the bone wall or *cortex*. Compact bone tissue is too thick to obtain its nutrition only from surface vessels and therefore relies on special vascular channels known as *Haversian canals*.

Figure 3. Ground section of human femur showing a typical Haversian canal surrounded by concentrically arranged lacunae and canaliculi. X300. (Courtesy of Dr. D. W. Fawcett from Bloom and Fawcett, 1975.)

Cancellous bone is a porous network of branching and anastomosing *trabeculae* of bone. This type of bone tissue fills the interior of the epiphyses and metaphyses with very little in the diaphyses. The thin trabeculae receive their nutrition from the surrounding blood vessels in the marrow spaces (see Fig. 4).

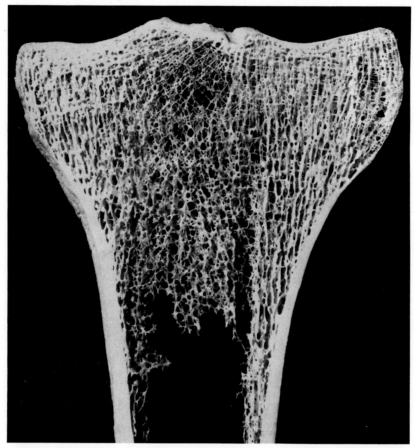

Figure 4. Longitudinal section of proximal tibia showing cancellous bone in the epiphyseal and metaphyseal regions. The cortex is composed of compact bone. (Courtesy of Dr. D. W. Fawcett from Bloom and Fawcett, 1975.)

BASIC BONE REACTIONS

Normal Bone Formation

Normal bone formation in the growing skeleton occurs by two different processes: intramembranous and endochondral. *Intramembranous ossification* is responsible for the formation of the frontal and parietal bones as well as enlargement of the other bones by subperiosteal bone apposition. In the embryo the mesenchymal tissue gives rise to a primitive connective tissue membrane. Small groups of osteoblasts develop from the membrane to produce nonmineralized matrix or *osteoid* which is subsequently mineralized to form woven bone trabeculae. Gradually these trabeculae form a latticework of osseous tissue, and the primitive connective tissue membrane condenses on the surface of the bone layer to form the periosteum. The periosteal osteoblasts lay down new bone in the latticework of trabeculae to eventually form a solid layer of bone. Concomitantly, the woven bone initially produced is replaced by lamellar bone.

Endochondral ossification is the major process of bone growth until fusion of the epiphyses occurs. By this process mesenchymal cells differentiate into chondroblasts which form a hyaline cartilage mass roughly corresponding to the shape of the bone in the adult skeleton. Cartilage cells in the center of this "cartilaginous model" enlarge, and the surrounding cartilage matrix is resorbed. Then the cartilage cells die and are invaded by rapidly growing blood vessels. The invading blood vessels contain mesenchymal cells which can differentiate into osteoblasts and hematopoietic cells. Thus, a primary center of ossification develops and spreads from midshaft towards both ends of the bone. At a later time secondary centers of ossification appear in the cartilaginous epiphyses. These centers expand until only a thin plate of hyaline cartilage separates the epiphysis from the shaft. This is called the *epiphyseal plate,* and continued growth of the cartilage followed by bone replacement provides for growth in length of the bone until adult stature is attained.

Normal Bone Resorption

The creation of properly proportioned bones requires bone resorption as well as bone formation. Moreover, bone is a living tissue necessitating constant renewal by bone resorption as well as bone formation.

Bone resorption or deossification is the dissolution of both the organic matrix (osteoid) and its mineral content. The *osteo-*

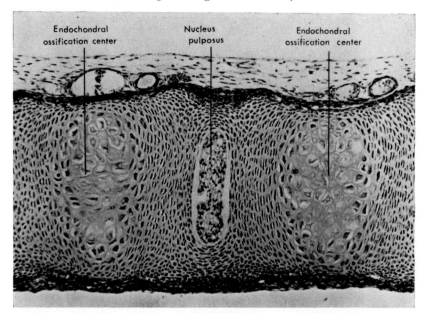

Figure 5. Photomicrograph of the cartilaginous vertebral column in a mouse embryo. In the center of each vertebra is an area of hypertrophied cartilage cells representing an early stage in the establishment of a center of endochondral ossification. (Courtesy of Dr. D. W. Fawcett from Bloom and Fawcett, 1975.)

clast appears to play the major role in bone resorption. This short-lived cell apparently arises through fusion of several mononuclear cells, but it is not known whether these cells were osteoblasts, osteocytes, macrophages, or less differentiated mesenchymal cells. Experimental evidence indicates that osteoclasts secrete both acidic substances which dissolve the bone mineral and lysosomal enzymes which depolymerize the organic matrix. This extracellular digestion releases minute bone fragments which are ingested by the osteoclasts and digested intracellularly.

BONE FORMATION AND RESORPTION IN PATHOLOGICAL CONDITIONS

Pathological conditions create an imbalance in the normal equilibrium of bone resorption and formation. Therefore, bone reacts to abnormal conditions by an increase or decrease in

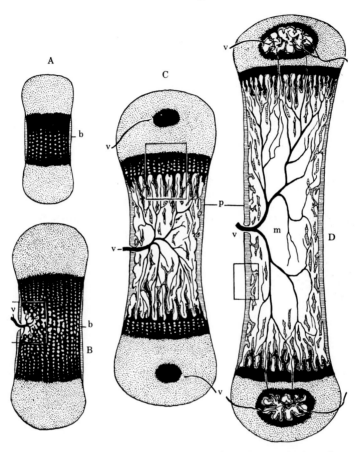

Figure 6. Diagrams of the ossification of a long bone: (A.) early cartilaginous stage; (B.) stage of eruption of the periosteal bone collar by an osteogenic bud of vessels; (C.) older stage with a primary marrow cavity and early centers of calcification in the epiphyseal cartilages; (D.) the condition shortly before birth with epiphyseal centers of ossification. Calcified cartilage in all diagrams is black; b, periosteal bone collar; m, marrow cavity; p, periosteal bone; v, blood vessels entering the centers of ossification. (Courtesy of Dr. W. F. Windle, 1955.)

the normal processes of bone formation, bone resorption, or a combination of the two processes at different locations in the bone. The repair of bone following a fracture is an excellent example of the response of bone tissue to abnormal or "supernormal" influences. (Refer to Chapter II for a discussion of bone repair.)

Mechanical Stress

Mechanical stress is one of the most important factors affecting bone architecture. Osteoblastic activity is in some way stimulated by increased tension or compression so that new lamellae are laid down according to the lines of increased stress. Abnormally high stresses may be exerted on bones in certain pathological conditions such as rickets. In such cases the osteoblasts produce osteoid in the areas of stress, but no calcium salts are present to harden the soft osteoid. The weight-bearing bones, such as the femur, become bowed. With healing, the thick layer of osteoid present along the concave side of the femur becomes mineralized, and even more new bone may be laid down in response to the abnormal stress on the deformed bone.

The absence of stress on the bone causes a decrease in osteoblastic activity. For example, the bones of a limb immobilized by paralysis or severe fracture become atrophied as osteoclastic resorption continues without new bone replacement. The bone tissue loss results in *osteoporosis*, a general term denoting a reduction in the amount of osseous tissue (mineralized osteoid) per unit of bone volume. Disuse atrophy due to the absence of normal mechanical stress is further discussed in Chapter VIII.

Blood Supply

Both bone formation and bone resorption are active cellular processes requiring an adequate blood supply. Thus, it is inaccurate to state that an increase in vascularity (*hyperemia*) always causes an increase in bone resorption only. For example, increased vascularity is found in the metaphysis of growing bone and in a healing bone fracture. Moreover, active bone formation and resorption often occur at the same site such as in a single bone trabecula. Increased blood supply thus favors both processes.

Of even more importance is insufficient blood supply or *ischemia* which results in the death of all bone cells in the affected area. Gradually blood vessels from the surrounding living tissue begin to invade the necrosed region, and mesenchymal cells of the vascular bud give rise to both osteoblasts and osteoclasts. Resorption of the dead bone and formation of new bone then takes place in the revascularized regions. The complete reorganization of a mass of dead bone depends both on the amount of dead tissue and the available blood supply. In certain anatomical regions, such as the head of the femur, the blood supply is very limited and aseptic bone necrosis will be long-lasting.

Inflammation

Inflammation is a normal response of tissue to an injurious agent such as an infection or fracture. The major feature of inflammation is the release of fluid, polymorphonuclear leukocytes, macrophages, and later mononuclear cells from the blood stream into the affected area. Blood vessel proliferation is also prominent. In purulent inflammations of bone, the increased pressure caused by the pus in the internal spaces of the bone leads to blood vessel occlusion and destruction. Thus, bone necrosis may be considerable in acute forms of inflammation where pus production is excessive.

New bone production is a prominent feature of inflammation. The exudative fluid and pus raise the periosteum from the bone, and this irritation of the periosteum stimulates exhuberant new bone formation. The increase in vascularity (granulation tissue) also favors the production of bone as well as the osteoclastic resorption of dead bone. In the final stages there is osteoclastic resorption of the primitive woven bone and replacement by highly structured lamellar bone.

Hormonal Imbalance

Hormones are very important in controlling bone growth, maintaining the equilibrium of bone resorption and formation, and mobilizing calcium salts from the bone to maintain a normal blood calcium level. No attempt will be made here to describe the response of bone to each type of hormone imbalance. In general, an increase or decrease of the hormone involved causes an increase or decrease in the normal physiological response of the bone to that hormone. For example, an increased secretion of somatotropic hormone (STH) from the anterior lobe of the pituitary causes an increase in bone growth resulting in giantism or acromegaly. Hyperparathyroidism is an increased secretion of parathyroid hormone which increases the number of osteoclasts. The result is a generalized resorption of bone or osteoporosis.

Neoplasia

Bone invasion by primary or secondary tumors usually causes both bone formation and bone resorption. Bone resorption is the main feature of most invasive tumors and appears to be mediated by a stimulation of the osteoclasts by the tumor cells. Bone formation is prominent in some types of tumors, particularly those that are slow-growing. Thus, one often finds osteosclerotic metastases of bone in slowly developing prostatic carcinoma or massive bone production by a meningioma slowly invading the cranium. In both instances the bone is produced by osteoblasts which have been stimulated to lay down bone. Most of the new bone produced is of the woven type, but lamellar bone may eventually replace it in the slower growing tumors.

Tumor cells of primary bone tumors originate from the bone tissue and therefore may directly cause bone formation and resorption. Bone destruction is prominent in giant-cell tumor (osteoclastoma) due to the proliferation of malignant osteoclasts. Similarly, bone production is considerable in osteosarcoma. It should be emphasized however, that much of the bone produced in oesteosarcoma is the result of stimulating normal osteoblasts in the endosteum and periosteum.

SUMMARY

This chapter has hopefully corrected the common misconception that bone is an inert structural framework impervious to the external environment. Instead, it is composed of living cells in a hard matrix which are quite sensitive to such influences as trauma, infection, mechanical stress, nutrition, and neoplastic growths.

With this general introduction to the gross anatomy of bone and its response to various pathological conditions, let us now examine specific types of gross bone pathology. The discussion will stress the gross features of bone pathology as these are of the greatest importance in diagnosing the specific disease entity from a dried bone specimen.

BONE AS A LIVING TISSUE BIBLIOGRAPHY

A. Bone Anatomy, Physiology, and Pathology

Aegerter, E. and J. Kirkpatrick: *Orthopedic Diseases*, 2nd ed., Saunders, Philadelphia, 1968.

Bass, W. M.: *Human Osteology: A Laboratory and Field Manual of the Human Skeleton*, Missouri Archaeological Society, Special Publications, Columbia, 1971.

Bloom, W. and D. W. Fawcett: *A Textbook of Histology*, 10th ed., Saunders, Philadelphia, 1975.

Enlow, D. H.: *Principles of Bone Remodelling*, Thomas, Springfield, 1963.

Frost, H. M.: *The Physiology of Cartilaginous, Fibrous, and Bony Tissue*, Thomas, Springfield, 1972.

Greenfield, G. B.: *Radiology of Bone Disease*, Lippincott, Philadelphia, 1969.

Hall, M. C.: *The Architecture of Bone*, Thomas, Springfield. 1966.

Hancox, N. M.: *Biology of Bone*, Cambridge Univ. Press, Cambridge, 1972.

Jaffe, H. L.: *Metabolic, Degenerative, and Inflammatory Diseases of Bones and Joints*, Lea and Febiger, Philadelphia, 1972.

Knaggs, R. L.: *Diseases of Bone*, Wood, New York, 1926.

Luck, J. V.: *Bone and Joint Diseases*, Thomas, Springfield, 1950.

McLean, F. C. and M. R. Urist: *Bone: Fundamentals of the Physiology of the Uhysiology of Skeletal Tissue*, Univ. of Chicago Press, Chicago, 3rd ed., 1968.

Putschar, W. G. J.: "General pathology of the musculo-skeletal system," in F. Buchner (ed.): *Handbuch der Allgemeinen Pathologie*, 2nd ed., Springer-Verlag, Berlin, Bd. III/2, 1960.

Ritvo, M.: *Bone and Joint X-ray Diagnosis*, Lea and Febiger, Philadelphia, 1955.

Shanks, S. C. and P. Kerley (ed.): *A Text-book of X-ray Diagnosis*. 4th ed., Lewis, London, vol. 4, 1969.

Weinmann, J. P. and H. Sicher: *Bone and Bones: Fundamentals of Bone Biology*, C. V. Mosby, St. Louis, 1947.

B. General Paleopathology

Ackerknecht, E. H.: "Palaeopathology; a survey," in A. L. Kroeber (ed.): *Anthropology Today,* pp. 120-127, Univ. of Chicago Press, Chicago, 1953.

Armelagos, G. J., J. H. Mielke, and J. Winter: *Bibliography of Human Paleopathology,* Research Reports Number 8, Dept. Anthropology, Univ. Mass., Amherst, 1971.

Brothwell, D. R.: *Digging up Bones,* British Museum (Natural History), London, 2nd ed., 1972.

――――――(ed.): *The Skeletal Biology of Earlier Human Populations,* Pergamon Press, Oxford, 1968.

Brothwell, D. R. and A. T. Sandison (ed.): *Diseases in Antiquity,* Thomas, Springfield, 1967.

Cockburn, T. A.: *The Evolution and Eradication of Infectious Diseases,* The Johns Hopkins Press, Baltimore, 1963.

Eckert, W. G.: *Paleopathology Bibliography,* Inform compilation, Wichita, 1970.

Goldstein, M. S.: "The palaeopathology of human skeletal remains," in D. R. Brothwell and E. Higgs (eds.): *Science in Archaeology,* Basic Books, New York, 1963.

――――――: "Human palaeopathology and some diseases in living primitive societies: A review of the recent literature," *Amer. J. Phys. Anthrop., 31*:285-294, 1969.

Hackett, C. J.: "Diagnosis of disease in the past," in E. Clarke (ed.): *Modern Methods in the History of Medicine,* Athlane Press, London, 1971.

Henschen, F.: *The History and Geography of Diseases,* Longmans, Green, and Co. London, 1966.

Jarcho, S.: "Some observations on diseases in prehistoric America," *Bull. Hist. Med., 38*:1-19, 1964.

――――――(ed.): *Human Palaeopathology,* Yale Univ. Press, New Haven, 1966.

Kerley, E. R. and W. M. Bass: "Paleopathology: meeting ground for many disciplines," *Science, 157*:638-644, 1967.

Moodie, R. L.: *Paleopathology: An Introduction to the Study of Ancient Evidences of Disease,* Univ. Illinois Press, Urbana, 1923.

Morse, D.: *Ancient Disease in the Midwest,* Illinois State Mus. Rep. Invest., #15, Springfield, 1969.

Pales, L.: *Paléopathologie et Pathologie Comparative,* Masson, Paris, 1930.

Ruffer, M. A.: *Studies in the Palaeopathology of Egypt,* edited by Roy L. Moodie, Univ. of Chicago Press, Chicago, 1921.

Sigerist, *A History of Medicine, vol. I, Primitive and Archaic Medicine,* Oxford Univ. Press, Oxford, 1951.

Wells, C.: *Bones, Bodies and Disease,* Thames and Hudson, London, 1964.

Chapter II

TRAUMA

TRAUMATIC LESIONS, particularly fractures, are frequently found in excavated skeletal material. Our interest in traumatic lesions lies not in their differential diagnosis, which is usually straightforward, but rather in the cultural significance and implications which a population study of such lesions may reveal. Traumatic lesions may be classified as fractures, crushing injuries, bone wounds caused by sharp instruments, and dislocations. Microtrauma in the growing bone induced by nutritional deficiency and disease is also discussed in this chapter.

FRACTURES

There are many types of fractures and these are determined by the direction and intensity of the causative force and the physical properties of the affected bone. A *complete* fracture refers to a fracture in which the two broken ends are completely separated. Such fractures may be transverse or oblique depending upon whether the direction of force was predominantly angular or rotatory. In an *incomplete* fracture, the two bone ends are still connected because the break is not complete. Such fractures often occur in the resilient bones of young children and are known as *"greenstick"* fractures. When the bone is weakened by any type of metabolic, infectious, or neoplastic disorder, it is much more vulnerable to fracture. Such fractures are termed *pathologic* and may be induced by very minor stresses on the weakened bones. A *comminuted* fracture occurs when the bone is broken into more than two fragments. Splintering of the bone is common in such cases and repair is slow.

When the fractured ends of the bone are exposed through the muscle and skin, the fracture is described as open or *compound*. Infection of the exposed bone frequently occurs producing a pyogenic osteomyelitis (see Chapter III). Simple fractures rarely become infected. Therefore, in an archaeological specimen of bone fracture complicated by osteomyelitis, the fracture was probably a compound one.

Pathogenesis of Fracture Healing

Fracture healing provides a good illustration of the reactions of bone discussed in the previous chapter. A fracture severs the rich supply of blood vessels in the medullary cavity, periosteum, and overlying muscle tissue. The extravasated blood forms a blood clot or hematoma at the fracture site. Opinion differs as to the function of the hematoma (see Collins, 1966). Its strands of fibrin may provide an important framework upon which the cellular elements proliferate. Osteoprogenitor cells of the periosteum, endosteum, and in the vascular spaces of the bone itself proliferate and modulate to osteoblasts laying down new, woven bone at a distance from the defect. In addition, cells from the surrounding connective tissue proliferate into the defect itself. These fibroblasts eventually transform into chondroblasts and osteoblasts by fibrocartilaginous and chondro-osseous metaplasia (Aegerter and Kirkpatrick, 1968). The proliferation and differentiation of osteogenic cells proceeds rapidly to form a primary cellular callus.

Vascularization of the callus begins within a few days after fracture. The vascular regeneration arises from budding of preexisting blood vessels in the soft tissues and medullary cavity. The hyperemia involves the bone as well as the hematoma. Calcium salts are released from dead bone fragments as well as the living bone and used in calcifying the osteoid callus. Cartilage produced in avascular areas of the defect is now replaced by bone through a process similar to endochondral ossification. This process may begin as early as two weeks following fracture (Blast et al., 1925).

Figure 7. Complete fracture of the left femur with wide separation of the broken ends. Strong bony union has occurred through callus formation to form a functional but shorter bone. (WAM 8060)

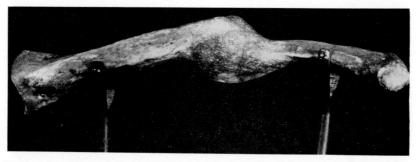

Figure 8. Fractured clavicle with large callus formed at site of break. Bone remodelling may continue over extended lengths of time to give the clavicle a nearly normal appearance. (WAM 5264)

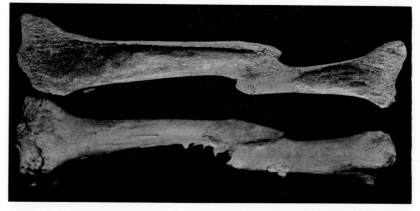

Figure 9. Longitudinal section of a fractured tibia showing the narrowing of the medullary cavity by new bone formation. (WAM 4776)

The bony callus joining the two bone ends forms rapidly and abundantly. Much of the exuberant callus is simply resorbed. The rest of this disorganized woven bone is gradually replaced by highly structured lamellar bone. The lamellar bone is laid down according to the functional lines of stress. This internal reorganization and remodelling will take place even in badly aligned fractures and may continue for months or even years after the injury (Todd and Iler, 1927). If the bone ends are aligned properly, the remodelling may completely obliterate all signs of the fracture.

Rate of Repair

The rate of repair of a fracture is modified by several factors such as age, type of fracture, degree of vascularization, amount of motion between the broken ends, and presence of infection. The capacity for growth of new tissue is greater in the infant than in the adolescent, and greater still than in the adult. Thus, union of a fractured femur by primary callus may take only one month in an infant, two months in a fifteen-year-old adolescent, and over three months in an adult (Watson-Jones, 1952; DePalma, 1970).

In spiral fractures, the marrow cavity is opened widely producing a larger vascular area for promoting tissue growth. Thus, union is more rapid in this type of fracture compared with horizontal fractures where medullary callus formation is limited. In any type of fracture where the blood supply is impaired to one or both fragments, the callus formation is delayed.

Union is more rapid if the broken ends are closely apposed than if there is a gap between them. Moreover, a gap often allows increased mobility between the two fragments. More cells differentiate into fibroblasts and chondroblasts when motion is increased. If this continues for a long period of time, the reparative process diminishes and the fibrocartilaginous tissue is not replaced by new bone. Fibrous or fibrocartilage union of the fractured bone ends is clinically termed *nonunion* (for paleopathological examples of nonunion fractures see Morse, 1969; Stewart, 1974).

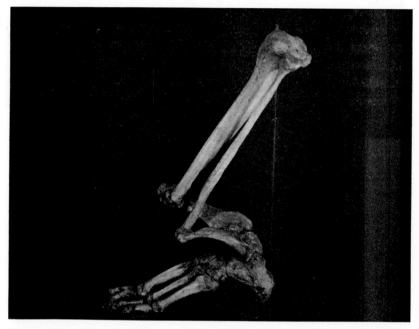

Figure 10. Lower left leg with fractured tibia and fibula. Inadequate immobilization has resulted in nonunion. Fibrous tissue connects the fractured ends with much deformity. (WAM 9813)

When the broken ends are very poorly aligned and motion is excessive, the proliferating cells differentiate primarily into chondroblasts. These cells produce a hyaline-like cartilage over the ends of the fractured bones. Other cells form synovial tissue which surrounds the bones and secretes lubricating fluid. Thus, a pseudarthrosis or false joint is created (see Fig. 11). Pseudarthrosis may affect any long bone and certain short bones. Some of the more common sites are the neck of the femur, middle one third of the clavicle, the humeral diaphysis, and the middle one third of the tibia.

Infection of the bone at the fracture site can seriously delay bone repair. If complete immobility is maintained, normal bone repair resumes after the infection subsides and fragments of dead bone (*sequestra*) are resorbed.

It is difficult to determine the age of a fracture in an archaeo-

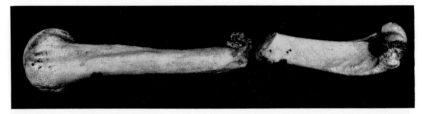

Figure 11. Right humerus with complete fracture that never reunited. The two broken ends have smooth surfaces and articulate with each other forming a pseudarthrosis. The osteophytic growth is produced by periosteal irritation. (WAM 5183)

logical specimen. If death occurs immediately following fracture, the broken ends of the bone will be sharp and lack any evidence of bone absorption or regeneration. Hyperemic or osteoclastic bone absorption and subperiosteal bone apposition may be grossly evident as early as two weeks following fracture. The lacunar osteocytes at the edge of the broken fragments have died and these are removed giving the bone edges a smooth surface. The subperiosteal woven bone forms at a distance from the bone defect on both sides. The primary cellular callus will not be present in the macerated specimen. Complete bony bridging of the fracture may take one month to over a year depending upon the factors discussed previously such as type of fracture and amount of motion between the bone ends.

The different types of fractures just described have been reported on many occasions in archaeological skeletal material. Although the bone pathology revealed by the specimens is usually not intrinsically important, the frequency and types of bone fracture in excavated skeletal series can provide important information about their lifestyle. To illustrate the possibilities of such an approach Table I lists the incidence of postcranial fracture in several American Indian populations from different time periods. More importantly, the populations represent different ways of life and the incidence of fracture varies accordingly. The highest incidence is found in the earliest population adequately represented—the Indian Knoll site. This same frequency of postcranial fracture continues through the Late Archaic period until the advent of the Woodland Tradition and permanent villages. The frequency of fracture is further decreased with the use of agriculture and development of large towns during the Mississippian period and continuing until historic times.

TABLE I

FREQUENCY OF POSTCRANIAL FRACTURE IN AMERICAN INDIAN POPULATIONS OF DIFFERENT PERIODS AND MODES OF LIFE*

Indian Population	No. Fractures No. Individuals	Percent	Source
Middle Archaic (4000-2000 B.C.)			
Indian Knoll, Ky.	57/521	10.7	Snow, 1948
Indian Knoll, Ky.	7/61	9.8	USNM series
Late Archaic (2000-1000 B.C.)			
Robinson Site, Tenn.	5/51	9.6	Morse, 1969
Morse Site, Ill.	6/62	9.7	Morse, 1969
Woodland (1000 B.C.-1000 A.D.)			
Tollifero Site, Va.	2/37	5.4	Hoyme and Bass, 1962
Steuben Site, Ill.	3/54	5.5	Morse, 1969
Klunk Site, Ill.	20/395	5.0	Morse, 1969
Mississippian (1000-1600 A.D.)			
Clarksville Site, Va.	3/77	3.9	Hoyme and Bass, 1962
Dickson Mounds, Ill.	11/331	3.3	Morse, 1969
Emmons Site, Ill.	1/83	1.2	Morse, 1969
Iroquois Tribe (1400 A.D.)			
Fairty Ossuary, Ont.	10/295	3.4	Anderson, 1964
Pueblo Indians (800-1800 A.D.)			
Pecos Pueblo, N. Mex.	19/503	3.8	Hooton, 1930
Texas Indians (800-1700 A.D.)			
Texas Sites	2/92	2.2	Goldstein, 1957

* There is a general pattern of decreasing incidence of postcranial fracture with respect to changing settlement patterns and means of obtaining food.

The frequencies of postcranial fracture listed in Table I are not entirely accurate for comparison with other skeletal series because of possible differences in the sex or age composition of the series. Further information might be gained by comparing the types of fracture and sites of fracture in the different populations.

CRUSHING INJURIES

Very severe comminuted fractures or crushing injuries may be caused by falls or blows from heavy objects. Common sites for such injuries are the bones of the hands, feet, and skull. Because of the extensive damage to the bone or bones involved, a large callus forms and may produce fusion and deformity of the splintered bones. Such fractures involving the skull may often become infected because the skull vault is easily exposed by trauma or soft tissue scalp infection can extend to the injured bone. Infection and the severity of the skull fractures may prevent complete bridging of the fracture, leaving a hole in the skull after the reparative process subsides. Such a defect may be confused with the healed opening left by surgical intervention (*trephination*), particularly since many trephinations were performed to alleviate compressed vault fractures (see Morse, 1969 plate 8).

BONE WOUNDS CAUSED BY SHARP INSTRUMENTS

Arrow and Spear Wounds

Bone wounds produced by arrows and spears may be diagnosed with certainty only when the projectile point remains in the bone. Such examples come from Mesolithic and Neolithic Europe and particularly the Woodland and Mississippian Cultures of the New World (see Janssens, 1970; Morse, 1969). Arrow wounds are most frequently found in the bones closest to the skin surface such as the vertebrae, ribs, and skull (see Wilson, 1901). If the bone tissue surrounding the projectile point shows any signs of healing, then the individual probably survived the injury for at least several weeks.

Scalping

Contrary to popular belief, scalping was by no means general among the Indians of North America in early times. The practice was originally confined to the eastern part of the continent, although cases have been reported in South America and British Columbia (see Friederici, 1907). In addition, scalping was

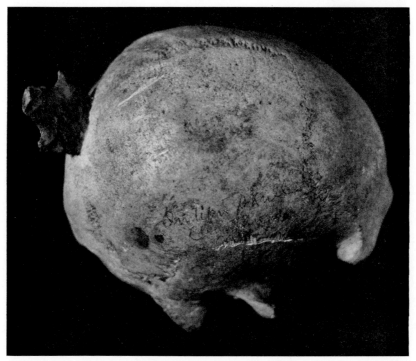

Figure 12. Projectile point imbedded in the right parietal of an unfortunate Indian from Chautauqua, New York. (Peabody Museum N/1393)

practiced by Scythians as described by Herodotus and even among the ancient Germans and French until at least 870 A.D. (Burton, 1864). Scalping apparently evolved from the older custom of keeping the entire head as a trophy. The spread of the scalping practice over central and western North America was a direct result of encouragement through scalp bounties offered by the French, British, and colonial governments (Mooney, 1900; 1910).

The operation of scalping was usually undertaken only after the victim had been slain. Different techniques were employed, and the area of the head which was scalped varied considerably (Nadeau, 1944). In general, the bones of the skull vault were rarely damaged if the incisions made into the soft tissues were by metal knives. However, when prehistoric scalping was carried

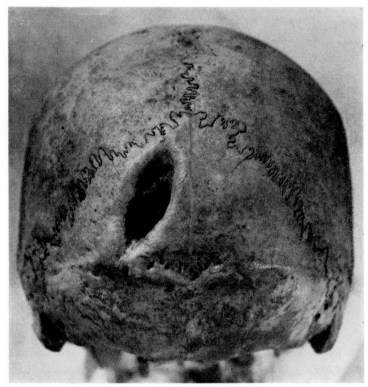

Figure 13. Female Indian from Mobridge, South Dakota. The large elliptical defect in the occipital bone is perhaps caused by intentional injury with a sharp implement. The rounding of the edges shows that bone healing took place. (USNM 325,342)

out using flint implements, the serrated edges produced numerous parallel incisions or scratches on the bone surface. This has been clearly demonstrated by experiments with flint implements on cadavers (Hamperl and Laughlin, 1959).

In some instances, scalping was performed on live individuals who then escaped or were released (Friederici, 1907). Besides having the scratch marks on their skulls, these individuals will exhibit bone changes resulting from the removal of the periosteum from the skull. Deprived of its periosteal blood supply, the outer table becomes necrotic and inflammatory granulation tissue then separates the necrotic area from the deeper layers of living bone. New bone is produced by these deeper layers and the wound gradually heals. The end result is a slightly depressed area of spongy bone corresponding to the area of scalping (Hamperl, 1967). Actual clinical information from a late nineteenth century victim of scalping confirms the sequence of bone changes described above (Reese, 1940).

Several possible examples of prehistoric scalping have been reported in North American skeletal material. One such case comes from the Middle Mississippian culture (1000-1500 A.D.) of Illinois (Neumann, 1940). In this adult skull, several cuts, apparently made by a flint knife, cross the skull vault in the frontal region and extend in a rough circle around the crown. Two similar cuts are present in the occipital bone. The appearance and location of these cuts correlate well with the usual method of scalping by freeing the scalp anteriorly, pulling it away from the skull, and then cutting it free in the occipital region. Judging from the crushed right side of the skull and fractured parietals, the individual certainly died immediately, and this explains the absence of any bone changes in the denuded vault area. Similar cut marks on the skull vault are present in three prehistoric and early historic skulls from Virginia (Hoyme and Bass, 1962).

Another possible case of scalping comes from Alabama during the Mississippian cultural period (Snow, 1941). In this skull of an adult female, a groove of osteitis completely encircles the top of the skull vault. The osteitis is limited to the groove implying that this was an unsuccessful scalping attempt. A fragmentary prehistoric Indian skull from Nebraska also exhibits a remarkably similar groove of osteitis in the frontal, parietals, and part of the occipital (Snow, 1942).

The skull of an Arikara Indian from South Dakota (1690-1720 A.D.) exhibits bone changes indicative of a healed scalping wound (Hamperl and Laughlin, 1959). At the back of the skull is an elliptical area of bone scarring measuring approximately 7 by 10 cm. The surface is uneven with many flattened depressions. A cross-section of the affected region reveals the thinned

Figure 14. Probable scalping in an adult Indian from Illinois. (A.) Superior view of the vault showing several cut marks crossing the frontal bone and extending around the crown.

outer table and normal inner table. The thin outer layer of bone is very dense because of the new bone formation.

The skull of a thirty-eight-year-old male from the Mississippian Culture of Arkansas exhibits similar evidence of osteitis and new bone formation following a possible scalping (Morse, 1973). A large oblong area measuring 8 by 10 cm at the top of the skull

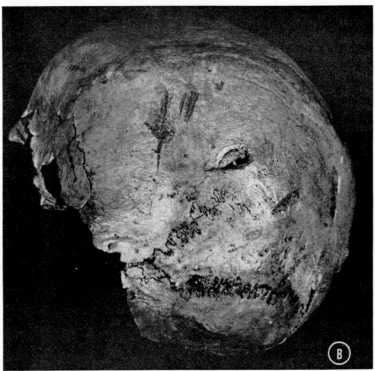

Figure 14B. Similar cuts in the parietal and occiptal region. (Courtesy of Dr. Dan Morse, 1969.)

is affected by irregular erosion and bone sclerosis without involving the inner table. The posterior portion of this area appears to have less evidence of complete healing (see Fig. 15).

Trephination (Trepanation)

Trephination is one of the first surgical procedures involving the skull of which we have any authentic record. The operation involves the removal of part of the skull vault without damaging the underlying meninges and brain. The first case of prehistoric trephination was discovered little more than a century ago by the anthropologist, E. G. Squier, in Peru. Since then well over one thousand specimens have been uncovered, accompanied by a voluminous literature concerning this fascinating operation by early man. Of particular note are three excellent reviews on the

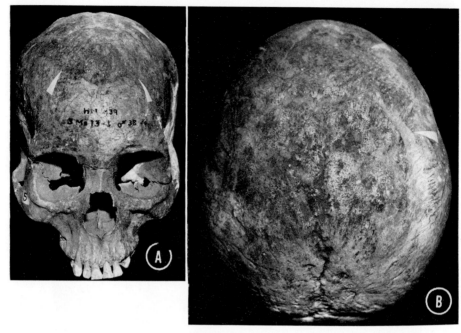

Figure 15. Possible healed scalping wounds in an adult male Indian from Arkansas. (A.) Frontal view of skull with arrows indicating sharply circumscribed area of osteitis. (B.) Superior view showing probable evidence of active infection in the posterior region. (Courtesy of Dr. Dan Morse, 1973.)

subject (Lisowski, 1967; Piggot, 1940; Stewart, 1958). Little can be added to these reports so the discussion here will be limited to a few general points.

Trephination has a remarkably wide distribution in both time and place. Although no cases have been reported from the Paleolithic, it certainly was widely practiced by the Neolithic period in Europe, particularly in France, England, Denmark, Germany, Italy, and Russia (Lisowski, 1967). Trephination continued in these areas well into the Middle Ages. In Asia the operation was well known, and numerous literary references to the procedure come from China and Japan in the early centuries A.D. Very few actual specimens have been found in Africa although trephination is still performed today by native medicine men in several parts of Africa (Margetts, 1967). Except for Melanesia, few definite specimens have been found in Oceania.

A similar paucity of cases occur in the massive amount of skeletal material from North America. In contrast, more trephined skulls have been found in Peru than in the rest of the world combined. The practice flourished in the Peruvian region between the fifth century B.C. and the fifth century A.D. (Stewart, 1958).

The dangerous operation of trephining the skull was performed for a variety of reasons as revealed by the specimens themselves, literary sources, and study of the procedure in primitive cultures today. One of the most important motives was the alleviation of intracranial pressure produced by compressed fractures of the skull vault complicated by edema. Many trephined skulls clearly reveal the cranial fracture near or at the site of the operation. This is particularly true of the Peruvian specimens. In other instances, trephination was performed to treat headaches, epilepsy, or other forms of mental illness (see Margetts, 1967). In such cases the opening of the skull allowed the release of imagined vapors, evil spirits, or even worms. The large number of trephined skulls in certain areas of France suggest a ritualistic reason for the operation. Other skulls from Europe were trephined postmortem to obtain rondels or amulets.

Several different techniques were employed to trephine the skull. The most common method of opening the skull was by scraping the bone with flint or metal blades. This resulted in a roughly circular opening with widely bevelled edges. In Peru, some trephinations were performed by drilling many small holes in a circular pattern and then cutting the remaining bone connections with a sharp instrument. Another method used in Peru but rarely elsewhere required four deep incisions to free a rectangular piece of bone. In several different regions the grooving method was employed. By repeatedly cutting a circular path, the groove deepened and eventually a large piece of bone was freed from the skull (see Fig. 16 concerning examples of pre-Columbian Peruvian trephination from the Tello Collection at the Peabody Museum.)

Although hindered by lack of anesthesia or antiseptic, the operation had an impressive recovery rate. Over 50 percent of two hundred fourteen cases examined showed complete healing and another 16 percent exhibited partial healing (Stewart, 1958).

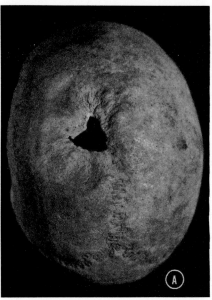

Figure 16A. Trephination of the left parietal performed by the scraping technique with a beveled flint instrument. The opening has widely beveled edges.

Figure 16B. Trephination of the frontal and right parietal performed by the grooving technique with a sharp instrument.

Figure 16C. Multiple trephinations of the vault.

Figure 16D. Two trephinations of the vault. One is almost completely healed by new bone bridging the defect. The other opening shows no evidence of bone repair.

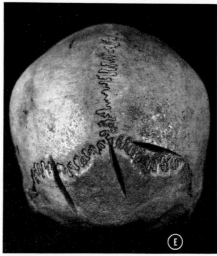

Figure 16E. Three straight incisions made into the occipital and parietal bones by a straight, wedge-shaped instrument.

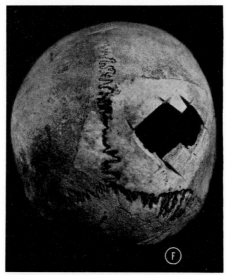

Figure 16F. Trephination of the right parietal made by eight straight incisions intersecting at right angles. Note the large angular area of sclerotic bone surrounding the wound which represents the maximum extent of the opening through the scalp.

Moreover, multiple trephinations may be found as vividly demonstrated by a Peruvian skull with seven well-healed trephine holes (Oakley et al., 1959). Encircling the healed margins of the opening is usually an osteoporotic pitting due to a septic osteitis or ischemic reaction (Stewart, 1956).

The differential diagnosis for trephinations may provide some difficulty as several conditions may closely resemble the trephine hole made by the scraping method. A sword cut which removes a small section of the skull may simulate an unhealed trephination. Similarly, holes made by a pick in the soft bone during excavation or postmortem localized erosion can be mistaken for a trephine hole (Brothwell, 1963). Such erosion is particularly liable to occur if the parietal bones are abnormally thin because of old age osteoporosis, congenital defect, or endocrine disorder. Enlarged parietal foramina, a developmental defect, may resemble a trephination (Lisowski, 1967). Benign and malignant tumors, such as benign bone cysts, metastatic carcinoma, and perhaps multiple myeloma, can cause lytic bone defects similar to unhealed trephine openings. Finally, infections of the bone, such as syphilis, tuberculosis, and a localized osteomyelitis, may occasionally produce openings in the skull. It is important to note that if only one or two skulls in a large amount of skeletal material have possible trephine holes, the diagnosis must remain tentative. Where practiced, trephination was usually performed on significant numbers of people—and the very nature of this operation must have required constant practice!

Sincipital T

Another type of primitive operation which should be briefly mentioned here is the extensive scarring of the skull vault, usually in the shape of a cross (sincipital T). One line runs anteroposterior along the sagittal suture, and the other groove intersects it at right angles along the parietal eminences. The lesion is produced by cauterizations of the skin affecting the periosteum. The damaged periosteum restricts the blood supply to the cauterized area producing ischemic necrosis often complicated by osteitis. This scarring may resemble cranial osteomyelitis.

The sincipital T was first described in 1895 by Manouvrier who noted the lesion in six female skulls from Neolithic France. Subsequently, cases have been described from Central Asia, Canary Islands, Africa, and Peru (Moodie, 1921; Sudhoff, 1908; Zaborowski, 1897). In almost all instances, the affected skulls are of females and small children suggesting a ritualistic significance such as an initiation rite. Manouvrier concluded that the operation was performed without change from Neolithic times to the surgeons of the Middle Ages who used cauterization for cases of epilepsy and insanity (see also MacCurdy, 1905).

Amputation

Amputations were certainly performed in various early populations, although the reason for the amputation in individual cases is hard to determine. War injuries from axes or swords may produce amputated limbs. Accidental trauma must be very severe to produce amputation and is therefore a less likely cause. Surgical amputation of diseased or injured limbs should be kept in mind, even in the most primitive populations. Finally, deliberate amputation may have been performed by some cultures as a punitive measure or as a means of recording the number of prisoners.

The bone changes subsequent to amputation are not grossly apparent unless the individual survived the insult for at least a week. The initial change is vascular erosion of the bone end and adjacent shaft. Further osteoporosis may occur later as a result of disuse atrophy. In cases where the marrow cavity is opened, endosteal callus forms after about two weeks to narrow the exposed end of the cavity. After several weeks to months, the endosteal callus forms a cap over the medullary cavity. A complete bony cap is not always produced, but in all cases there is a rounding and smoothing of the stump. Finally, osteophytes may form at the bone end producing complications. In some cases the osteophytes may create a firmer stump by fusing the freed ends of the tibia and fibula or radius and ulna (See Barber, 1929; 1930; 1934).

Changes may occur in other bones not directly involved in the amputation as a result of disuse atrophy (See Chapter VIII). For example, an amputated humerus leads to ultimate reduction in size of the scapula and clavicle. In contrast, very little atrophy of these bones is found in amputations of the forearm (Todd and Barber, 1934).

Very few cases of amputation have been described in archaeological remains. One very typical case comes from Egypt during the 9th Dynasty or about 2100 B.C. (Brothwell and Møller-Christensen, 1963a). The specimen consists of a right ulna and a fragmentary radius which is fused to the remaining distal portion of the ulna. An estimated 10 cm of both radius and ulna have been lost through amputation. The bone ends are well rounded and partially capped by callus. Inflammatory pitting is still evident on both shafts suggesting that the loss of the limb was not of extremely long duration. Another amputation reported in Egypt comes from the Ptolemaic period (330-30 B.C.). The mummy of an elderly male has an amputation several cm above the wrist. An artificial limb complete with digits was added by the embalmers (Gray, 1966).

A possible amputation of the right hand is present in a forty-two-year-old male from the late Hopewell Culture of Illinois (500 B.C.-500 A.D.). The distal ends of the radius and ulna are fused with enlargement and rounding of the distal radius (Morse,

Figure 17. Possible amputation of the right hand in an adult male from the late Hopewell Culture of Illinois. The distal ends of radius and ulna are fused with enlargement and rounding of the distal radius. (Courtesy of Dr. Dan Morse, 1969.)

1969). A fragmentary fused left ulna and radius from the Mayan ruins of Altar de Sacrificios in Guatemala may be due to amputation (Saul, 1972). The specimen comes from an adult male of the Boca or Jimba phases dating 800-1000 A.D. A possible amputation of the distal third of an ulna is briefly reported in pre-Columbian skeletal remains from Mexico (Hurtado, 1970). A midshaft amputation of a left humerus has been reported in early historic skeletal remains from Hawaii (Snow, 1974). A pseudarthrosis would create similar bone changes and must also be considered a possible cause.

An interesting case of multiple amputation has been described in the remains of a forty- to fifty-year-old man from Britain in the 7th century A.D. The left forearm was severed just above the wrist, and the right leg was amputated 5 cm above the ankle. Osteophytes firmly unite the distal ends of both the radius and ulna and the tibia and fibula. The stump ends are rounded and relatively smooth. Inflammatory changes are slight and very localized. The presence of multiple amputation and remarkably good healing without infection suggest a deliberate surgical amputation of these limbs for therapeutic or punitive reasons (Brothwell and Møller-Christensen, 1963b).

Another very similar case of multiple amputation has been reported in a Slavonic skeleton from the 9th century A.D. of Czechoslovakia (Vyhnánek et al., 1966). In this skeleton the distal ends of the right forearm and left leg exhibit the bone changes typical of amputation.

It has recently been suggested that some cases described as amputations might be the result of pseudarthrosis or extensive bony callus formation following fracture (Stewart, 1974). For example, F. Wood Jones illustrated bony bridging between the radius and ulna in two ancient Egyptian specimens of healed fractures (Jones, 1910 plates 65 and 69; see also Hrdlička, 1899). However, the bony bridging occurred proximally leaving the distal ends of the bones intact and unchanged. Except for the fragmentary specimen described by Saul, the specimens illustrated by Jones bear little resemblance to the cases of amputation described in archaeological or clinical material. Pseudarthrosis following nonunion of fractures may produce bony fusion and rounding very similar to that occurring in amputation. However, it is significant that in none of the cases reported above were any bones distal to the affected limbs recovered.

DISLOCATIONS

Dislocations can only be diagnosed if the bone or bones remain dislocated for an extended length of time allowing bone modifications to take place. As long as the joint remains dislocated, the articular cartilage cannot be nourished by the synovial fluid. In some cases, the blood supply to the subchrondral bone is also restricted. The cartilage therefore begins to degenerate, and arthritic changes may be produced in areas of bone-on-bone friction. Atrophic changes occur if the bones are immobilized by the dislocation.

Dislocations occur most frequently in adults after ossification of the epiphyses and before the onset of senility. In younger individuals a violent force causes epiphyseal separation rather than joint dislocation. In elderly people, a similar force results in a fracture of the brittle bone.

Dislocations of the fingers and toes can easily be reduced and therefore bone changes are rare. Dislocations of the vertebrae usually cause immediate death by severing the spinal cord. Dislocations of the shoulder are very common, but most are easily reduced. However, when the humeral head remains displaced from the glenoid fossa for several weeks, bone changes are produced. The glenoid fossa fills with granulation tissue and the head of the humerus is bound by scar tissue. Atrophy of the humerus subsequently takes place. In some cases, a secondary fossa forms on the scapular wing for the displaced humerus (see Fig. 18).

Dislocations of the hip are commonly the only type of dislocation found in ancient skeletal material. These dislocations may be either congenital or traumatic. Congenital dislocation of the hip varies greatly in incidence among different populations today. In general, girls are affected five times more often

Figure 18. Left scapula showing the bone changes of a dislocated shoulder. A new articulation with the humerus has formed on the anterior surface of the scapular wing near the axillary border. From an early Eskimo from Morton Bay, Alaska. (USNM 346,205)

than boys. Prenatal relaxation of the joint capsule at the hip appears to be the basic lesion. Limitation of abduction by restrictive clothing or by actual binding of the legs in adduction by Lapps and certain American Indian tribes may be responsible for the later onsets of permanent dislocations in these groups (Caffey, 1972). Congenital dislocations are usually unilateral and characterized by an underdeveloped, shallow acetabulum. With continued dislocation, the head of the femur becomes flattened and later mushroom-shaped. The neck of the femur may assume a more anterior position in relation to the lower end of the femur.

In advanced cases, a shallow secondary acetabulum may develop on the wing of the ilium (Shands, 1957).

Traumatic hip dislocation may closely resemble congenital hip dislocation except the primary acetabulum will be fully developed. When a hip has been dislocated posteriorly for several years, a new acetabulum forms on the posterior aspect of the ilium as a result of functional adaptation. Such an individual walks with a peculiar limp because the new joint has formed in a plane more posterior than normal and muscle relations are changed.

Figure 19. Dislocation of the right hip. The acetabulum is partially filled with irregular bone growth and a new articulation for the femur has formed superiorly. An early Eskimo from Kodiak Island, Alaska. (USNM 372,897)

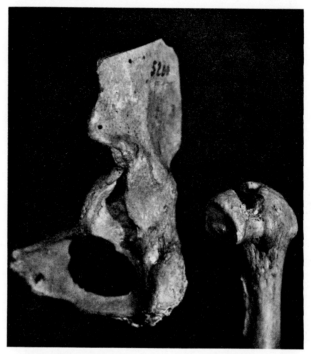

Figure 20. Dislocation of the left hip of many years duration. The old acetabulum is partially obliterated by the new articulation. (WAM 5200)

Paleopathology

Several specimens of congenital or traumatic dislocation have been reported, all affecting the hip joint. Congenital dislocations of the hip have been found in Neolithic France and the Early Iron Age of Greece (Pales, 1930; Angel, 1946). Several cases come from Anglo-Saxon and Medieval times in Britain (Brothwell, 1961; Wells, 1964). A bilateral hip dislocation has been described in a fifteen-year-old girl from early burials in Hawaii (Snow, 1974). Another bilateral hip dislocation comes from Late Archaic times in Illinois (around 2000 B.C.). The left innominate is still preserved showing the shallow primary acetabulum in this ten-year-old child (Morse, 1969).

An unusual case, probably secondary to trauma, comes from the Bronze Age of Britain (Ashley Montagu, 1931). The femoral head has been displaced downwards onto the acetabular notch with the formation of a new socket in this position. The individual is an adult male, and the dislocation certainly occurred many years before death. The femoral head is freely movable within the new socket, but osteoarthritic lipping prevents the withdrawal of the femur in the dried bone specimen.

TRANSVERSE LINES (HARRIS'S LINES, GROWTH ARREST LINES)

In paleopathology one is often limited to a picture of the individual at the moment of death, with little idea about the varying health conditions of this individual as he progressed through life. However, the study of the transverse lines in the long bones from such an individual may provide some important evidence concerning his health history and that of the entire skeletal population in general.

General Appearance

Transverse lines must be visualized roentgenographically or by sectioning the bone. They form symmetrically throughout the skelton. The term "transverse" actually applies only to the long and short tubular bones with transverse growth plates. In the round and flat bones such as the calcaneus, ilium, and ischium, the lines follow the contour of the epiphyseal growth plate. The transverse lines are most common in the distal tibia followed in decreasing frequency by the proximal tibia, distal femur, distal radius, and metacarpals. The bones of the pelvis and scapula are rarely demonstrably affected (Garn et al., 1968).

Transverse lines range in thickness from a millimeter to more than a centimeter. They are thickest at the ends of the bones which grow more rapidly such as the sternal ends of the ribs and both ends of the femur and tibia. Due to remodelling, the older lines deeper in the shaft are thinner and less distinct.

When macerated bones are sectioned and examined, the transverse lines are shown to be layers or lattices of horizontally arranged bone strands corresponding to remnants of the bony

Figure 21. Transverse lines in the distal femur of a juvenile from the Early Horizon of California (5000-2000 B.C.). (A.) Roentgenogram showing 23 transverse lines. (B.) Sectioned femur showing the latticework of bone trabeculae corresponding to the transverse lines of the roentgenogram. (Courtesy of Dr. Henry McHenry, 1968.)

plates formed in early infancy and childhood (Park, 1964; McHenry, 1968). The trabeculae are much thicker peripherally than centrally. The lines deeper in the shaft (those formed earlier) are more transversely oriented than the more recently formed lines near the ends of the bone. These obliquely oriented lines usually do not completely traverse the diameter of the bone. When sectioned they are shown to be strands of bone within the marrow cavity and are more numerous at the medial surface of the distal femur.

Figure 22. Sectioned femur of an adult Archaic Indian from Kentucky. The latticework of bone trabeculae corresponds to a transverse line in an anteroposterior roentgenogram. (Courtesy of Dr. Stanley Garn et al., 1968.)

Mechanism of Formation

The mechanism by which transverse lines form has been thoroughly worked out through experimental studies in young rats (Park 1954; 1964; Park and Richter, 1953). Dietary deficiency or disease primarily affects the epiphyseal cartilage plate which becomes reduced in thickness. The immature cartilage cells of this plate are resistant to penetration by the capillaries and osteoblasts, and therefore, longitudinal growth is arrested. The thin layer of mature cartilage cells immediately beneath the plate are replaced by numerous osteoblasts which form a thin layer of bone termed the *primary stratum*.

When normal nutrition is restored or when the disease passes, the osteoblasts in the primary stratum recover immediately and begin laying down bone on the inferior surface of the primary stratum because the superior surface remains impenetrable by immature cartilage cells. The trabeculae of the thickened stratum are arranged horizontally rather than vertically, and this forms the transverse line seen in the roentgenogram. After four to nine days, the cartilage cells of the epiphyseal plate are ready for osteoblastic invasion, and the bony trabeculae replacing this cartilage are arranged in the normal vertical direction.

Thus, there are two factors involved in the formation of a transverse line. The first is growth arrest by disease or nutritional deficiency, resulting in the production of the primary stratum and its horizontal arrangement. The second is the recovery factor responsible for the thickening of the primary stratum so that it is visible as a transverse line.

Etiology

Transverse lines were first intensely studied by Asada in 1924 and Harris in 1933, and their possibilities in paleopathology have just recently been applied. Transverse lines represent periods of renewed or increased growth following inhibited growth of the bone. Many types of diseases and nutritional deficiency may produce transverse lines in the growing bone. Such childhood diseases as measles, whooping cough, influenza, laryngitis, chicken pox, and pneumonia are capable of producing these lines (Harris, 1933). Transverse lines will form in experimentally starved rats, rabbits, and dogs (Asada, 1924; Harris, 1933; Acheson, 1959). Rats deprived of vitamin A also develop transverse lines (Wolbach, 1947). Protein deficiency produces the lines in pigs (Platt and Stewart, 1962). Humans suffering from kwashiorkor, a protein-calorie deficiency, will also develop transverse lines (Jones and Dean, 1959). Thus, transverse lines are formed in the growing bone whenever the individual is subjected to severe stress such as starvation or disease. The lines may even form in prenatal infants if the mother is suffering from malnutrition. They may also form immediately after birth when the infant shifts from placental oxygenation and nutrition to that of the lungs and alimentary tract (Sontag and Harris, 1938).

Transverse Lines in Infancy

Transverse lines are not produced by every disease or nutritional stress during the growing period. In a longitudinal study of several hundred children using medical records and tibial roentgenograms, the transverse lines followed a disease episode approximately 25 percent of the time (Gindhart, 1969). Transverse lines may develop and then vanish over a period of months so that some of the lines may have been missed by the six-month examination intervals. In many cases, however, bone growth continues even at the expense of previously formed bone and transverse lines do not develop (Garn et al., 1968).

Of the 201 subjects followed from one month of age through early adulthood, all but two of the subjects developed transverse lines. New lines appeared with greatest frequency in girls at 2.5 years and boys at 2.0 years, coinciding with peaks in the occurrence of the common childhood diseases. Boys had approximately twice as many lines per individual as the girls in all age groups. This latter finding may support the belief that boys are more vulnerable to environmental insults or may reflect the greater rate of subperiosteal apposition and linear growth of bone in boys (Garn et al., 1968; Gindhart, 1969).

Transverse lines are not left permanently in the bone but are gradually removed by endosteal and subperiosteal cortical re-

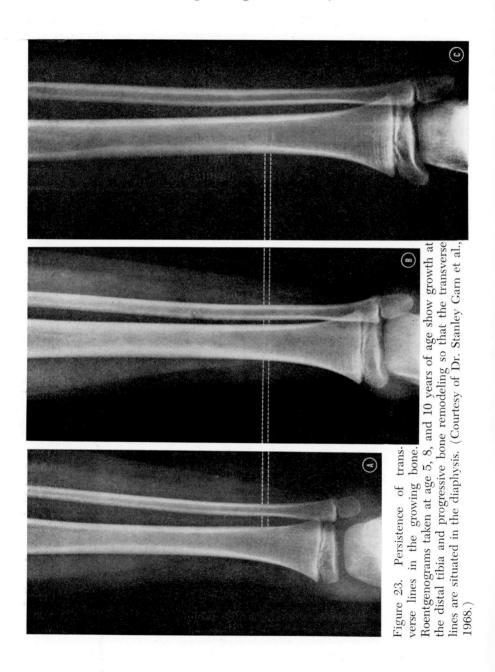

Figure 23. Persistence of transverse lines in the growing bone. Roentgenograms taken at age 5, 8, and 10 years of age show growth at the distal tibia and progressive bone remodeling so that the transverse lines are situated in the diaphysis. (Courtesy of Dr. Stanley Garn et al., 1968.)

modelling (Garn and Schwager, 1967). The lines appear to endure longer in females, but almost all lines disappear within ten years (Gindhart, 1969). The lines that do persist into adulthood are those that were laid down very early in life (Garn and Schwager, 1967).

Transverse Lines in Adults

Although nearly 100 percent of the children examined in the longitudinal study exhibited transverse lines, only 24.3 percent of 136 adults aged 25 to 50 years exhibited the lines (Garn and Schwager, 1967). Females were affected twice as often as males probably reflecting the increased bone remodelling in males. Between fifty and sixty years of age, the rate of loss of cancellous bone and resorption at the endosteal surface are at a maximum and this continues until death. Thus, an incidence of only 11.8 percent with transverse lines was found in 110 adults aged 51 to 86 years. It is significant that the width of the medullary cavity is greater in older subjects who do not exhibit transverse lines (see also Garn et al., 1967).

Transverse lines were first described as "growth arrest lines" implying that individuals with several of these lines may have a shorter stature than normal. However, the term "growth arrest" is correct only so far as the cartilage component is concerned (Park, 1964). The bone stratum representing the transverse line is actually the result of recovery from the disease or nutritional stress and signifies renewed or even increased growth. In support of this, no significant difference in final adult stature was noted between subjects with many transverse lines and those with few lines (Gindhart, 1969).

Transverse Lines in Paleopathology

The technique of transverse line analysis has only recently been applied to ancient human populations. Such studies can provide important information on the amount of morbidity suffered by individuals during their growing years. The recovery factor discussed above is important when interpreting the results of transverse line studies in skeletal populations. A transverse line will not form unless the individual recovers from the illness or obtains adequate nutrition. Thus, a population chronically malnourished will not exhibit as many transverse lines as a population with seasonal periods of starvation and good nutrition. This may explain the large number of lines per individual in California Indians from the Early Horizon (5000-2000 B.C.). The thirty-four adult femurs examined showed an average of over eleven lines per femur giving strong support to an existence of seasonal starvation during the winter and early spring followed by several months of good nutrition during the salmon runs (McHenry, 1968).

Transverse line studies may provide valuable information in comparative studies of health conditions among various skeletal populations. These studies are particularly useful when comparing populations from the same locality but from different time periods, especially if their modes of nutrition are thought to differ. For example, the fewer number of transverse lines per femur found in California Indians from the Late Horizon as compared to the Early Horizon supports other evidence that this group subsisted on stored acorns during the winter months (Baumhoff, 1963 cited by McHenry, 1968). In a similar study of Saxon, Medieval, and Bronze Age populations in England, the Saxons had a significantly higher number of lines per distal tibia than the other groups (Wells, 1961; 1967). Ancient Peruvian skeletal remains from the coastal areas had a higher frequency of transverse lines than did individuals from mountain cultures, perhaps due to a disease such as malaria which is present along the coast but not in the mountains (Allison et al., 1974).

By measuring the distance of a transverse line from the end of the bone, one may also determine at what approximate age these lines were formed and thereby compare different populations according to the amount of morbidity during given age groups (see Garn et al., 1968). Morbidity measurements may also be determined in a similar manner to compare different sexes within the same age group and population. For example, in a Middle Saxon population, girls were found to have a higher percentage of morbidity (more transverse lines) occurring in the

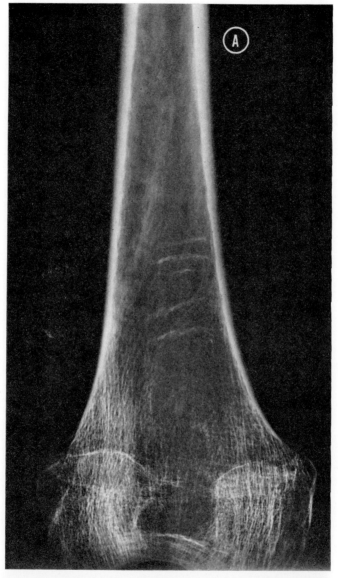

Figure 24. Transverse lines in the distal femora of Archaic Indians from Kentucky. (A.) Most of the lines are obliquely oriented and are grouped disproportionately at the medial surface, reflecting the regional differences in bone remodeling.

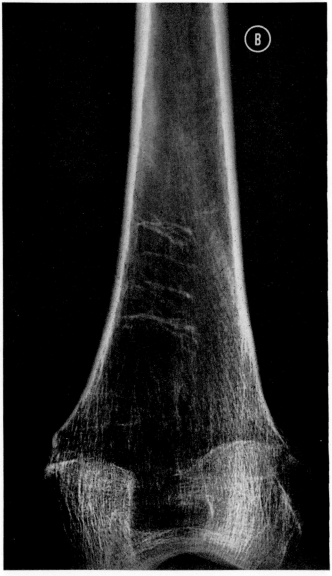

Figure 24B. Figures 24A and 24B show good correspondence between the left leg (A.) and the right leg (B.) of a single Indian.

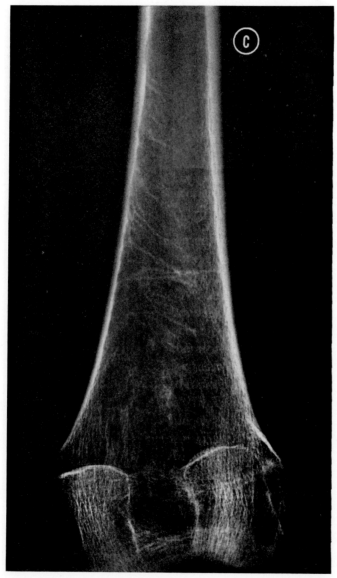

Figure 24C. Transverse lines in left leg of Archaic Indian from Kentucky.

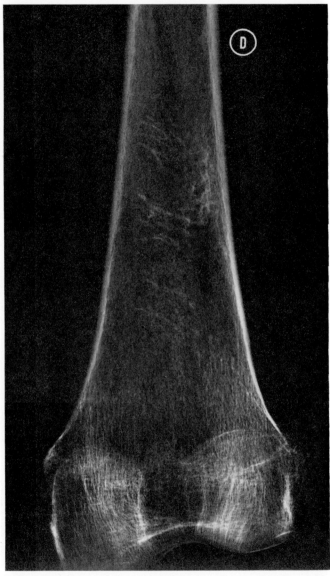

Figure 24D. Transverse lines in right leg of Archaic Indian from Kentucky. (Courtesy of Dr. Stanley Garn et al., 1968.)

early years of life. This may be explained in terms of the social structure of the Saxons with emphasis placed on the male warriors and thus the infant and young boys had a higher status than the young girls (Wells, 1967).

There are of course limitations and dangers in the application of transverse line analysis to skeletal populations. At present it is impossible to determine whether a particular transverse line was produced by a disease of short duration such as measles or the result of chronic malnutrition. Furthermore, comparisons between skeletal populations must take into account age-associated loss of transverse lines. Otherwise, inaccurate interpretations may result from comparison of two series widely differing in age composition. Nevertheless, transverse line analysis provides a new and promising means of studying the skeletal biology of early human populations.

TRAUMA BIBLIOGRAPHY

Acheson, R. M.: "Effects of starvation, septicemia, and chronic illness on the growth cartilage plate and metaphysis of the immature rat," *J. Anat.*, 93:123-130, 1959.

Aegerter, E. and J. Kirkpatrick: *Orthopedic Diseases*, W. B. Saunders, Philadelphia, 2nd ed., 1968.

Allison, M. J., D. Mendoza, and A. Pezzia: "A radiographic approach to childhood illness in precolumbian inhabitants of southern Peru," *Amer. J. Phys. Anthrop.*, 40:409-415, 1974.

Anderson, J. E.: "The people of Fairty," *Natl. Mus. Canada Bull.*, #193, 1963.

Angel, L.: "Skeletal change in ancient Greece," *Amer. J. Phys. Anthrop.*, 4:69-97, 1946.

Asada, T.: "Ueber die Entstehung und pathologische Bedeutung der im Röntgenbild des Röhrenknochens am Diaphysenende zum Vorschein Kommenden 'parallelen Querlinienbildung,'" *Mitteil. Med-Fakult. Kyushu Univ.*, 9:43-95, 1924.

Ashley Montagu, M. F.: "Dislocation of the femur upon the acetabular notch in a pre-Roman Briton," *J. Bone Joint Surg.*, 13A:29-32, 1931.

Barber, C. G.: "Immediate and eventual features of healing in amputated bone," *Ann. Surg.*, 90:985-992, 1929.

———: "The detailed changes characteristic of healing bone in amputation stumps," *J. Bone Joint Surg.*, 12:353-359, 1930.

———: "Ultimate anatomical modification in amputation stumps," *J. Bone Joint Surg., 16*:394-400, 1934.

Baumhoff, M. A.: "Ecological determinants of aboriginal California populations," *Univ. Calif. Publ. Amer. Archeol. Ethnol., 49*:155-236, 1963.

Blast, T. H., W. E. Sullivan, and F. D. Geist: "The repair of bone," *Anat. Rec., 31*:255-280, 1925.

Brothwell, D. R.: "The palaeopathology of early British man," *J. Roy. Anthrop. Inst., 91*:318-344, 1961.

Brothwell, D. R. and V. Møller-Christensen: "Medico-historical aspects of a very early case of mutilation," *Danish Med. Bull, 10*:21-27, 1963a.

———: "A possible case of amputation dated to c. 2000 B.C.," *Man, 63*:192-194, 1963b.

Burton, R.: "Notes on scalping," *Anthrop. Rev., 2*:49-52, 1864.

Caffey, J.: *Pediatric X-Ray Diagnosis*, vol. II, Yearbook Medical Publ., Chicago, 6th ed., 1972.

Collins, D. H.: *Pathology of Bone*, Butterworth, London, 1966.

DePalma, A. F.: *The Management of Fractures and Dislocations*, W. B. Saunders, Philadelphia, 1970.

Friederici, G.: "Scalping in America," *Annual Report of the Smithsonian Institution* for 1906, pp. 423-438, 1907.

Garn, S. M., C. G. Rohman, B. Wagner, and W. Ascoli: "Continuing bone growth throughout life," *Amer. J. Phys. Anthrop., 27*:375-378, 1967.

Garn, S. M., F. N. Silverman, F. P. Hertzog, and C. G. Rohman: "Lines and bands of increased density. Their implication to growth and development," *Med. Radiog. and Photog., 44*:58-88, 1968.

Gill, E.: "Examination of Harris's lines in recent and fossil Australian aboriginal bones," *Curr. Anthrop., 9*:215, 1968.

Gindhart, P. S.: "The frequency of appearance of transverse lines in the tibia in relation to childhood illnesses," *Amer. J. Phys. Anthrop., 31*:17-22, 1969.

Goldstein, M. S.: "Skeletal pathology of early Indians in Texas," *Amer. J. Phys. Anthrop., 15*:299-312, 1957.

Gray, P. H. K.: "A radiographic skeletal survey of ancient Egyptian mummies," *Excerpta Medica Internat. Congress Series, 120*:35-38, 1966.

———: "Radiography of ancient Egyptian mummies," *Med. Radiog. and Photog., 43*:34-44, 1967.

Hamperl, H.: "The osteological consequences of scalping," in D. R. Brothwell and A. T. Sandison (ed.): *Diseases in Antiquity*, Thomas, Springfield, 1967.

Hamperl, H. and W. S. Laughlin: "Osteological consequences of scalping," *Hum. Biol., 31*:80-89, 1959.

Harris, H. A.: *Bone Growth in Health and Disease,* Oxford Univ. Press, London, 1933.

Hooton, E. A.: *The Indians of Pecos Pueblo,* Yale Univ. Press, New Haven, 1930.

Hoyme, L. E. and W. M. Bass: "Human skeletal remains from the Tollifero (Ha 6) and Clarksville (Mc 14) sites, John H. Kerr Reservoir Basin, Virginia," *Bureau of Amer. Ethnol. Bull., 182*:329-400, 1962.

Hrdlička, A.: "A new joint-formation," *Amer. Anthrop., 1*:550-551, 1899.

Hurtado, E. D.: "Pre-Hispanic osteopathology," in R. Wauchope (ed.): *Handbook of Middle American Indians,* vol. 9, Univ. of Texas Press, Austin, 1970.

Janssens, P. A.: *Palaeopathology,* Humanities Press, New York, 1970.

Jones, F. W.: "Pathological report," *Archaeological Survey of Nubia Bull.,* #2, plates 65 and 69, 1910.

Jones, P. R. M. and R. F. A. Dean: "The effects of kwashiorkor on the development of the bones of the knee," *Pediatrics, 54*:176-184, 1959.

Lisowski, F. P.: "Prehistoric and early historic trepanation," in D. R. Brothwell and A. T. Sandison (ed.): *Diseases in Antiquity,* Thomas, Springfield, 1967.

MacCurdy, G. G.: "Prehistoric surgery—a neolithic survival," *Amer. Anthrop., 7*:17-23, 1905.

Manouvrier, L.: "Le T sincipital: curieuse mutilation crânienne néolithique," *Bull. Soc. Anthrop. Paris, 6*:357-360, 1895.

Margetts, E. L.: "Trepanation of the skull by medicine-men of primitive cultures, with particular reference to present-day native East African practice," in D. R. Brothwell and A. T. Sandison (ed.): *Diseases in Antiquity,* Thomas, Springfield, 1967.

Marshall, W. A.: "Problems in relating radiopaque transverse lines in the radius to the occurrence of disease," *Symp. Soc. Hum. Biol., 8*:245-261, 1966.

McHenry, H.: "Transverse lines in long bones of prehistoric California Indians," *Amer. J. Phys. Anthrop., 29*:1-18, 1968.

Møller-Christensen, V.: "Skelefrund fra St. Jorgensbjaerg Kirke, Roskilde," *Med. Forum, 14*:97-111, 1961.

Moodie, R. L.: "A variant of the sincipital T in Peru," *Amer. J. Phys. Anthrop., 4*:219-222, 1921.

Mooney, J.: "Myths of the Cherokee," *Nineteenth Annual Report of the Bureau of American Ethnology,* pp. 208-209, 1900.

———: "Scalping," *Bureau of Amer. Ethnology Bull., 30*:482-483, part 2, 1910.

Morse, D.: *Ancient Disease in the Midwest,* Illinois State Mus. Rep. Invest., #15, 1969.

———: "Pathology and abnormalities of the Hampson skeletal collection," in D. F. Morse (ed).: *Nodena,* Arkansas Archeological Survey Publ. Research Series, #4, 1973.

Nadeau, G.: "Indian scalping technique in different tribes," *Ciba Symposia,* 5:1677-1681, 1944.

Neumann, G. K.: "Evidence for the antiquity of scalping from central Illinois," *Amer. Antiq.,* 5:287-289, 1940.

Oakley, K. P., W. Brooke, A. R. Akester, and D. R. Brothwell: "Contributions on trepanning or trephination in ancient and modern times," *Man,* 59:287-289, 1959.

Pales, L.: *Paléopathologie et Pathologie Comparative,* Masson, Paris, 1930.

Park, E. A.: "Bone growth in health and disease," *Arch. Dis. Child., 29:* 269-291, 1954.

———: "The imprinting of nutritional disturbances on the growing bone," *Pediatrics, 33:*815-862, 1964.

Park, E. A. and C. P. Richter: "Transverse lines in bone, the mechanism of their development," *Bull. Johns Hopkins Hosp.,* 41:364-388, 1953.

Piggott, S.: "A trepanned skull of the Beaker period from Dorset and the practice of trepanning in prehistoric Europe," *Proc. Prehist. Soc.,* 6:112-132, 1940.

Platt, B. S. and R. J. C. Stewart: "Transverse trabeculae and osteoporosis in bones in experimental protein-calorie deficiency," *Brit. J. Nutrition, 16:*483-495, 1962.

Reese, H. H.: "The history of scalping and its clinical aspects," *Yearbook of Neurology, Psychiatry and Endocrinology,* pp. 3-19, 1940.

Saul, F. P.: "The human skeletal remains of Altar de Sacrificios," *Papers Peabody Mus. Archaeol. Ethnol.,* vol. 63, #2, 1972.

Shands, A. R.: *Handbook of Orthopaedic Surgery,* C. V. Mosby, St. Louis, 5th ed., 1957.

Snow, C. E.: "Anthropological studies at Moundsville," *Alabama Mus. Nat. Hist. Mus. Paper, 15:*55-57, 1941.

———: "Additional evidence of scalping," *Amer. Antiq.,* 7:398-400, 1942.

———: "Indian Knoll skeletons of site Ohc, Ohio County, Kentucky," *Univ. Ky. Rep. Arch.,* 4:371-554, 1948.

———: *Early Hawaiians, An Initial Study of Skeletal Remains from Mokapu, Oahu,* Univ. Ky. Press, Lexington, 1974.

Sontag, L. W. and L. M. Harris: "Evidence of disturbed prenatal and neonatal growth in bones of infants aged one month, II. Contributing factors," *Amer. J. Dis. Child,* 56:1248-1255, 1938.

Stewart, T. D.: "Significance of osteitis in ancient Peruvina trephining," *Bull. Hist. Med.,* 30:293-320, 1956.

———: "Stone age skull surgery," *Annual Report of the Board of Regents of the Smithsonian Instiution* for 1957, pp. 469-491, 1958.

――――: "Nonunion of fractures in antiquity, with descriptions of five cases from the New World involving the forearm," *Bull. N.Y. Acad. Med., 50*:875-891, 1974.

Sudhoff, K.: "Le T-sincipital Neolithique," *Bull. Soc. Franc. D'Hist. Med., 7*:175-179, 1908.

Todd, T. W. and C. G. Barber: "The extent of skeletal change after amputation," *J. Bone Joint Surg., 16*:53-64, 1934.

Todd, T. W. and D. H. Iler: "The phenomena of early stages in bone repair," *Ann. Surg., 88*:715-736, 1927.

Vyhnánek, L., M. Stloukal, and J. Kolář: "A case of amputation in an old Slavonic skeleton from the 9th century A.D." *Cas. Lek. Cesk., 105*: 296-297, 1966.

Watson-Jones, R.: *Fractures and Joint Injuries,* Vol. I., Williams and Wilkins, Baltimore, 1952.

Wells, C.: "A new approach to ancient disease," *Discovery, 22*:526-531, 1961.

――――: "Hip disease in ancient man. Report of three cases," *J. Bone Joint Surg., 45B*:790-791, 1963.

――――: "A new approach to palaeopathology: Harris's lines," in D. R. Brothwell and A. T. Sandison (ed.): *Diseases in Antiquity,* Thomas, Springfield, 1967.

Wilson, T.: "Arrow wounds," *Amer. Anthrop., 3*:513-531, 1901.

Wolbach, J. B.: "Vitamin A deficiency and excess in relation to skeletal growth," *J. Bone Joint Surg., 29*:171-192, 1947.

Zaborowski: "Le T sincipital. Mutilation des crânes néolithiques, observée en Asie Centrale," *Bull. Soc. Anthrop. Paris, 8*:501-503, 1897.

Chapter

NONSPECIFIC INFECTION: PYOGENIC OSTEOMYELITIS

OSTEOMYELITIS IS A general term describing infection of the bone by various kinds of microorganisms. Strictly defined, it refers to inflammation of the marrow cavity, but in general use, osteomyelitis also includes periostitis (inflammation of the periosteum) and osteitis (inflammation of the bone). Microscopic examination supports this grouping because localization of the inflammatory process rarely occurs, there being some degree of involvement of all these anatomical components.

Under normal circumstances bone is not exposed to the external environment, and pathogens must reach the bone indirectly through the bloodstream. Infection of the bone or bones in this manner is known as *hematogenous osteomyelitis.* The primary focus of infection may have been a furuncle, skin lesion, or subcutaneous abcess. In certain cases the bone may become infected directly after being exposed by severe trauma such as a compound fracture or burn. These primary bone infections may in turn create secondary foci by hematogenous spread to other bones. Depending upon the virulence of the microorganism and the resistance of the host, osteomyelitis may be an acute or chronic infection and the gross pathology will differ accordingly.

Osteomyelitis is almost always caused by infection with a microorganism capable of producing pus, hence the name pyogenic or *suppurative osteomyelitis.* Numerous reports confirm that almost 90 percent of cases of pyogenic osteomyelitis are caused by *Staphylococcus aureus.* The remaining 10 percent are due to pus-producing organisms, such as streptococci, pneumococci, meningococci, and occasionally salmonella or colon bacilli (see Knaggs, 1926; Wilensky, 1934; and Waldvogel et al., 1971).

Thirty-eight cases of hematogenous osteomyelitis and pyogenic osteomyelitis by direct infection were examined at the Smithsonian Institution, the Army Medical Museum (now part of the Armed Forces Institute of Pathology), the Peabody Museum of Archaeology and Ethnology, and the Warren Anatomical Museum at Harvard Medical School. This data along with the clinical findings of various workers will be used in this chapter to describe acute and chronic osteomyelitis according to its pathogenesis, gross morphology, and site of involvement in the skeleton.

ACUTE OSTEOMYELITIS

Acute osteomyelitis is a destructive invasion of the bone, periosteum, and marrow by pyogenic bacteria. The mode of infection may be hematogenous or by direct infection of bone.

Acute Hematogenous Osteomyelitis

Acute hematogenous osteomyelitis occurs most frequently between the ages of three and fifteen years, which is the period of most active bone growth. In most series, 80 to 90 percent of cases are under fifteen years of age (Trendel, 1904; Speed, 1922; and Winters and Cahen, 1960). Occasionally, hematogenous osteomyelitis may occur in infants below the age of one year. In 63 percent of these infantile cases, the causal organism is *Streptococcus* rather than *Staphylococcus* (Green and Shannon, 1932). Acute hematogenous osteomyelitis is rare among adults, who are more likely to have the chronic form of infection (Zadek, 1938).

In acute hematogenous osteomyelitis among children, males predominate over females by a ratio of 3 or 4 to 1. This sex differential is smaller among infants and greater among adults (Jaffe, 1972). Increased trauma and soft tissue infection is responsible for the greater number of cases in males. Minor trauma causes a minute hemorrhage and microthrombi which lower the resistance to bacteria of the region involved.

Location of Lesions

Any part of the skeleton may become a focus of infection, but the major location is in a long bone metaphysis (see Fig. 25). There are several reasons for the increased susceptibility of the metaphyses to infection. Between the ages of three and fifteen years when most cases occur, the metaphyseal capillaries adjacent to the growth plate turn down in acute loops and reach a system of large sinusoidal veins which are connected with the venous network of the medullary cavity (see Fig. 26). This region of slow blood flow provides an ideal environment for the pyogenic bacteria, particularly *Staphylococcus aureus* (Trueta, 1957, 1959). Moreover, the afferent loop of the metaphyseal capillary lacks phagocytic lining cells, and the efferent loop contains functionally inactive phagocytic cells incapable of resisting a septicemia (Hobo, 1921-22). The capillary loops adjacent to the epiphyseal growth plate are nonanastomosing branches of the nutrient artery, and any obstruction by bacterial growth results in small areas of avascular necrosis (Morgan, 1959).

The incidence of osteomyelitis in a given long bone metaphysis is in direct proportion to the rate of growth at that metaphysis and to the size of the bone (Phemister, 1928b; Putschar, 1960). The more rapidly growing end of a bone is more susceptible to the infection. Thus, the distal end of the femur is affected more often than the proximal end. The proximal ends of the tibia and humerus are involved more frequently than the distal ends of these same bones. Table II lists the major sites of acute hematogenous osteomyelitis according to various authors. The long bones of the lower extremities are involved much more frequently than those of the arms. In adults, the shafts may be affected as often as the metaphyses.

As a result of the hematogenous spread of infection, multiple bone lesions occur in 5 to 25 percent of all cases of acute osteomyelitis (Green et al., 1956; Winters and Cahen, 1960). Two bones are involved in most of these cases, but as many

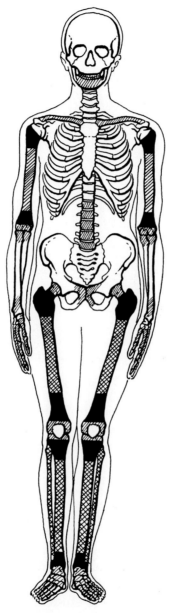

Figure 25. Skeletal distribution of acute hematogenous osteomyelitis. The solid black areas indicate the most frequent sites; checked areas mark common sites; and diagonal lines indicate occasional sites.

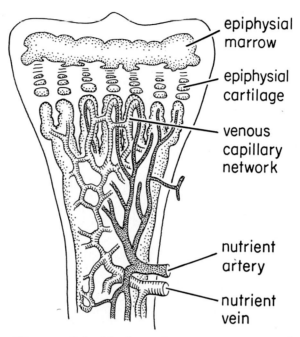

Figure 26. Diagram of the blood supply in a growing bone showing the capillary loops adjacent to the epiphyseal growth plate. (Modified from Hobo, 1921-22.)

as five or even ten bones may become infected (Trendel, 1904; Wilensky, 1934). The multiple foci may all be due to spread from the soft tissue focus or may arise from a secondary site of infection in a bone.

Pathogenesis

The pathogenesis of a typical case of acute hematogenous osteomyelitis is as follows: Pyogenic microorganisms enter a metaphysis through a nutrient artery and then set up a focus of infection in the metaphyseal sinusoidal veins. The epiphyseal growth plate usually prevents extension of the infection into the epiphysis and joint space. Instead, the infection spreads laterally and perforates the thin metaphyseal cortex. The pus lifts the periosteum which is only loosely anchored in the growing bone except where it is firmly adherent at the epiphyseal

TABLE II
DISTRIBUTION OF ACUTE HEMATOGENOUS OSTEOMYELITIS IN THE SKELETON

	Trendel 1904	Wilensky 1934	Gilmour 1962	Huebler 1927-28	Shandling 1960	Pyrah and Pain 1933	Hilpert 1914	Green et al. 1956	Winters and Cahen 1960
femur	39.7	23.1	36.1	27.1	33.4	36.6	34.5	18.8	43.0
tibia	30.2	15.3	22.4	34.1	35.0	35.1	36.9	18.2	29.2
humerus	8.0	9.3	7.0	6.9	7.8	6.4	11.3	13.0	11.1
radius	3.4	3.2	1.8	1.3	2.1	5.3	3.3	3.2	5.5
fibula	2.8	3.5	7.7	4.3	6.3	10.0	4.4	6.5	5.5
ulna	2.7	0.9	1.2	0.9	1.2	1.1	3.0	6.4	1.4
mandible	1.8	5.2	—	2.2	—	0.4	1.2	—	—
metatarsals	1.2	1.4	1.2	0.0	0.9	0.8	0.0	3.9	—
calcaneus	1.2	2.8	5.0	0.0	2.8	1.5	1.2	2.6	—
pubis	1.1	3.5	3.3	0.0	2.4	—	0.6	1.3	—
ilium	1.1	0.6	2.5	5.6	0.6	0.8	1.2	4.5	—
cranium	1.0	11.8	3.3	0.0	0.9	—	0.0	3.9	1.4
vertebrae	0.6	2.8	2.2	0.0	—	0.4	0.6	—	—
phalanges	0.5	8.9	0.6	0.0	0.3	0.4	0.0	6.5	—
metacarpals	0.5	0.3	0.3	6.9	—	—	1.2	—	—
patella	0.1	1.1	1.2	0.0	—	—	0.6	—	—
other	4.1	6.3	4.2	0.0	6.3	1.2	0.0	11.2	2.9
TOTAL %	100.0	100.0	100.0	100.0	100.0	100.0	100.0	100.0	100.0
TOTAL CASES	1279	346	328	316	300	262	178	99	66

line (Starr, 1922). A subperiosteal abcess is formed which raises the periosteum over large areas of the cortex (see Fig. 27).

The infection provokes an acute suppurative inflammation. The bacteria, drop in pH, proteolytic enzymes from leukocytes, internal pressure produced by localized edema, and ischemia all contribute to the necrosis of bone tissue and absorption of trabeculae (Waldvogel et al., 1971). Extension of the infection to the surrounding bone, particularly the medullary cavity, occurs through the Haversian and Volkmann canals which have been enlarged by the bone absorption process.

The elevation of the highly vascular periosteum and the occlusion of the Volkmann and Haversian systems interfere with the blood supply to the cortex, and this results in the death of those parts of the shaft which are deprived of their blood supply. The dead areas are separated from the living bone by the action of granulation. These cortical and sometimes cancellous portions of dead bone are called *sequestra* and vary in size from small fragments to the entire shaft of the bone (see Fig. 28). Deeper areas of the bone may also sequestrate when their blood supply is cut off by thrombosis in branches of the nutrient artery or compression of these vessels by inflammatory exudate (Larsen, 1938).

The elevation of the periosteum and infection of the subperiosteal space induces exhuberant new bone formation which eventually encloses the diseased portion of the shaft more or less completely. This subperiosteal new bone mass is called an *involucrum* and may be perforated by one to several crater-like openings or cloacae through which pus drains into the soft tissues (see Fig. 29). If the infection subsides and the sequestra absorbed or extruded through the draining sinuses, the massive involucrum rapidly narrows down and is gradually replaced by mature lamellar bone. With further healing the trabeculae and Haversian systems again align themselves in the lines of functional stress.

Gross Morphology and Roentgenographic Appearance

The gross morphology and roentgenographic appearance of acute hematogenous osteomyelitis reflect the bone destruction,

Nonspecific Infection: Pyogenic Osteomyelitis 67

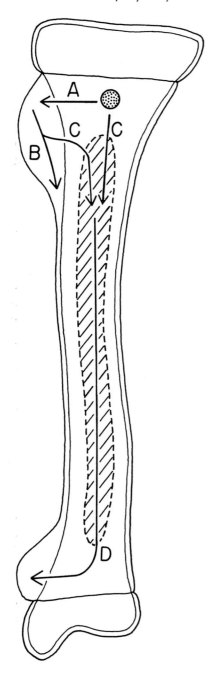

Figure 27. Route of infection in hematogenous osteomyelitis: (A.) Lateral spread of infection from primary focus in metaphysis and perforation of the thin metaphyseal cortex. (B.) Formation of subperiosteal abscess as pus raises the periosteum from the cortex. (C.) Infection reaches medullary cavity from the metaphyseal focus or from the subperiosteal abscess via nutrient artery canal. (D.) Spread of infection down medullary cavity to the other metaphysis and formation of a late subperiosteal abscess. (Modified from Knaggs, 1926.)

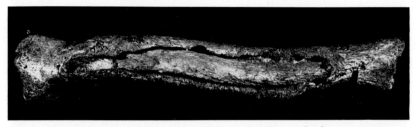

Figure 28. Pyogenic osteomyelitis of the right tibia. The large sequestrum shown here is a shaft of dead bone 22 cm long. New bone is growing around this sequestrum to form an involucrum. (WAM 1291)

subperiosteal new bone formation, and blood-deprived bone sequestration discussed in the pathogenesis of this disease. Bone destruction is the first manifestation and appears in the roentgenogram as irregular areas of increased translucency. The destruction extends from the original focus in the metaphysis along the shaft and invades the cortex. These rarefied areas are not sharply defined and show a tendency to coalesce. They are often elongated and aligned in the long axis of the bone, especially those in the cortex (see Fig. 30).

Between the areas of bone destruction are regions of normal bone density which are sequestra. In more chronic cases the sequestra are more radiopaque and better defined. The sequestra are usually smooth in outline due to partial absorption and may be partially or completely surrounded by the living bone (involucrum) which is somewhat less opaque.

New bone formation is readily apparent even in less severe cases with subperiosteal bone apposition limited to the metaphyseal area. In more severe cases when the periosteum is elevated from large areas of the cortex, bone apposition occurs over most of the bone forming an involucrum. The layers of periosteal bone sometimes form sequentially to give a lamellated appearance resembling Ewing's sarcoma. More often, the bone of the involucrum is bulky and very irregular in outline with no alignment of the trabeculae to the lines of stress. Several cloaca usually perforate the involucrum and are easily visible in the gross specimen (see Fig. 29).

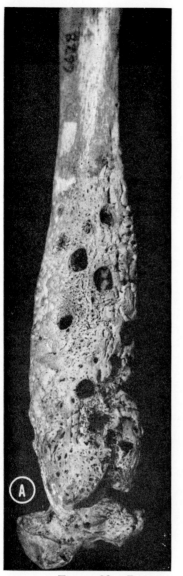 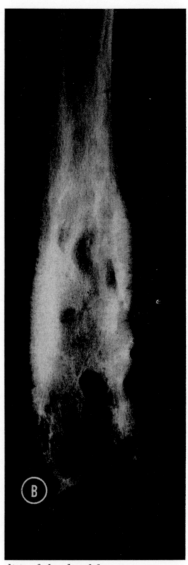

Figure 29. Pyogenic osteomyelitis of the distal femur.

Figure 29A. Several large cloaca have formed for pus drainage externally. Bone formation is irregular and composed of primitive or woven bone. The distal epiphysis is almost completely separated from the metaphysis and has become involved by the infection.

Figure 29B. Roentgenogram of the femur revealing the extent of new bone formation. (WAM 6628)

70   *Paleopathological Diagnosis and Interpretation*

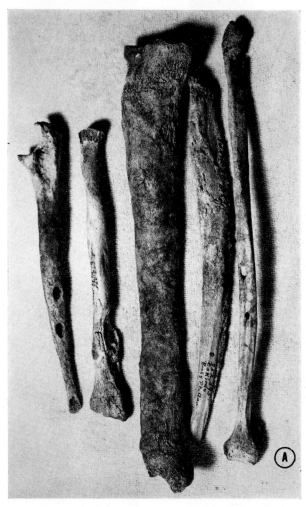

Figure 30. Hematogenous osteomyelitis in an adult male Eskimo. (A.) Bones shown are from right to left: right ulna, right radius, left tibia, right fibula, and left fibula.

Even in well-healed cases, pyogenic osteomyelitis generally leaves bone changes which are visible throughout the person's life. The bone architecture remains somewhat irregular with areas exhibiting varying degrees of sclerosis and sharply defined cavities lacking bone matrix.

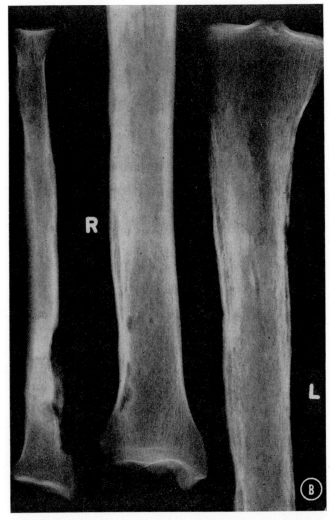

Figure 30B. Roentgenogram of right radius and both tibiae showing areas of destruction in the metaphyses and spread to the diaphyses of these bones.

Pyogenic arthritis occurs in about 15 percent of all cases of acute hematogenous osteomyelitis and usually involves the hip or knee. Infection and subsequent destruction of the epiphyses and articular surfaces occurs primarily in adults and in infants

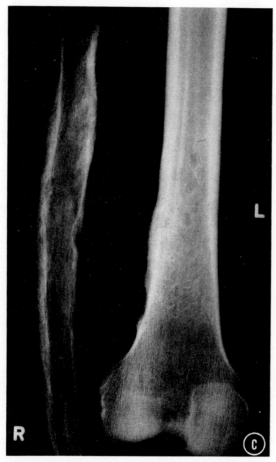

Figure 30C. Roentgenogram of right fibula and left femur. The fibula is greatly enlarged by subperiosteal new bone formation. (USNM 352,395)

below the age of one year. This unusual distribution may be explained by the anatomical arrangement of the vascular system in these two age groups. The pathogenesis and bone changes of pyogenic arthritis will be discussed in a separate section.

Acute Pyogenic Osteomyelitis by Direct Infection

Pyogenic osteomyelitis by direct infection may occur at any age. However, the highest percentage of cases occur in adults in contrast to the younger age group in hematogenous osteomyelitis. This is probably due to the higher incidence of long bone fractures, chronic soft tissue infections, and periodontal infections in older individuals (Wilensky, 1934). Compound fractures are the major cause of direct infection of the bone. The tibia and femur are by far the major sites of fracture and subsequent infection. Fractures of the mandible with exposure of bone externally or into the oral cavity are also frequently infected.

Pathogenesis and Gross Morphology

In compound fractures of the long bones, the broken portions of the diaphyses are exposed, and clotting of extravasated blood occurs in the broken ends. Deprivation of blood supply to segments of bone tissue bordering the fracture may result in an aseptic necrosis very similar in gross appearance to a localized osteomyelitis. Pyogenic bacteria will grow in the extravasated blood clot and fractured surfaces of the bone. The inflammatory response of the bone is due to the bacterial infection and in part is a reparative effort.

The gross morphology of pyogenic osteomyelitis by direct infection is similar to acute hematogenous osteomyelitis with minor differences. The initial bone infection is often in the shaft rather than the metaphysis of the long bone. Rapid extension of the infection due to increased internal pressure of the pus does not occur because the open fractured ends provide drainage of fluid. However, there may be spread of the bacteria to the metaphyses of the involved bone and indeed, hematogenous spread of the bacteria to other bones may also take place. In such cases, the marked features of severe bone fracture and frequent bone deformity following callus formation should provide a clue to the initial site of infection.

Sequestration due to infection or to the splintering of bone and involucrum formation are prominent features of osteomyelitis following compound fracture. Due to the presence of the large fragments of dead bone, the infection usually develops into a chronic pyogenic osteomyelitis, and the involucrum (callus) surrounding these portions becomes riddled with cloacae for pus drainage. The callus formation is considerable and unevenly distributed around the bone. The chronic infection may prevent proper union of the broken ends.

CHRONIC OSTEOMYELITIS

Chronic Pyogenic Osteomyelitis

Many cases of acute pyogenic osteomyelitis become chronic infections for a variety of reasons. The most frequent cause of chronicity is the presence of sequestra which are too large to be extruded through a draining sinus. Such wounds fail to heal and may continue to suppurate for decades. Many chronic cases are due to persistence of a low-grade infection in the bone and adjacent soft tissues or extension of the infection into a joint. The organism may remain viable in small abcesses and in sequestra for months or years causing flare-ups of osteomyelitis at irregular intervals. Epithelization of the soft tissue and bone sinus tracts further retards the healing of the disease. Such draining sinuses may persist for decades, and one reported case continued for sixty-two years (Milgram, 1931). In .5 percent of the chronic cases, malignant changes may occur in the sinus epithelium giving rise to an epithelioid carcinoma (Benedict, 1931).

The bone changes reflect the chronicity of the pyogenic osteomyelitis. Old sinuses are slowly plugged by granulation tissue, and new sinuses form as the pus production causes a build-up of internal pressure. This process continues for years until the involucrum becomes riddled with cloacae. Bone regeneration accompanying osteomyelitis appears as variously sized nodules near areas of destruction or by bone granulation followed by thickening of the cortex. With formation of the involucrum and cortical thickening, the infected bone often becomes greatly enlarged (see Fig. 32).

Brodie's Abscess

Brodie's abcess represents a special type of chronic pyogenic osteomyelitis in which the infection is walled off by granulation,

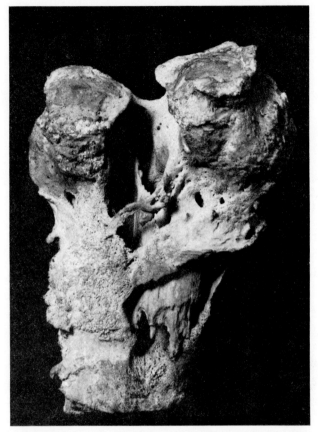

Figure 31. Chronic pyogenic osteomyelitis affecting the distal end of the right femur. The articular surfaces are roughened by erosion and irregular new bone formation. Note the large sequestrum which was formerly the shaft in this thirty-five-year-old man who had the infection for eighteen years. (WAM 1290)

which collagenizes to form a fibrous capsule. The capsule gradually becomes osteosclerotic, and the fluid within it is often sterile. The abscess develops in the cancellous tissue of the metaphysis of a long bone and most commonly in the distal end of the tibia. Other common sites are the distal metaphyses of the radius and femur. The abscess is usually small, measuring 1 to 3 cm in diameter (Brailsford, 1938).

The diagnosis of Brodie's abscess in archaeological material

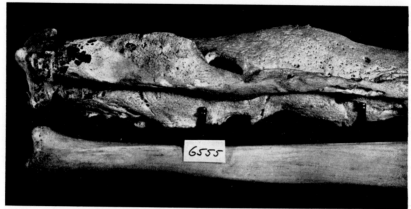

Figure 32. The left tibia with ankylosed fibula shows the gross enlargement of the bone as a result of bone regeneration in pyogenic osteomyelitis. Beneath the diseased bone is a normal tibia for comparison. (WAM 6555)

depends on the use of radiology or gross sectioning, because the bone may appear nearly normal externally. Brodie's abscess accounted for 2.8 percent of 346 cases of pyogenic osteomyelitis (Wilensky, 1934). The abscess generally occurs in adolescents between the ages of fourteen and twenty years. The abscess will become permanent unless there is surgical intervention and therefore may often be found in adults who have had the lesion for years. In such cases the abscess may be in the diaphysis due to further growth of the bone.

In the roentgenogram the abscess often appears as a translucent oval area with the long axis of the defect aligned with the axis of the long bone (Shanks and Kerley, 1950). The well-defined margin exhibits osteosclerosis, and in some cases, a small sequestrum may be present. The cortex may be increased in width and density although periosteal reaction is not present. The differential diagnosis with benign bone cyst is generally easy (see Chapter X). Benign bone cysts are usually much larger and have cortical thinning and expansion.

Chronic Nonsuppurative Osteomyelitis

(Sclerosing Osteomyelitis of Garré)

Sclerosing osteomyelitis is a rare form of chronic osteomyelitis characterized by a lack of pus formation and absence of an acute

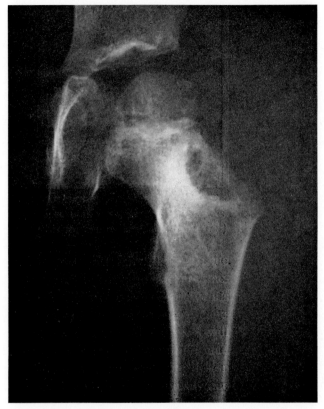

Figure 33. Brodie's abscess of the right femoral metaphysis in a young child. (Courtesy of Dr. E. B. D. Neuhauser)

phase. It generally affects the shafts of the tibia and femur in adults. A causative microorganism cannot be isolated in many cases, and the lesion may be produced by interference with the blood supply to the bone.

Instead of producing localized areas of necrosis, the infection spreads extensively throughout a segment of bone inciting osteoblast activity. Small abscesses may be seen in the roentgenogram, but the new bone formation is by far the major feature. The cortex is increased in width and density with resultant narrowing of the medullary cavity. A fusiform expansion of the shaft results. These bone changes are very similar to a syphilitic osteitis of a long bone.

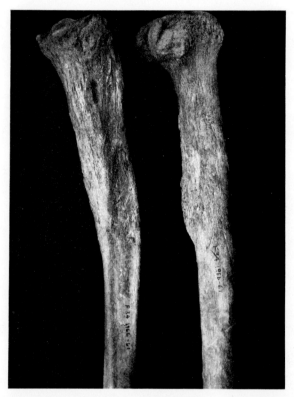

Figure 34. Possible nonsuppurative osteomyelitis also known as sclerosing osteomyelitis of Garré. The right tibia and left femur exhibit irregular thickening and a healed sinus tract in the tibia. The skull of this twenty-year-old male from the Mississippian Culture of Illinois has a circumscribed lesion containing a sequestrum in the left parietal. (Courtesy of Dr. Dan Morse, 1969.)

PYOGENIC ARTHRITIS

Pyogenic arthritis is a complication of pyogenic osteomyelitis occurring in about 15 percent of all cases (see Beekman, 1928; Pyrah and Pain, 1933; and Gilmour, 1962). It occurs most frequently in infants and adults and is the most common cause of infectious arthritis. The joints most often involved are the knee and hip which comprise up to 70 percent of all cases (Heberling, 1941). The ankle and shoulder joints together account for most of the other cases. Multiple joint involvement occurs very infrequently.

The usual mode of entry into the joint is direct spread of the infection from the metaphyseal focus. One of several routes may be taken, and these routes vary according to the age of the individual (see Trueta, 1957, 1959). In infants below the age of one year, vessels from the metaphysis penetrate the epiphyseal growth plate providing access to the joint for pyogenic bacteria. In children between the ages of one and sixteen years the metaphyseal capillaries no longer penetrate the growth plate, and this explains the infrequency of joint lesions in this age group. In adults there is fusion of the epiphysis with the metaphysis, and vascular connections are established between the two regions. The microorganisms may thus enter the joint through erosion of the articular cortex and cartilage. In large diarthrodial joints, a metaphyseal infection need only pierce the thin cortex to enter the joint space. For this reason an infection which originates in the neck of the femur almost always results in pyogenic arthritis because the neck lies within the joint space (Phemister, 1928a).

Gross Pathology

Upon entering the joint space, the infection causes pus production and destruction of the articular cartilages by proteolytic enzymes from dead leukocytes. The destruction is most rapid at sites of contact of the articular surfaces. With destruction of the articular surfaces, the process becomes chronic, and arthritic bone changes occur. Sequestra formation is rare and bony ankylosis is frequent. If joint motion is preserved in spite of the cartilage destruction, degenerative arthritis with its marginal osteophytes and subchondral sclerosis slowly develops (Luck, 1950).

DIFFERENTIAL DIAGNOSIS

The typical, well-advanced case of osteomyelitis is easy to recognize. However, when dealing with early or atypical cases, a differential diagnosis must be used to separate pyogenic osteomyelitis from the conditions outlined in the following discussion.

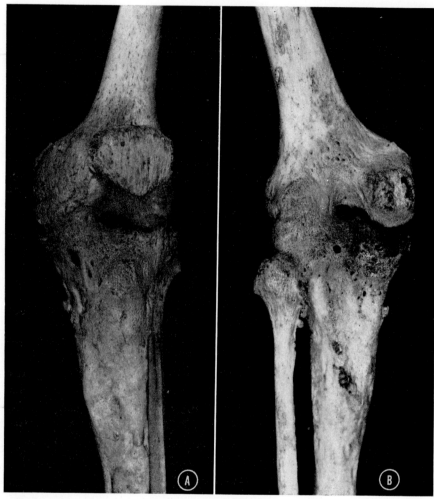

Figure 35. Pyogenic arthritis of left knee. (A.) Anterior view of left knee with complete ankylosis from pyogenic arthritis. (B.) Posterior view. (WAM 1205)

Skeletal tuberculosis often produces abscesses similar to a Brodie's abcess but lacks the dense sclerosis surrounding the lesion. Diaphyseal involvement in tuberculosis is not as common as in pyogenic osteomyelitis. Sequestra formation is infrequent but usually involves cancellous tissue rather than the cortex. Such sequestra are generally smaller than in pyogenic osteomyelitis. Cortical thickening through subperiosteal bone apposition occurs infrequently in tuberculosis while such new bone formation is characteristic of pyogenic osteomyelitis. Finally, joint involvement is more frequent in tuberculosis than in osteomyelitis (see Chapter V).

Tertiary syphilis most commonly results in subperiosteal deposition of hard dense bone and narrowing of the medullary cavity which may closely resemble the rare lesion of sclerosing osteomyelitis. In pyogenic osteomyelitis, new bone formation is more excessive and irregular in outline than in syphilis. Furthermore, the cranium is rarely involved by osteomyelitis while this is a frequent site of syphilitic gummata. Multiple bone lesions occur more frequently in syphilis. Sequestra formation usually involves cancellous bone rather than the cortex.

Subperiosteal ossifying hematomas caused by trauma or scurvy may resemble a localized pyogenic osteomyelitis. New bone formation occurs wherever the periosteum is raised by the imprisoned blood. The absence of metaphyseal cancellous or cortical destruction in the roentgenogram provides the major differentiating feature of pyogenic osteomyelitis.

Benign bone cysts resemble Brodie's abscess but lack bony sclerosis. Their differential diagnosis is discussed in the section on Brodie's abscess. Osteoid-osteoma and Brodie's abscess may produce identical bone lesions and are difficult to differentiate (see Chapter X).

Osteosarcoma and Ewing's sarcoma may simulate pyogenic osteomyelitis in the early stages of their growth. However, with few exceptions archaeological material will always show the end stages of these malignant tumors which appear markedly different from pyogenic osteomyelitis.

Paget's disease is characterized by multiple bone involvement, particularly thickening of the cranium. Bowing deformity of the long bones may also occur.

PALEOPATHOLOGY

Periostitis and osteomyelitis together account for the great majority of pathological changes found in early human and animal bones (Sigerist, 1961). Indeed, a classic example of chronic pyogenic osteomyelitis has been described in *Dimetrodon*, a primitive reptile from the Permian deposits of 250 million years ago (Moodie, 1923). The spinous processes of *Dimetrodon* are extremely long and vulnerable to injury. A compound fracture occurred near the base of the spinous process and became chronically infected. Similar cases of periostitis and osteomyelitis have been described in several dinosaurs from the Jurassic and Cretaceous of Wyoming (Moodie, 1923) and in Pleistocene remains of cave bears (Virchow, 1895; Abel, 1924), bison (Moodie, 1923), lions and wolves (Moodie, 1926). The characteristic sinus formation found in many of these specimens strongly suggests pyogenic bacteria as the causal organisms, although it is impossible to ascertain if the bacteria were staphylococci.

Periostitis and osteomyelitis are common pathological lesions in early human populations. In the New World these nonspecific infections may be found in Indian populations of every archaeological horizon and geographical location (*cf.* Whitney, 1886; Hooton, 1930; Goldstein, 1957; Roney, 1959; Hoyme, 1962; Morse, 1969; Ragir, 1972). In the Old World, periostitis and osteomyelitis are also commonly found in archaeological remains, particularly from Neolithic times onward (Ruffer, 1921; Pales, 1930; Brothwell, 1963). Periostitis and osteomyelitis may be exacerbated and indeed induced by poor hygiene and malnutrition as shown by the rise of these diseases in British burial grounds during the industrial revolution (Wells, 1973). Changes in diet and shift to sedentary agriculture produced a similar increase in nonspecific infections of bone among pre-Columbian Maryland Indians (Hoyme and Bass, 1962).

A thorough knowledge of the gross bone morphology of pyogenic osteomyelitis in all of its pathogenetic stages and forms is obviously of great importance to the paleopathologist. A differential diagnosis between nonspecific osteomyelitis and the specific disease entities such as tuberculosis and syphilis is very difficult but crucial to a better understanding of the history of these diseases. The following chapter will attempt to present the evidence for the antiquity of specific bone infections together with descriptions of their gross morphology.

OSTEOMYELITIS BIBLIOGRAPHY

Abel, O.: "Neuere studien über Krankheiten fossiler Wirbeltiere," *Verhandl. d. zool. botan. Ges. Wien,* 73:104, 1924.

Beekman, F.: "Acute hematogenous osteomyelitis," *Ann. Surg.,* 88:270-296, 1928.

Benedict, E. B.: "Carcinoma in osteomyelitis," *Surg. Gynec, and Obst.,* 53:1-11, 1931.

Brailsford, J. F.: "Brodie's abscess and its differential diagnosis," *Brit. Med. J.,* 2:119-123, 1938.

Brothwell, D. R.: *Digging up Bones,* British Museum (Natural History), 1963.

Gibson, A.: "A clinical study of the pathology of osteomyelitis," *Canadian Med. Assn. J.,* 12:844-848, 1921.

Gilmour, W. N.: "Acute hematogenous osteomyelitis," *J. Bone Joint Surg.,* 44B:841-853, 1962.

Goldstein, M. S.: "Skeletal pathology of early Indians in Texas," *Amer. J. Phys. Anthrop.,* 15:299-312, 1957.

Green, M., W. L. Nyhan, and M. D. Fousek: "Acute hematogenous osteomyelitis," *Pediatrics,* 17:368-382, 1956.

Green, W. T. and J. C. Shannon: "Osteomyelitis in infants. A disease different from osteomyelitis of older children," *Arch. Surg.,* 32:462-493, 1936.

Heberling, J. A.: "A review of 201 cases of suppurative arthritis," *J. Bone Joint Surg.,* 23A:917-922, 1941.

Hilpert: *Contribution to the Question of Acute Osteomyelitis,* Inaugural Dissertation, Freigurg, 1914.

Hobo, T.: "Zur Pathogenese der akuten haematogen Osteomyelitis, mit Berucksichtigung der Vitalfarbungslehre," *Acta Sch. Med. Univ. Kioto,* 4:1-29, 1921-22.

Hooton, E. A.: *Indians of Pecos Pueblo,* Yale Univ. Press, New Haven, 1930.

Hoyme, L. E. and W. M. Bass: "Human skeletal remains from Tollifero (Ha 6) and Clarksville (Mc 14) sites, John H. Kerr Reservoir Basin, Virginia," *Bureau of Amer. Ethnol. Bull.,* 182:329-400, 1962.

Huebler: "Acute osteomyelitis in childhood," *Wien. Med. Wchnschr.,* 77:1456, 1927-28.

Jaffe, H. L.: *Metabolic, Degenerative and Inflammatory Disease of Bones and Joints,* Lea and Febiger, Philadelphia, 1972.

Knaggs, R. L.: *Diseases of Bone,* William Wood and Co., New York, 1926.

Larsen, R.: "Intramedullary pressure with particular reference to massive diaphyseal bone necrosis," *Ann. Surg.,* 108:127-140, 1938.

Luck, J. V.: *Bone and Joint Disease,* Thomas, Springfield, 1950.

Milgram, J. E.: "Epithelization of cancellous bone in osteomyelitis," *J. Bone Joint Surg.*, *13*:319-324, 1931.

Moodie, R.: *Paleopathology: An Introduction to the Study of Ancient Evidences of Disease*, Univ. of Illinois Press, Urbana, 1923.

——————: "Pleistocene examples of traumatic osteomyelitis," *Ann. Med. Hist.*, 8:413-418, 1926.

Morgan, J. D.: "Blood supply of growing rabbit's tibia," *J. Bone Joint Surg.*, *41B*:185-203, 1959.

Morse, D.: *Ancient Disease in the Midwest*, Illinois State Mus. Rep. of Inves., 15, 1969.

Pales, L.: *Paléopathologie et Pathologie Comparative*, Masson and Co., Paris, 1930.

Phemister, D. B.: "The pathology and treatment of pyogenic osteomyelitis." *Penn. Med. J.*, *32*:52-57, 1928a.

——————: "Unusual forms of osteomyelitis," *North-west Med.*, 27:460-466, 1928b.

Putschar, W. G. J.: "General pathology of the musculo-skeletal system," in F. Buchner (ed.), *Handbuch der Allgemeinen Pathologie* Bd. III/2, Springer-Verlag, Berlin, 2nd ed., 1960.

Pyrah, L. N. and A. B. Pain: "Acute infective osteomyelitis. A review of 262 cases," *Brit. J. Surg.*, *20*:590-601, 1933.

Ragir, S.: "The early horizon in central California prehistory," *Contrib. of the Univ. of Calif. Arch. Res. Facility*, #15, 1972.

Roney, J. G.: "Paleopathology of a California archaeological site," *Bull. Hist. Med.*, *33*:97-109, 1959.

Ruffer, M. A.: *Studies in the Palaeopathology of Egypt*, R. L. Moodie (ed.), Univ. of Chicago Press, Chicago, 1921.

Shandling, B.: "Acute hematogenous osteomyelitis; a review of 300 cases treated during 1952-59," *South Afr. Med. J.*, *34*:520-524, 1960.

Shanks, S. C. and P. Kerley (ed.): *A Text-book of X-ray Diagnosis*, vol. 4, H. K. Lewis and Co., London, 1950.

Sigerist, H. E.: *A History of Medicine I. Primitive and Archaic Medicine, Medicine*, Oxford Univ. Press, London, 2nd ed., 1961.

Speed, J. S.: "An analysis of 160 cases of osteomyelitis with end results," *Southern Med. J.*, *15*:721-726, 1922.

Starr, C. L.: "Acute hematogenous osteomyelitis," *Arch. Surg.*, *4*:567-587, 1922.

Steinbach, H. L.: "Infections of bones," *Seminars in Roentgenology*, *1*:337-369, 1966.

Trendel: "Contribution to our knowledge of acute infectious osteomyelitis and its sequelae," *Beitr. z. klin. Chir.*, *41*:607-675.

Trueta, J.: "The normal vascular anatomy of the human femoral head during growth," *J. Bone Joint Surg.*, *39B*:358-394, 1957.

———————: "The three types of acute hematogenous osteomyelitis," *J. Bone Joint Surg., 41B*:671-680, 1959.

Virchow, R.: "Knochen vom Hohlenbaren mit krankhaften Veranderungen," *Zeit. f. Ethnol., 27*:706-708, 1895.

Waldvogel, F. A., G. Medoff, and M. N. Swartz: *Osteomyelitis,* Thomas, Springfield, 1971.

Wells, C.: "The palaeopathology of bone disease," *The Practitioner, 210*:384-391, 1973.

Whitney, W. F.: "Notes on the anomalies, injuries, and diseases of the bones of the native races of North America," *Eighteenth and Nineteenth Annual Reports of the Trustees of the Peabody Museum, 3*:433-448, 1886.

Wilensky, A. O.: *Osteomyelitis: Its Pathogenesis, Symptomatology, and Treatment,* Macmillan, New York, 1934.

Winters, J. L. and I. Cahen: "Acute hematogenous osteomyelitis," *J. Bone Joint Surg., 42A*:691-704, 1960.

Zadek, I.: "Acute osteomyelitis of the long bones of adults," *Arch. Surg., 37*:531-545, 1938.

Chapter **IV**

TREPONEMA INFECTION: SYPHILIS, BEJEL, AND YAWS

THE PALEOPATHOLOGY, origin, and epidemiology of the four treponemal infections, venereal syphilis, nonvenereal syphilis (bejel), yaws, and pinta, have been sources of considerable controversy and disagreement. A complete discussion of all the different theories and opinions regarding these four diseases or syndromes would be both lengthy and confusing. However, a brief review of the problem together with an introduction to the vast literature on the subject is presented here to stress the importance of historical, biological, and epidemiological data in addition to paleopathological evidence.

INFECTIOUS DISEASE: SPECIFIC INFECTIONS

Infectious disease has been a major natural selective factor in human evolution and has deeply influenced the development and decay of civilizations. It is extremely unfortunate that only a few of the many important diseases leave their mark on the skeleton. Their distribution and antiquity may be ascertained from the skeletal material while many other diseases leave no trace. Ancient art and literature have increased our knowledge of some of these diseases, but many gaps in our present-day knowledge remain unfilled.

One method of attacking the problem of determining a disease's age and distribution is outlined in an excellent book on the evolution of infectious disease (Cockburn, 1963). This method involves constructing vertical arrays of related microorganisms, close examination of the parasite's life cycle, and consideration of limiting factors such as climate and minimum population density required of the host. Such a biological approach has useful applications in paleopathology for studying diseases which may or may not affect the bone. Let us now examine some of the specific diseases which cause a bone reaction or have bone involvement.

THEORIES RELATING TO THE SPREAD OF SYPHILIS

Columbian Theory

Many authorities have maintained that Christopher Columbus and his crew carried syphilis to Europe from the New World on their return to Spain in 1493 (Williams, 1927; Pusey, 1933; Zinsser, 1935; Harrison, 1959; Crosby, 1969). The introduction of a totally new disease would explain its rapid spread to every part of Europe in less than two years as shown by proclamations and personal accounts of the period. However, it is significant that the New World is not mentioned as the source of syphilis until some thirty years after Columbus' return. Moreover, the major impetus for regarding the New World as the cradle of syphilis was the discovery of guiac, a kind of wood, as a remedy for the disease. According to the reasoning of that period, "God always provides a remedy where he inflicts a disease." Therefore, since guiac came from the New World, then syphilis must have originated there as well.

Pre-Columbian Theory

This theory holds that syphilis was present in Europe long before the return of Columbus and his crew. Its advocates cite descriptions of diseases resembling syphilis in the literature prior to Columbus, including possible ancient Roman and Greek references (Buret, 1891; Sudhoff, 1926; Holcomb, 1930, 1934, 1935; Hudson, 1961b, 1963b, 1964). They maintain that venereal syphilis was present in Europe but was not distinguishable from among a number of diseases which were grouped under the term "leprosy." This is supported by the many references to "venereal leprosy" in the 13th and 14th centuries and not long afterward to "hereditary leprosy," (Hudson, 1961b). Leprosy is never spread sexually nor transmitted congenitally, but both features are prominent in syphilis. Furthermore, the Chinese and Arabs have used mercury rubs and inhalations for thousands of years to treat "leprosy." In the 12th and 13th centuries the Crusaders returned from the Holy Land with "Saracen ointment" containing mercury to treat the increasing numbers of "lepers." Mercury is ineffective against true leprosy but has been the mainstay in treating syphilis until the discovery of salvarsan and later penicillin.

During and after the Crusades and innumerable pilgrimages to the Holy Land (1096-1221), "leprosy" rapidly spread to every part of Europe (Gramberg, 1959). Some 19,000 "leprosy" asylums or lazarets were established in Europe (Buret, 1891). Their main function was not to cure or care for the "lepers," but to isolate them and thus prevent the spread of such a contagious disease. True leprosy is hardly contagious and requires an incubation period of three to ten years. On the other hand, syphilis may be spread sexually or in areas of poor hygiene by simple body contact. Its incubation period is only a few days or weeks rather than years. In Iraq today, there are isolated villages where people with nonvenereal syphilis (bejel) are said to have "leprosy," (Hudson, 1958a).

The "Epidemic" Around 1500

Proponents of the Columbian theory point to the so-called epidemic of syphilis around 1500 as evidence of a virulent disease striking people never before exposed to it. Descriptions of the disease by the physicians of that time include the three stages associated with syphilis as it is known today (Zimmerman, 1934, 1935). However, another phase of the disease was described with prolonged illness consisting of fever, sweating, diarrhea, and jaundice and often culminating in coma and death (Holcomb, 1930). Such symptoms do not occur in syphilis but are prominent features of several acute epidemic diseases such as typhus, typhoid, cholera, dysentery, and influenza. These conditions had not yet been recognized as separate diseases and were probably a major factor in the "epidemic" of syphilis.

The invention of movable type in 1450 produced a tremendous upsurge in scientific thought and work through easier exchange of ideas and information. This may account for the recognition around 1500 of venereal syphilis as a disease distinct from leprosy, and in later years the epidemic diseases were defined as well (Castiglioni, 1941). While more and more was being written on venereal syphilis around 1500, it is very significant indeed that "leprosy" rapidly became less common. Indeed, one physician left his chapter on "venereal leprosy" untouched and simply changed the title to morbus gallicus, the current name for syphilis (Holcomb, 1934).

Thus, the "epidemic" of syphilis around 1500 was more likely caused by identification of syphilis as a disease distinct from other diseases such as leprosy. Knowledge of syphilis as a separate disease quickly spread to all parts of Europe where the disease had already been present either in its venereal or nonvenereal form for centuries.

Some medical historians agree that syphilis was present in both the Old and New Worlds before the return of Columbus but assert that his crew introduced a more virulent strain causing the epidemic (Castiglioni, 1941; Singer and Underwood, 1962; Jarcho, 1964). This combination of elements from both the Columbian and pre-Columbian theories is possible but hardly likely. As mentioned above, the epidemic of syphilis was largely caused by the recent recognition of syphilis as a separate disease. Furthermore, an increased prevalence of venereal syphilis may be explained by historical events unrelated to the return of Columbus.

For example, papal proclamations of 1490 and 1505 abolished all "leper" asylums of the Order of St. Lazarus, thus releasing an untold number of syphilitics all over Europe (Holcomb, 1934). In late 1494 the army of King Charles VIII of France beseiged Naples, and after its fall his army was stricken by syphilis. The disease forced the partial disbandment of his largely mercenary army in 1495 and 1496, resulting in the dispersal of infected soldiers all over Europe. The syphilis present in Naples did not come from Columbus' crew via Spanish reinforcements to the town as popularly supposed. Syphilis had been in Italy at least since the Crusades and probably earlier.

Finally, beginning March 30, 1492, somewhere between 160,000 and 400,000 Jews were expelled from Spain (Holcomb, 1934). The syphilis these homeless and wandering people carried to all parts of Europe did not originate in the New World. It had been present in Spain for over a thousand years in venereal and nonvenereal form during occupation by the Saracens and their Negro slaves and allies (Hudson, 1963a). For centuries both Spain and Portugal imported slaves by land from the savannahs and rain forests of sub-Saharan Africa where nonvenereal syphilis and yaws flourish. Beginning in 1492, increasing numbers of slaves were imported by the easier sea route, thus providing Spain with a constant source of treponemal infection long before the discovery of the New World.

Unitarian Theory—The Biological Evidence

The most recent and provocative theory regarding the origin and spread of syphilis is the Unitarian theory (Butler, 1936; Hamlin, 1939; Hudson, 1958a, b; 1963a, 1965a, b). Hudson has been the most ardent supporter of this theory and his articles are required reading for all students interested in the history of syphilis. The major features of this theory and the supportive evidence will be briefly presented here.

The Unitarian theory states that the four treponemal infections: pinta, yaws, nonvenereal syphilis (bejel), and venereal syphilis are not separate diseases caused by different organisms. Instead, they represent syndromes caused by slightly different strains of one organism, *Treponema pallidum*. The treponeme will produce all of the syndromes under the environmental and sociological conditions appropriate to each. The four syndromes collectively form the disease called *treponematosis*.

The Four Syndromes

PINTA. This type occurs in tropical America from Mexico to Ecuador. It occurs mostly in dark-skinned people and produces white patches on the skin, particularly the palmar surfaces. The internal organs are not involved.

YAWS. Yaws occurs in rural populations of the humid tropics. The ulcerating lesions of yaws affect both skin and bones, but cardiovascular and neurologic manifestations are rare.

NONVENEREAL SYPHILIS. This type of syphilis is often referred to as *endemic syphilis* or by many local names such as bejel, dichuchwa, sibbens, and button scurvy depending on its geographic locality. This syndrome is similar to yaws but today is found only in warm, arid climates. Cardiovascular lesions occur with some frequency, but involvement of the central nervous system is rare. Both yaws and endemic syphilis are usually acquired in childhood, but these mild infections may last for many years. World surveys have revealed that the number of human beings affected by endemic syphilis is far greater than the number affected by venereal syphilis (Hudson, 1961a).

VENEREAL SYPHILIS. Venereal syphilis has no climatic restrictions but is not found where another form of treponematosis predominates. It is acquired by adults through sexual intercourse and may be transmitted congenitally by passage of the treponeme through the placenta of an infected mother into the fetus. Venereal syphilis may affect any tissue of the body. The skin is involved initially, followed by internal organs such as bone, liver, heart, and brain.

THE BIOLOGICAL AND EPIDEMIOLOGICAL EVIDENCE

The treponemes, also known as spirochetes, of the four syndromes cannot be differentiated from each other by any known test. They are morphologically identical with a common antigenic structure which differs only quantitatively. Inoculated rabbits exhibit cross-immunity between syphilis and yaws (Turner and Hollander, 1957). In man, partial cross-immunity exists between syphilis (both venereal and nonvenereal), yaws, and pinta (Cannefax, 1967). This evidence strongly supports the idea of four different strains of treponeme rather than different species.

Gross and microscopic examination of diseased human tissue from all four syndromes cannot distinguish one form of treponematosis from the other. Pinta may produce marked skin discoloration, but this is also found in the other syndromes only to a lesser degree. Thus, all pathologic differences are merely quantitative with considerable overlapping between the syndromes. The four syndromes of treponematosis produce a pathological gradient extending from the cutaneous manifestations of pinta to the ulcers of yaws involving both skin and bone, to similar lesions of endemic syphilis affecting the skin, bone, and cardiovascular system, and finally to the lesions of venereal syphilis affecting all of the organs just mentioned in addition to the nervous system.

Climate

Treponema pallidum is a very delicate organism easily killed when exposed to oxygen or dessication. The strains of treponeme have adapted to particular environments, and therefore, one strain predominates in a given climate. In hot, humid climates such as the rain forests of Africa and the jungles of South America, the human skin is kept continually moist. This allows the treponeme to live and produce lesions all over the body surface as are found only in yaws. In drier climates such as the desert or savannah, the human skin is kept dry, and the treponeme can exist only in the mouth, armpits, and crotch where there is sufficient moisture. This distribution of lesions constitutes the syndrome of endemic syphilis. Venereal syphilis has no climate restrictions but is more common in temperate areas where individuals are heavily clothed. The clothing prevents nonvenereal spread by body-to-body contact as in yaws and endemic syphilis leaving only sexual intercourse as a means of transmission.

Living Conditions

Personal hygiene and living conditions are also important

determining factors in treponematosis. Endemic syphilis is limited today mainly to the hot, dry areas of the world relatively untouched by civilization. Overcrowding in nomadic huts, communal use of eating and drinking utensils, and scanty clothing among the children provide a great amount of body contact for transmission of the treponemes. Endemic syphilis also existed in more temperate regions in past centuries and even in recent years where people were living under very unsanitary conditions. Thus, one finds the sibbens of Scotland, button scurvy of Ireland, radesyge of Norway, saltfluss of Sweden, and frenga of Bosnia (Grin, 1935; Willcox, 1960; Morton, 1967). All of these foci of endemic syphilis have disappeared as living conditions improved —and this is where venereal syphilis enters the scene. With better hygiene and an improved standard of living, the cycle of childhood infection with endemic syphilis is broken, leaving only sexual intercourse as a means of transmission in adults.

There are many examples showing transition of one syndrome of treponematosis into another due to alterations in the climate or living conditions. When people in the hot and humid lowlands infected with yaws move to the cooler mountains, they quickly lose the generalized lesions of yaws and develop the restricted lesions of endemic syphilis (Hudson, 1965a). The sibbens or endemic syphilis of Scotland, which first appeared in the middle of the 17th century, was blamed on Cromwell's army carrying venereal syphilis. The venereal syphilis changed to endemic syphilis under the unhygienic conditions of the times (Morton, 1967).

Thus, every human population or subpopulation has the kind of treponematosis that is appropriate to its physical environment and socio-cultural status. With alterations of the environment, such as by migration or with changes in the living conditions, there is a concomitant shift in the form of treponematosis affecting that population. This conclusion has important implications concerning the history and paleopathology of venereal syphilis. Wherever groups of people improved their level of hygiene and standard of living, their endemic syphilis disappeared and was replaced by venereal syphilis as the dominant form of treponematosis. This could have happened many times and in many places in both the Old and New World (see Hudson, 1963a, 1965a).

THE PALEOPATHOLOGICAL EVIDENCE

The paleopathological literature on syphilis and other forms of treponematosis is considerable but disappointing. The dating of many specimens is inexact or even unknown in several cases. In most instances, the specimen consists of a solitary bone or bone fragment because the rest of the skeleton had deteriorated or was left behind as having no scientific value! The diagnosis of syphilis, when based on a single bone exhibiting nonspecific periostitis or osteoperiostitis, is very difficult and vulnerable to error—and yet this occurs quite often in the literature.

Moreover, until recently the epidemiology of syphilis in early human populations has been ignored. Instead of describing isolated cases of possible syphilis, it is important to determine the prevalence of such lesions in the entire skeletal series. So many individuals have "swollen" long bones resembling syphilis in certain archaeological series that some workers believe a nutritional deficiency or disturbance of some unknown type must be the cause (Hoyme and Bass, 1962; see also Brues, 1966). Determining the age of the affected individuals is important, and all infant and adolescent remains should be closely examined for evidence of congenital syphilis.

The cultural and climatic context must also be considered to determine the likely form of treponematosis affecting the population. Except for pinta which does not involve bone, the other syndromes produce nearly identical bone lesions—and yet venereal syphilis is almost invariably considered the cause of such lesions in American Indian skeletal material. It is certainly possible and indeed likely that venereal syphilis existed in the large urban centers and town and temple communities of such cultures as Inca, Aztec, Pueblo, Mississippian, and perhaps Adena-Hopewell (see Willey, 1960). However, the small populations representative of the hunter-gatherers and small village farmers were more likely affected by endemic (nonvenereal) syphilis. The scarcity of possible cases of congenital syphilis supports this idea, because nonvenereal syphilis is not transmitted congenitally. Although endemic syphilis primarily affects children, the disease is chronic so that the bone lesions will often be present in adults. Similar lesions in adolescents should also be found but few have been reported in American skeletal material.

It is not the purpose of this chapter to review the entire paleopathological literature on yaws and syphilis in the Old and New Worlds. However, the more important articles are listed here for the reader's information.

New World

The first major report on syphilis in pre-Columbian American Indian remains was made by Joseph Jones in 1876. This important report on aboriginal remains exhibiting syphilitic lesions from several sites in Tennessee and Kentucky stimulated considerable interest in paleopathology on both sides of the Atlantic. A number of workers disagreed with Jones (Whitney, 1883; Hyde, 1891; Morgan, 1894) but several others reported new finds supporting Jones (Bruhl, 1890; Ashmead, 1896; Lamb, 1898; Gann, 1901). Many of the specimens cited came from the so-called "Mound-builders" of southeastern United States (Jones, 1876; Lamb, 1898; Orton, 1905; Means, 1925; Denninger, 1935; Wakefield et al., 1937; Haltom and Shands, 1938; Snow, 1945). These Indians built elaborate earthworks enclosing burial mounds and earthworks constructed in the shape of animals. These mounds are associated with the Adena-Hopewell culture (800 B.C.-200 A.D.).

Other cases of pre-Columbian syphilis were reported from Peru and Central America (Ashmead, 1896; Gann, 1902; Eaton, 1916; MacCurdy, 1923; Tello and Williams, 1930; Krumbhaar, 1936; Goff, 1953, 1963, 1967; Stewart, 1956; Anderson, 1965; and Saul, 1972). Five skulls with extensive osteitis characteristic of syphilis have been reported from Puerto Rico and are dated 1400-1600 A.D. (Gejvall and Henschen, 1971). Many of the remaining possible examples of pre-Columbian syphilis have been described from excavations of the Pueblo Indian dwellings in New Mexico (Hooton, 1930; Denninger, 1938; Cole et al., 1955).

The early literature on pre-Columbian syphilis in the New World was excellently reviewed by Herbert U. Williams in 1932 and 1936. He also described new specimens from the Hopewell culture of Ohio, Aztec culture of Mexico, and Basketmaker culture of Arizona.

Possible cases of syphilis have been reported from several other areas and time periods. The Indian Knoll site in Kentucky, a Late Archaic site dated around 3300 B.C., has yielded several possibly syphilitic skeletons (Snow, 1948). Nine individuals from a Middle Horizon site in California (around 500 B.C.) exhibited periostitis suggestive of syphilis (Roney, 1959, 1966). A brief description of syphilis involving assorted long bones and a skull is found in a report on dental conditions among prehistoric Indians of northern Alabama (Rabkin, 1942). No date is mentioned for the skull although the material described in the report spanned Archaic to early historic periods. A solitary skull, probably from the Nodena phase of the Mississippian culture, exhibits a sclerotic periosteal reaction with pitting involving the left half of the frontal bone and closely resembles syphilis (Morse, 1973). Two skeletons from Mississippian sites in Illinois (1000-1400 A.D.) have lesions highly characteristic of syphilis (Morse, 1967).

Several possible examples of syphilis come from precontact sites in Oklahoma and Texas (Goldstein, 1957; Brues, 1958, 1959). Several sites in Florida have yielded likely specimens of syphilis (Hrdlička, 1922; Snow, 1962; Bullen, 1972). They range in date from an Archaic site around 3300 B.C. to several Weeden Island Complex sites around 850 A.D. Postcontact examples of syphilis have been described in Aleut and Eskimo skeletons from Alaska (Holcomb, 1940).

The paleopathological evidence of pre-Columbian syphilis in the New World continues to accumulate in a rather piecemeal fashion. The paucity of lesions and questionable dating and diagnosis of specimens has caused some authorities to doubt the presence of syphilis in the New World before European contact (Hrdlička, 1932; Stewart, 1940; 1973). The author agrees with

T. Dale Stewart's call for a general survey of the New World material, particularly newly excavated material, to establish the prevalence and distribution of venereal and nonvenereal syphilis in well-dated specimens.

Old World

The paleopathological evidence of syphilis in the Old World prior to 1500 A.D. is remarkably scanty. The early literature has been reviewed and provides few likely specimens (Pales, 1930; Williams, 1930). The earliest possible cases of syphilis have come from Asia, particularly Siberia (Rokhlin and Rubasheva, 1938). They consist of several tibiae, a radius and ulna dated 1000-800 B.C., two tibiae dated 500-200 B.C., and three skulls dated 100-700 A.D.

Important negative evidence is provided by the absence of any syphilitic lesions in the massive Egyptian series totalling some 25,000 skeletons (Smith and Jones, 1910), although several possible specimens have recently been described (Hussein, 1949). Excavation of several medieval leprosaria cemeteries has failed to reveal clear evidence of syphilis before 1500 (Andersen, 1969; Møller-Christensen, 1969).

The skull of an adult female from London, England presents the classic "worm-eaten" appearance of syphilis (Morant and Hoadley, 1931). The skull was excavated from the churchyard of St. Mary Spittle which was in use from 1197-1537 A.D. A skeleton with very characteristic syphilitic lesions has been described in Denmark and dates from the late Middle Ages or about 1500 A.D. (Møller-Christensen, 1958). Two possible cases of syphilis have been reported from the Baltic coast of Lithuania and Sweden and dating around 1600-1700 A.D. (Gejvall, 1960; Derums, 1965). Of similar date are four syphilitic skulls from among 5300 examined in the Catacombs of Paris (Møller-Christensen and Jopling, 1964). The author's acquaintance with the European and Asian literature on early specimens of syphilis is not extensive, but it seems clear that very few cases have been reported before 1500 A.D.

VENEREAL SYPHILIS

Congenital Syphilis

Venereal syphilis may be either congenital or acquired. Congenital syphilis is transmitted *in utero* by treponemes invading the fetus from the placenta of an infected mother. The disease is easily transmitted as shown by an attack rate of 84 percent in the offspring of infected mothers (Fournier, 1899). The treponemes cross the placenta after the third or fourth month of gestation and spread through the fetal bloodstream to virtually every bone in the body. They may be found particularly at the sites of active endochondral ossification such as the metaphyses of the long bones.

Frequency of Skeletal Involvement

The bone manifestations of congenital syphilis may be divided into early and late categories. Early congenital syphilis includes the skeletal lesions which appear from birth to three or four years. The majority of the cases of late congenital syphilis occur between five and fifteen years of age. Almost all cases of early congenital syphilis have some degree of bone involvement (see McLean, 1931; Parmalee and Halpern, 1935). However, in about 50 to 75 percent of the cases, the bone changes are so slight or will heal without treatment leaving no evidence of the disease in the skeletal remains. Late congenital syphilis is roughly as frequent as early congenital syphilis, but bone lesions are not as common. Only 7.2 percent of 1010 cases of late congenital syphilis had bone involvement (Cole, 1937). However, unlike early congenital syphilis, these bone lesions are chronic and produce bone deformities which persist for many years.

Early Congenital Syphilis

The osseous lesions of early congenital syphilis in order of frequency are: osteochondritis, periostitis, and diaphyseal osteomyelitis. *Osteochondritis,* also known as *metaphysitis,* often occurs during the first six months of life and in most cases affects several bones bilaterally. The tibia is most frequently involved followed by the ulna, radius, femur, humerus, and fibula (McLean, 1931). Osteochondritis first appears as a widened band of increased calcification in the metaphysis which may not be preserved in the archaeological specimen. In more severe cases, this zone of provisional calcification becomes quite irregular producing a "saw-toothed" metaphysis. The formation of granulation tissue makes the metaphyseal-epiphyseal border even more irregular and decreases the amount of normal spongiosa. The epiphysis may be displaced or even crushed down into the defective ends of the shaft. The presence of osteochondritis alone is not pathognomonic of syphilis and may be produced by other infections of the newborn such as tuberculosis and pyogenic osteomyelitis (see Caffey, 1939; Evans, 1940).

Periostitis is often found in early congenital syphilis as a reactive and reparative response to the infection. New bone is laid down over the affected region and the cortex becomes thickened. Occasionally the periosteum covers the entire area with one thick layer or several layers of fiber bone. This may gradually be converted into dense, lamellar bone ensheathing the cortex. Such a massive periosteal reaction, known as "periosteal cloaking," is usually associated with syphilitic osteomyelitis of the diaphysis and was present in eight of 102 cases of congenital syphilis (McLean, 1931).

Diaphyseal osteomyelitis occurs in about 50 percent of all cases of congenital syphilis and usually begins in the metaphyses. The cortex may be focally destroyed or even thickened due to periosteal new bone. When foci of osteomyelitis are present at the proximal ends of both tibiae on their medial aspects, this is highly suggestive of congenital syphilis and is known as Wimberger's sign (see Fig. 36).

The Skull in Early Congenital Syphilis

Involvement of the skull in early congenital syphilis is relatively infrequent, being found in about 5 to 10 percent of all cases involving the skeleton. In nearly all cases there is prominent involvement of the postcranial skeleton. Two types of lesions may affect the skull, a necrotizing osteitis or a hypertrophic

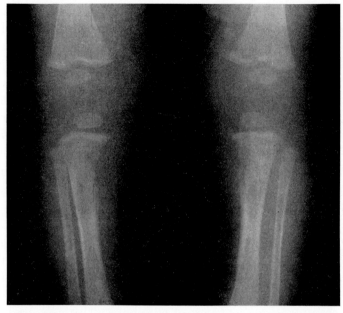

Figure 36. Clinical roentgenogram of early congenital syphilis. The bilateral osteomyelitis affecting the medial portions of the proximal tibiae is known as Wimberger's sign. (Courtesy of Dr. E. B. D. Neuhauser.)

periostitis. The necrotizing osteitis may involve both tables of the skull and form sequestra. The hypertrophic periostitis of the skull has caused a great deal of confusion in the literature as discussed below. It is rarely found in uncomplicated early and late congenital syphilis.

Congenital Syphilis and Rickets

Congenital syphilis and rickets affected large numbers of infants during the middle part of the 19th century when the bone changes of these disorders were first adequately described. Moreover, the severity of the congenital syphilis in these infants produced nutritional disturbances leading to a high incidence of associated rickets. Thus, it is not surprising to find confusion in the early literature regarding the bone changes in both diseases. For example, Jules Parrot described in detail the "rachitic phase" of congenital syphilis with particular emphasis on the formation of prominent bony eminences on the frontal and parietal bones due to subperiosteal bone apposition (Parrot, 1879; see also Barlow, 1879). With this misconception, Parrot diagnosed congenital syphilis in several ancient skulls from Ecuador and Peru simply because the cranial vaults exhibited prominent, circumscribed areas of bone overgrowth. These lesions are more likely the result of rickets, iron deficiency anemia, or one of the congenital anemias (see Chapters VII and VIII).

Even today there remains some confusion regarding the exact etiology of the "hot-crossed bun" appearance of the cranium. Most current textbooks avoid the problem by allotting only a few sentences to the bone changes of congenital syphilis because of its low incidence today. Jaffe briefly states that the cranial bossing is common in rickets but may be found in congenital syphilis rarely. Jeans and Cooke state that such cranial changes are entirely due to rickets (Jeans and Cooke, 1930). Perhaps the best discussion is found in Taylor's classic monograph of 1875 in which he presents the differential diagnosis between congenital syphilis and rickets. Taylor notes the much greater frequency of cranial bossing in rickets than in syphilis and concentrates instead on the markedly different postcranial bone changes in the two diseases.

Late Congenital Syphilis

Early congenital syphilis affecting the bone is a very severe disease with a mortality in the younger age group of almost 50 percent (McLean, 1931). In many cases however the bone changes may spontaneously heal and even disappear after a few years. The syphilitic infection remains latent and may flare up again, particularly in children who had a mild or clinically unrecognized case of congenital syphilis in early infancy. The skeletal manifestations are known as late or tardive congenital syphilis in these children who mainly range from five to fifteen years of age. Unlike early congenital syphilis, the bone lesions in this older age group are usually chronic and the deformities persist for many years.

The bone lesions of late congenital syphilis tend to resemble

TABLE III

BONES AFFECTED IN 49 CASES OF LATE CONGENITAL
SYPHILIS INVOLVING BONE*

One tibia	32	Tibia, femur and skull	1
Two tibias	4	Tibia, humerus and skull	1
Tibia and femur	2	Ulna and fibula	1
Tibia and finger	1	Skull	1
Tibia and ulna	1	Radius	1
Tibia, ulna and radius	1	Ulna	1
Tibia, ulna and skull	1	Humerus	1

* Of the 49 cases, the tibia was involved in 43, the ulna in 5, the skull in 4, the femur in 3, the humerus in 2, the radius in 2, the fibula in 1, and the finger in 1. (From Jeans and Cooke, 1930.)

those of acquired syphilis in their gross appearance and distribution in the skeleton (see Table III). The two main types of lesions are hyperplastic osteoperiostitis and gummatous osteomyelitis. As Table III illustrates, the tibia is by far the most frequently involved followed by the ulna, skull, and femur. The great preponderance of cases involving the tibia is perhaps related to the increased exposure to trauma of this bone. Other evidence indicates that the tibia and other bones frequently involved are the favorite sites for treponemes during the latent stage and subsequent activation in both congenital and acquired syphilis.

Hyperplastic osteoperiostitis primarily involves the diaphyses of the long bones rather than the metaphyses as in early congenital syphilis (see Pendergrass and Bromer, 1929). The tibia is frequently affected and often bilaterally to produce the well known *saber-shin tibia*. Saber-shin tibiae were reported in 43 percent of 202 cases of late congenital syphilis (Stokes, 1934). The resemblance of the tibia to a cavalry saber is due to the marked subperiosteal apposition of new bone on the anterior surface (shin) of the tibia. The new bone is laid down parallel to the cortex and tends to remain distinct from it (see Fig. 37). The thickening produced by the new bone is usually fusiform, involving the middle third of the shaft. The normally sharp anterior aspect of the tibia becomes rounded. In addition, new bone is laid down endosteally causing a narrowing of the

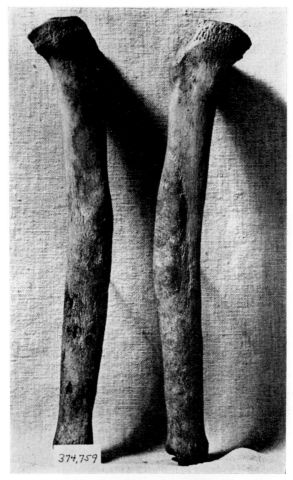

Figure 37. Late congenital syphilis in an adolescent Aleut from Kodiak Island, Alaska. Lateral view of the tibiae shows the apposition of bone along the anterior margins and blunting of the normally sharp anterior surface. The right tibia reveals that this new bone has not completely merged with the underlying cortex. Destructive lesions perforating the left tibia are sites of destructive gummata. (USNM 374,759)

medullary cavity. The original cortex may remain normal or be involved with gummatous osteomyelitis. After several years, the initially porous periosteal and endosteal bone is replaced by very dense, sclerotic bone which is tightly fused to the cortex (see Fig. 38).

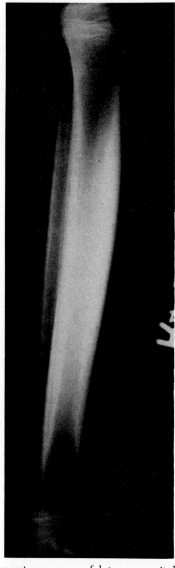

Figure 38. Clinical roentgenogram of late congenital syphilis showing the typical saber-thin tibia. (Courtesy of Dr. E. B. D. Neuhauser.)

In some cases, the tibia is truly bowed forward and flattened laterally. This is caused by accelerated tibial growth during epiphyseal irritation which is hindered at two fixed points by the unaffected fibula (Jaffe, 1972). This bowing deformity differs

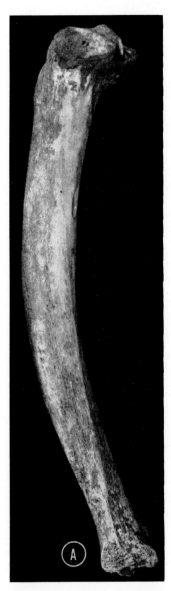 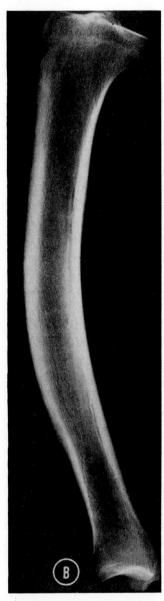

Figure 39. Late congenital syphilis in a young adult. (A.) Lateral view showing true bowing of the tibia has occurred as well as apposition of new bone anteriorly. (B.) Roentgenogram of the tibia. (WAM autopsy specimen)

from that occurring during rickets in which the weakened tibia becomes bowed laterally due to the pressure of weight-bearing.

Gummatous osteomyelitis or osteitis is often present in addition to the osteoperiostitis of the long bones. The osteomyelitis is generally diffuse producing dense and irregular trabeculae and an uneven cortical border as shown by roentgenogram (Shanks and Kerley, 1950; see also Pendergrass et al., 1930). Although the alteration in cancellous structure is diffuse, the main destruction of bone is patchy and localized to the sites of the gummata. The osteomyelitis is chronic and unlike pyogenic osteomyelitis, sequestrum and sinus formation rarely occur. The major bones affected are the tibia and other long bones. The gummatous osteitis involving the skull occurs less frequently but is diagnostically important and will be discussed further under acquired syphilis.

Destruction of the bony and cartilaginous elements of the nose may occur in early and late congenital syphilis to produce a "saddle nose." This occurs in 17.5 to 30 percent of all cases of late congenital syphilis (Ong and Selinger, 1922; Stokes, 1934). Such nasal destruction is nonspecific and must be accompanied by other bone changes to warrant a diagnosis of congenital syphilis.

The *dental stigmata* of late congenital syphilis are important but not pathognomonic signs of this disease. In some series, the reported incidence is as high as 32 percent of 202 cases, 40 percent of 463 cases, 43 percent of 480 cases, and 45 percent of 254 cases of late congenital syphilis (Stokes, 1934; Jeans and Cooke, 1930; Fournier, 1899; Putkonen, 1962). In the latter series, vitamin D prophylaxis prevented rachitic dental changes from confusing the picture and one notes that the incidence of dental stigmata is still high. The dental stigmata are associated with characteristic bone lesions in about 50 percent of all cases, and in such instances the diagnosis is very reliable.

The permanent dentition and in some instances the deciduous molars are the only teeth affected, evidently because these elements form after the treponemes have invaded the fetus. The exact pathogenesis of the dental stigmata remains uncertain and

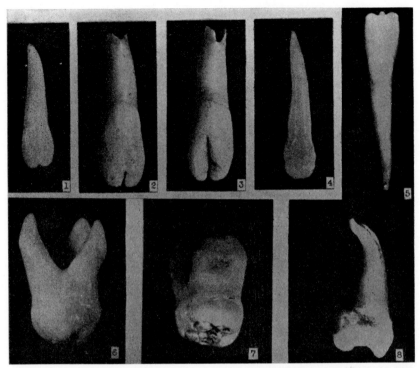

Figure 40. Dental stigmata of congenital syphilis. 1. Maxillary lateral incisor with notch in incisal margin and rounding of incisal angle. 2. Labial surface of fissured maxillary lateral incisor. 3. Lingual surface of fissured maxillary lateral incisor. 4. Shallow depression in incisive and lateral margins of mandibular incisor. 5. Tubercle projecting from incisive notch of mandibular incisor. 6. Maxillary first molar with small occlusal surface and rounded angles. 7. Maxillary first molar with changes typical of a mulberry molar. 8. Maxillary supernumerary tooth showing deformity characteristic of congenital syphilis. (Courtesy of Dr. R. V. Bradlaw, 1953.)

has been reviewed elsewhere (see Karnosh, 1926; Thoma, 1950; Bradlaw, 1953; Putkonen, 1962). The major teeth involved are the maxillary incisors and all first molars. The mandibular incisors are less frequently affected.

Classic Hutchinson's teeth are notched and narrowed incisors, especially the permanent central incisors of the upper jaw (see Hutchinson, 1863, plate I). These incisors are barrel-shaped with convergence of both lateral margins towards the cutting surface. The amount of notching varies widely: In some there is a distinct crescentic notch at the cutting edge, in some there is only a depression on the anterior surface of the tooth immediately above the cutting edge, and in some both of these alterations are present. In many cases no notching is apparent at all, but the convergence of the lateral margins occurs to give the teeth a "screw-driver" shape. Any changes in the teeth are not necessarily bilateral.

The altered first molar in congenital syphilis is known as a mulberry molar, bud molar, dome-shaped molar, Moon's molar, or Fournier's tooth. This molar is considerably smaller than the normal first molar, particularly in the mesiodistal length of the crown. The occlusal surface is rough and irregular with several small knobs representing atrophic cusps. Enamel hypoplasia similar to that seen in rickets may also be present.

In Putkonen's series, 45 percent of the cases had Hutchinson's teeth of the upper central incisors. Of this group, 40 percent had associated mulberry molars. In contrast, 22 percent of the entire series had mulberry molars and 80 percent of this group had associated Hutchinson's teeth. Again, it should be stressed that the presence of dental stigmata is not pathognomonic of congenital syphilis unless associated with characteristic bone lesions such as the saber-shin tibia.

Acquired Syphilis

Of the many studies concerning the antiquity of syphilis and its existence in pre-Columbian America or the Old World, few have approached the problem from an epidemiological perspective. The great majority of reports have been descriptions of isolated specimens exhibiting lesions suggestive of syphilis. None of these reports sought to determine the prevalence of such lesions within the skeletal population and few even attempted to place the individual accurately in time. The following three sections discuss the incidence of bone lesions in cases of acquired syphilis, the probable prevalence of syphilis endemic to early human populations, and the location of osseous syphilitic lesions in the skeleton. Only with this as a background can one intelligently seek to make a paleopathological diagnosis of syphilis using knowledge of gross pathology which will be discussed later.

Incidence of Bone Lesions in Acquired Syphilis

The available literature gives very little information as to the actual frequency of bone involvement in acquired syphilis. Part of this is due to the undoubted existence of unsuspected and symptomless bone lesions which are noted only in radiographic studies of syphilitic patients and resulting in high frequencies of bone involvement. On the other hand, many early studies of syphilis noted bone involvement only when there was demonstrable pain in the area and the resultant frequencies are therefore low. Unfortunately, all studies after 1943 must be suspect because the use of penicillin combined with early systematic and intensive antiluetic treatment has resulted in a dramatic decrease in secondary and tertiary syphilis (cf. King and Catterall, 1959).

The frequencies of bone lesions in acquired syphilis listed in Table IV are from sources predating the use of penicillin and should therefore provide the most accurate information for osseous syphilitic lesions in early human populations.

TABLE IV

FREQUENCIES OF BONE LESIONS IN CASES OF ACQUIRED SYPHILIS

9.5% of 10,000 cases	(McKelvey and Turner, 1934)
11.5% of 399 cases	(Vondelehr et al., 1936)
12.2% of 473 cases	(Bruusgaard, 1929)
14.9% of 314 cases	(Symmers, 1916)
17.0% of 2,231 cases	(Keyes, 1908)
20.8% of 544 cases	(Whitney, 1916)
23.6% of 165 cases	(Wile and Senear, 1916)

It is important to note that about 65 percent of all syphilitic individuals die with no anatomical evidence of syphilis (Bruusgaard, 1929; Rosahn, 1946). As Table IV indicates, the frequency of bone lesions in cases of acquired syphilis ranges between 10 and 20 percent. Perhaps the most reliable figures are those given by Bruusgaard and Vonderlehr et al. who followed a series of untreated patients over many years.

Prevalence of Syphilis in Early Human Populations

The next problem is determining the prevalence of syphilis in early human populations in order to estimate how many individuals in a skeletal series will exhibit syphilitic lesions if syphilis is indeed endemic to the population. Such an estimate is hampered by the possibility that the treponema organism may have been more virulent in earlier times and has gradually become attenuated over a span of several hundreds or thousands of years (Jarcho, 1966). Population size, contact with other populations, and cultural restrictions (or lack of them) concerning promiscuity would also greatly affect the epidemiology of syphilis.

There are many studies of the prevalence of syphilis in the preantibiotic era with widely varying results due to differences in the means of detection and sample selection. For reviews of this confusing and often statistically biased literature see Whitney, 1915; Hazen, 1922 p. 23; Jeans and Cooke, 1930 p. 86; Hudson, 1932-33 p. 452; Stokes, 1934 p. 68; Shattuck, 1938. Nevertheless, one obtains a rough estimate of 5 percent for the prevalence of syphilis in a civilized and largely urban adult population prior to the use of penicillin. Multiplying this number by the 10 and 20 percent frequencies of bone lesions in all cases of syphilis, one obtains the frequency of osseous syphilitic lesions to be expected in a skeletal series representing that adult population. Using the numbers above, this prevalence ranges from .5 to 1 percent. This range is a very rough estimate and further research is obviously necessary to more accurately predict the frequency of osseous syphilitic lesions, particularly in primitive human populations.

A word must be said concerning the prevalence of syphilis in small populations, especially since many early human populations were small in size. With a population composed of several extended families, the number of births per female was often quite high (see Acsádi and Nemeskéri, 1970). Although syphilitic females may have several miscarriages due to the disease, a large proportion of the infants will develop the osseous lesions of early and late congenital syphilis. A representative skeletal series should present evidence of such lesions.

As discussed in a previous section, other forms of treponematosis should be considered as well depending upon the geographic location and the standards of personal hygiene within the population. Indeed, nonvenereal syphilis (bejel) is much more likely to affect small, primitive populations than venereal syphilis. This is supported by a survey of 322 Bosnian villages which revealed that the prevalence of nonvenereal syphilis was inversely proportional to the population (see Table V).

TABLE V

PREVALENCE OF NONVENEREAL SYPHILIS IN A SURVEY OF 322 BOSNIAN VILLAGES*

Village Population	Percent Infected
Less than 200	22.5
200 to 400	16.2
400 to 600	12.6
Over 600	9.1

* The prevalence is inversely proportional to population size. (From Grin, 1952.)

Nonvenereal syphilis produces bone lesions nearly identical to venereal syphilis. Although bone involvement is not as frequent in this predominantly cutaneous disease, the prevalence of nonvenereal or endemic syphilis is much greater. For example, nonvenereal syphilis may infect 25 to 75 percent of the individuals in the small towns of Yugoslavia and in the small nomadic populations of the Near East (Grin, 1952; Hudson, 1958). With a frequency of bone involvement in only 3.6 percent of 9,052 cases (computed from Hudson, 1958), one should expect to find osseous involvement in about 0.9 to 2.7 percent of the total population—which is slightly higher than the estimated frequency for venereal syphilis.

In any case, these low percentages obtained for the frequency of osseous syphilis (venereal or nonvenereal) may explain why few cases of syphilis have been reported in archaeological material on both sides of the Atlantic. Such a paucity of cases has been cited as evidence against the existence of pre-Columbian syphilis in the New World, but in fact it may simply reflect the low frequency of osseous syphilis and the statistical inadequacy of the small skeletal series to exhibit this incidence.

Location of Osseous Lesions in Acquired Syphilis

Literature before the use of penicillin provides fairly adequate information concerning the frequencies of osseous lesions in the different bones of the skeleton. Slight differences in incidence may be noted in Tables VI and VII, but part of this apparent disagreement is due to differences in the presentation of the data. For example, some workers regard the bilateral involvement of the tibia as one lesion while others say two and still others give no indication at all. In any case, the bones most often affected are the tibia, frontal and parietal bones of the cranium, the nasal-palatal region, sternum, clavicle, vertebrae, femur, fibula, humerus, ulna, and radius in that approximate order. As Tables VI and VII indicate, the frontal and parietal bones, bones of the nasal-palatal region, and tibia are by far the most often involved when compared with the rest of the skeleton.

TABLE VI
THE LOCATION OF BONE LESIONS IN 302 CASES OF OSSEOUS SYPHILIS

Bone	I.*	II.*	Bone	I.	II.
tibia	88	119	fibula	3	5
nasal	24	56	maxilla	3	3
parietal	10	35	calcaneus	3	3
frontal	11	29	humerus	3	3
palate	10	28	temporal	0	3
sternum	4	12	occipital	1	2
clavicle	3	11	patella	2	2
ulna	0	10	ilium	2	2
femur	3	9	radius	2	2
ribs	5	9	metatarsals	2	2
mandible	3	9	scapula	2	2
metacarpals	3	8	tarsals	1	1
phalanges	6	8	sacrum	1	1
TOTAL:	170	343	TOTAL:	25	31
Solitary lesions		195			
Multiple lesions		70			
Bilateral lesions		37			
CASES		302			

* Column I. gives the number of cases in which only one bone is involved. Column II. gives the number according to the location of all bone lesions observed. A total of 374 lesions were observed in a total of 302 cases. (Data from Keyes, 1908.)

TABLE VII
STUDIES OF THE LOCATION OF BONE LESIONS IN OSSEOUS SYPHILIS

Bone	Percent	Bone	Percent	Bone	Lesions	Bone*	Lesions
cranium	42.0	nasal	29.6	humerus	28	tibia	34
tibia	26.0	tibia	23.4	femur	26	clavicle	17
sternum and ribs	10.0	palate	23.4	tibia	18	cranium	15
clavicle	7.5	sternum	7.8	clavicle	16	fibula	15
vertebrae	6.0	maxilla	6.2	radius	10	femur	8
femur	4.5	frontal	1.6	scapula	4	humerus	5
fibula	4.0	parietal	1.6	sternum	3	rib	4
humerus	3.5	vertebrae	1.6	vertebrae	3	ulna	3
forearm and hand	2.5	scapula	1.6	ulna	2	scapula	3
other	4.0	ulna	1.6	patella	1	facial	3
		patella	1.6			mandible	2
TOTAL:	100.0	TOTAL:	100.0	TOTAL:	111	radius	1
Data from 239 cases. (Stokes et al., 1944)		Data from 64 cases. (Jullien, 1879)		Data from 49 cases. (Gangolphe and Perret, 1896)		sternum	1
						vertebrae	1
						other	5
						TOTAL:	117
						Data from 67 cases (Kampmeier, 1964)	

* nasal-palatal lesions excluded

Gross Pathology of Osseous Syphilis

The following observations on the gross pathology of osseous syphilis are based on the extensive medical literature and on sixty-eight cases of dried bone syphilis studied at the Smithsonian Institution, Army Medical Museum, Harvard University's Peabody Museum, and the Warren Anatomical Museum at Harvard Medical School. Special emphasis is placed on syphilitic lesions of the cranium because many skeletal series consist only of the skulls and associated mandibles. This is reflected by the fact that thirty-seven of the sixty-eight cases examined consisted only of the cranium. The cranium is also very important because the syphilitic cranial lesions are highly characteristic of syphilis (Williams, 1932; Virchow, 1856; 1896).

Venereal syphilis is acquired through sexual intercourse. The treponeme is then carried to the site of infection by the blood, but bone infection may not occur for several years after the initial infection. Osseous syphilis is classified as tertiary although

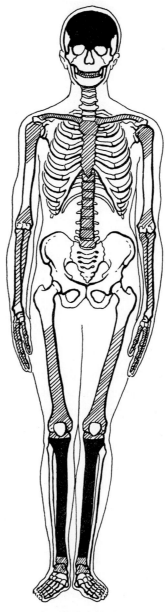

Figure 41. Skeletal distribution of venereal syphilis. The solid black areas indicate the most frequent sites and diagonal lines mark occasional sites.

subperiosteal lesions may be found in secondary syphilis (Hazen, 1921).

Periostitis

Syphilis is a blood-borne infection and the walls of the arterioles are the primary sites of involvement. The metaphyses of the long bones are well vascularized and it is in these areas that the initial periostitis begins. The bones most often affected are the tibia, fibula, clavicle, radius, and ulna. The periostitis is characterized by formation of subperiosteal bone as a reaction to the periosteal inflammation. The subperiosteal bone apposition is usually limited and circumscribed as shown in many cases exhibiting thickening of the medial third of the clavicle or the distal portion of the tibia (Fig. 42). However, when the entire periosteum becomes inflamed, the subperiosteal response results in thickening and even deformation of the involved bone. This is especially true in the tibia, clavicle, radius, and ulna. Apparently the increased incidence of syphilitic involvement in these bones is due to the greater incidence of constant irritation and minor trauma to these more exposed bones. Several cases have been described in which syphilitic osseous lesions followed soon after major trauma to the bone of a person infected with syphilis. Minor but constant trauma to the bones mentioned may produce similar conditions for syphilitic periosteal inflammation (Keyes, 1908).

The new bone deposited on the original cortex is loosely connected to it at first but gradually merges with the underlying cortex as the fiber bone is replaced by dense lamellar bone (Fig. 43). In cases where this new bone layer is thick, it may substitute for the original cortex which may be gradually resorbed or rapidly destroyed by a syphilitic medullary gumma. The gumma sometimes invades the subperiosteal new bone and destroys it as well.

Osteitis and Osteoperiostitis

Osteitis and osteoperiostitis are extensions of the disease from the periosteum to the bony tissue Thus, the bones mentioned

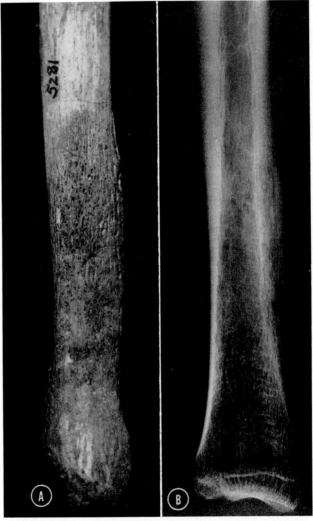

Figure 42. Syphilitic periostitis of the distal tibia. (A.) The gross specimen shows a localized periostitis affecting the medial aspect of the right tibia. (B.) The roentgenogram reveals the subperiosteal new bone and increased thickening of the cortex proximally. (WAM 5281)

above are most often affected as well as the frontal, parietal, and nasal-palatal bones. There is thickening and increased density which is usually not uniform but appears as roughened and irregular areas of bone on the shafts of the long bones and

Figure 43. Syphilitic periostitis affecting the entire left tibia and fibula. Externally, the bones appear irregularly swollen. The longitudinal section reveals the complete remodeling of woven bone into dense, lamellar bone. The cortex is subsequently thickened both externally and internally. (WAM 10363)

cranium. The medullary cavities of the long bones are greatly narrowed by the cortical thickening, particularly in the tibia and the femur may be involved as well (see Figs. 44 & 45). The medullary cavities may also be partially obstructed by the thickened trabeculae of close-meshed spongy bone. Some destruction is evident in the long bones and cranium by pitting on the surface. This is the result of the localized inhibition of the blood supply and not necessarily due to destruction from gummata. Sequestrum formation in the long bones is unusual at this stage.

The bone regeneration reaction of syphilis is interesting because very few infections provoke the formation of such an excess of bone concomitant with destructive granulation produced by gummata. The subperiosteal bone regeneration is quite irregular and spicules of new bone project at right angles from the shaft of the bone forming a fine matrix of trabeculae, particularly in the fibula. A similar radial pattern of subperiosteal spiculation is also found in osteosarcoma, a malignant tumor which will be described in a later section. Indeed syphilitic lesions of the limb

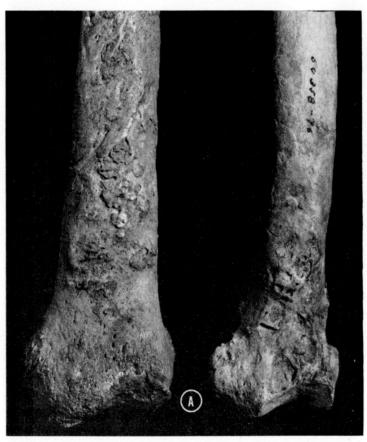

Figure 44. Distal portions of two left femora with osteitis and osteoperiostitis. (A.) Externally, the bones appear roughened by the irregular growths of bone.

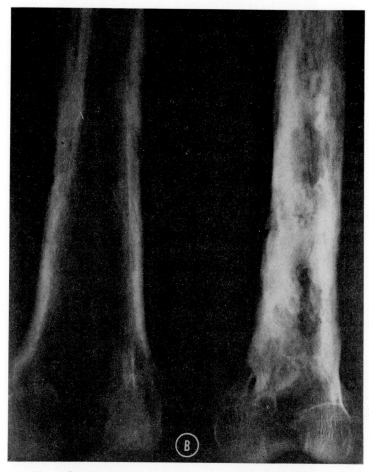

Figure 44B. The roentgenogram reveals a greater amount of new bone formed in the femur on the right. (WAM 1224 and PM 60358-16)

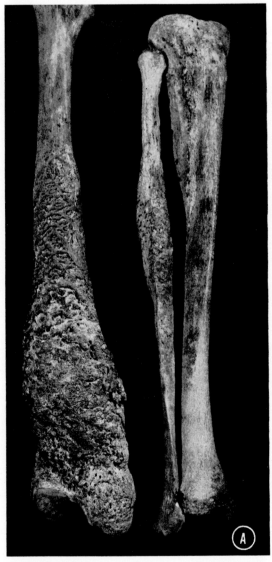

Figure 45. Syphilitic osteoperiostitis in the right femur, tibia, and fibula. (A.) The distal femur is greatly expanded by irregular bone growth. The fibular shaft shows the fusiform swelling often found in this bone.

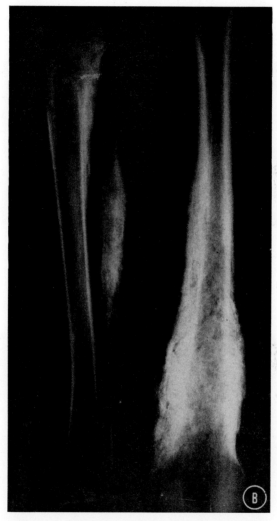

Figure 45B. The roentgenogram reveals slight involvement of the tibia laterally. The femoral epiphysis is practically unaffected by the extensive osteoperiostitis. (WAM 1224)

bones may so closely resemble the malignant osteosarcoma, that some have been mistakenly amputated in an effort to save the individual from the tumor (Stokes et al., 1944).

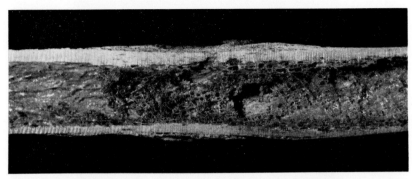

Figure 46. Longitudinal section of a femoral shaft showing the beginning stages of osteoperiostitis. The periosteal new bone has not completely merged with the underlying cortex. Endosteal new bone and coarsening of the spongy trabeculae may be seen internally. (Peabody Museum N/1267)

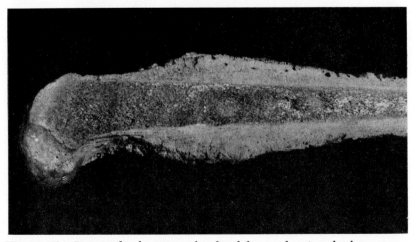

Figure 47. Longitudinal section of a distal femur showing the later stages of osteoperiostitis. The metaphysis is the site of greatest involvement with fusiform swelling. Much of the new bone is lamellar. The spongy trabeculae are thickened and coarse. (WAM 1224)

Gummatous Osteomyelitis

Gumma formation is the combined result of thrombosed blood vessels restricting the blood supply to the bone and to the toxic products of treponeme degeneration. The necrotizing reaction may be surrounded by a zone of collagenized connective tissue, and the central portion of the lesion undergoes a degenerative process that transforms it into a moderately firm, amorphous mass called a *gumma*. Inoculation studies indicate that gumma formation is a hyperimmune reaction to reinfection (Grin, 1952; Magnuson et al., 1956). The reinfection with treponemes or other possible factors may cause a reactivation of residual treponemes in a sensitized individual, thus upsetting the host-parasite relationship.

Gummata vary greatly in size. They are often small and discrete, but may coalesce to form a large mass filling the entire medullary cavity. The periosteal gummata are usually much smaller and fewer in number than those occurring in the medullary cavity. Most gummata begin in the metaphysis but extensive involvement of the entire marrow cavity soon results with both proliferative and degenerative changes. All bone is destroyed in the space occupied by the gumma but the surrounding bone often becomes very sclerotic. As the gumma resorbs by the action of leukocytes and macrophages, some new bone may be found in the necrotic cavity.

The secondary involvement of bone from soft tissue gummata may result in osteophytic overgrowths as well as sieve-like formation from gummatous destruction along the Haversian canals. Sequestra may form but even in this advanced stage of infection, they occur infrequently. When sequestra do form, they are found in the cancellous bone of the metaphyseal and epiphyseal regions. Portions of the skull may sequestrate when multiple small areas of gummatous destruction become confluent.

Radiologically, the long bones will reveal a marked thickening of the cortex, roughened along the external border, generally smooth along the medullary canal, and interspersed with radiolucent gummatous destruction (See Figs. 49 & 50).

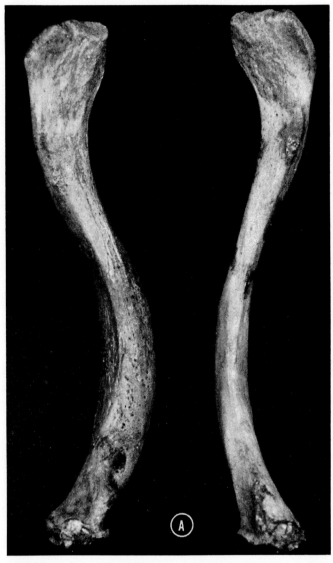

Figure 48. Gummatous osteomyelitis affecting the left clavicle. (A.) The medial portion of the clavicle appears swollen compared to the nearly normal right clavicle. An area of gummatous destruction has perforated the cortex near the medial end.

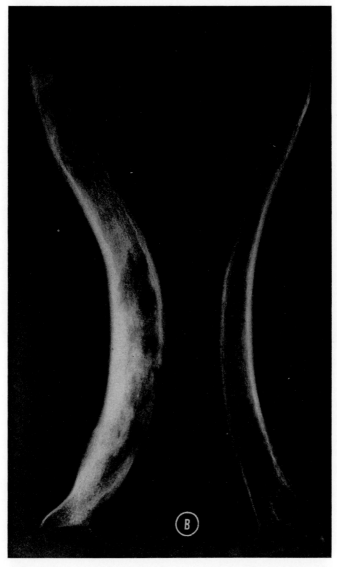

Figure 48B. The roentgenogram shows pronounced new bone formation interspersed with lytic areas occupied by gummata. (PM N/1267)

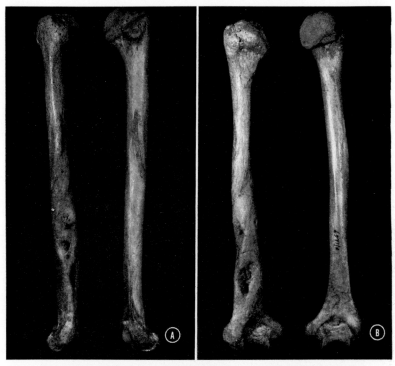

Figure 49. Gummatous osteomyelitis affecting the left humerus compared to the normal right humerus. (A.) Lateral view showing swelling of the distal portion and several perforations of the cortex. (B.) Posterior view reveals the extent of new bone growth.

Vertebral Syphilis

Both congenital and acquired syphilis may affect the vertebral column, but the bone changes are often very slight (see Whitney and Baldwin, 1915). The cervical vertebrae are most frequently involved, and disease rarely affects more than two vertebrae (Ziesché cited by Hazen, 1921). The periosteum of the vertebral body is usually the initial site of involvement with both necrosis and later new bone formation. The vertebral body may become rounded and larger than normal in size. Large bony spurs sometimes develop between the vertebrae closely resembling degenerative osteophytosis.

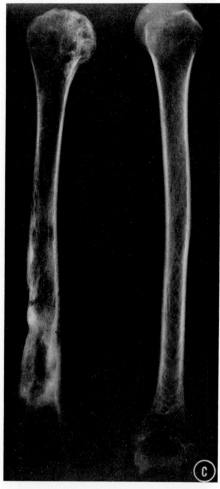

Figure 49C. The roentgenogram reveals the generalized involvement of the humerus including the humeral head. (PM N/1267)

Cranial Syphilis

Several case studies have found the cranium to have the highest incidence of osseous syphilitic lesions (Von Zeissl, 1886; Stokes et al., 1944). Of the thirty-seven crania which I examined, all thirty-seven exhibited frontal lesions, twenty-five had lesions on one or both parietals, five presented lesions on the occipital,

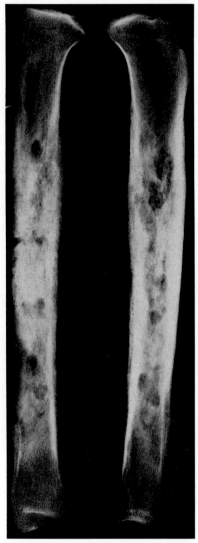

Figure 50. Roentgenogram of a sectioned left tibia showing syphilitic gummatous destruction originating in the medullary cavity and perforating the cortex. (PM N/1267, roentgenogram courtesy of Dr. C.W. Goff.)

and only two had lesions on one or both temporal bones. Similar frequencies have been noted by others (Hazen, 1921; Stokes et al., 1944). Unlike other series which report a high frequency of nasal-palatal involvement, only eight cases (21.7 percent) of such lesions were found and many were in very advanced cases of cranial syphilis. Perhaps this difference is due to an inclusion by other workers of cases exhibiting cartilaginous destruction of the nasal area and destruction of the soft palate which are not present in dried bone material.

Erosion of the cranial vault by gummata has long been regarded as highly characteristic of acquired syphilis (Virchow, 1856, 1896; Williams, 1932). This gummatous destruction begins almost invariably on the external surface of the calvarium by extension of the infection from the soft tissues of the pericranium. There is extensive bone resorption and this destruction follows the paths of the small blood vessels entering the cranium from the pericranium. The frontal bone is frequently the first bone involved. Soon multiple, scattered areas of destruction occur and coalesce into a much larger area of destruction.

Two apparently opposed processes are evident in each area of calvarial erosion (see Virchow, 1856; Coulson, 1869). In the center of the lesion, the destruction produces a depression reaching down to the spongy part of the diploë. While the destructive process is going on in the cranial depression, a regenerative process takes place around the circumference laying down new bone which gradually becomes very sclerotic. When the gummy matter is finally resorbed, the stellate lesion characteristic of cranial syphilis remains (see Figs. 51 & 52).

Further progression of the erosion increases the extent and depth of the depression. As the destruction deepens, new bone is deposited peripherally around the edge of the depression, and the borders acquire a folded or wrinkled appearance. If there is subsequent healing at this stage, the wrinkles appear to emanate from the initial stellate lesions scattered about the frontal and parietal bones. The new bone gradually merges with the surface of the cranium and becomes very sclerotic. The erosive and reparative processes may completely obliterate the diploë.

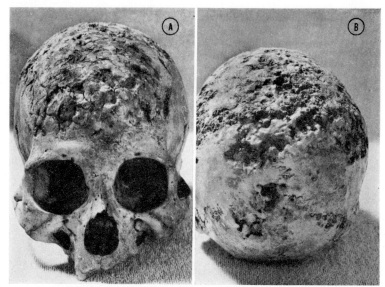

Figure 51. Cranial syphilis in a female Eskimo from St. Lawrence Is., Alaska (500-1700 A.D.). (A.) Frontal view showing the irregular growth of new bone which gives the vault a wrinkled or stellate appearance. (B.) Posterior view showing greater destruction of bone where the infection is still active. Sclerotic clumps of bone are evident in the midparietal region.

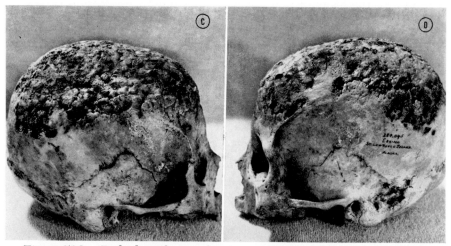

Figure 51C. Right lateral view showing progression of the infection from the frontal bone posteriorly to the parietals. (D.) Left lateral view. (USNM 280,095)

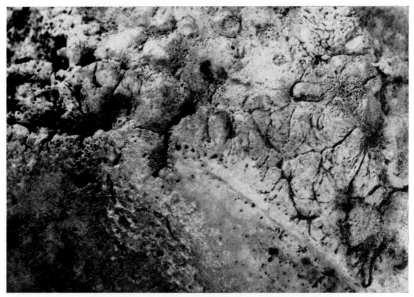

Figure 52. Close-up view of the right frontoparietal region in the previous figure. This illustrates several important features of cranial syphilis. To the right, are the stellate scars formed by new bone production around foci of gummatous destruction. To the left, the gummatous destruction and bone necrosis are more extensive producing a "worm-eaten" appearance in the cranium. Note also the relative lack of syphilitic lesions below the temporal line. (USNM 280,095)

When the calvarial erosion is very deep and extensive, areas of the vault are cut off from their blood supply. Necrosis of these areas is also influenced by the rapid progress of the gummatous granulation and the frequent occurrence of a secondary pyogenic infection (MacFarlane and Rannie, 1958). The extensive bone necrosis combined with gummatous erosion may cause large sequestra to form in the cranium. In the final stages, perforation of the entire thickness of the calvarium occurs. The overall picture of the cranium often reminds one of a "worm-eaten" appearance.

Syphilis of the Joints

Joint lesions constitute about 12 percent of all cases of osseous syphilis (Stokes et al., 1944). The three major types of syphilitic

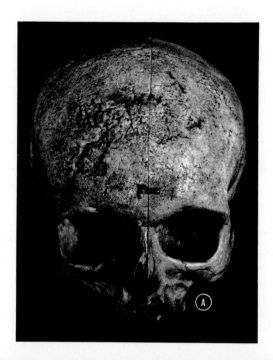

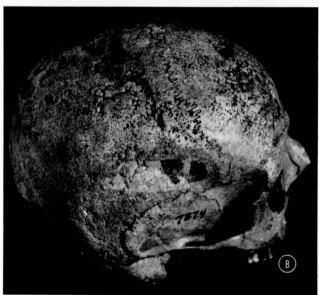

Figure 53. (A.) Frontal and (B.) lateral views showing the stellate scars of cranial syphilis. Note the area of active destruction in the right frontal region. (WAM 7874)

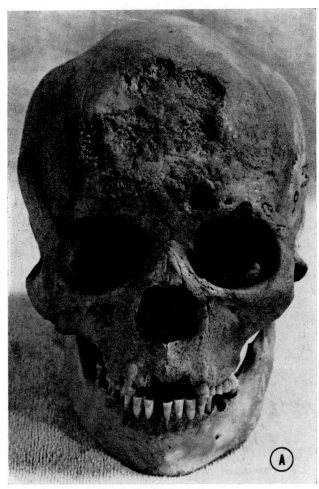

Figure 54. Cranial syphilis. (A.) Female Eskimo from Kagamil Island, Alaska and probably of early historic date. Syphilitic lesions have coalesced to affect a large portion of the frontal and nasal bones. Stellate scars are evident and a few foci of destruction are located more posteriorly. (USNM 377,819)

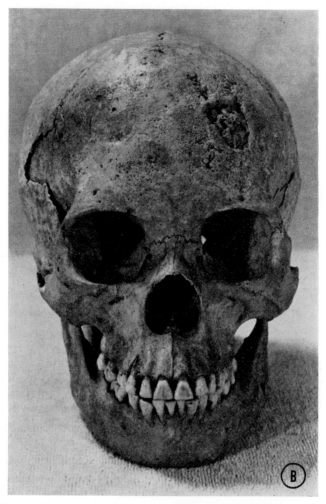

Figure 54B. Female Indian from Arkansas. The large lesion in the frontal bone has a definite border of new bone surrounding the destruction. Stellate scars are evident around the periphery. Similar lesions are present in both parietals. The left side of the jaw is greatly thickened and perforation resembling a sinus is present. This specimen presents the problems typical of many archaeological specimens: lack of accurate dating and absence of postcranial remains. The lesions are strongly suggestive of syphilis, but hematogenous osteomyelitis must also be considered. (USNM 263,413)

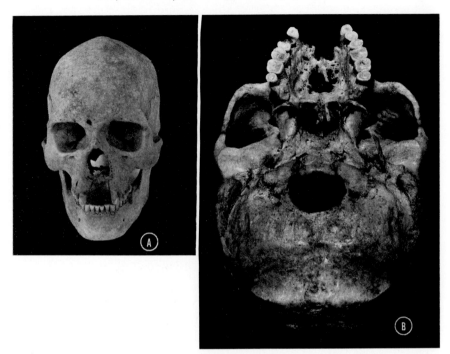

Figure 55. Adult male Indian of the Mississippian Culture in Illinois (about 1200-1400 A.D.). (A.) Frontal view shows considerable destruction of the nasal area, including the anterior alveolar area. The margins of the nasal orifice are rounded by bone repair following an inflammatory reaction. The vault bones are slightly affected. (B.) Basal view shows more clearly the anterior alveolar destruction and irregular perforation of the hard palate. Postcranial bones with characteristic syphilitic lesions include both tibiae, right femur, right ulna, both humeri, and left clavicle. (Courtesy of Dr. Dan Morse, 1969.)

arthritis are: Clutton joints in late congenital syphilis, gummatous arthritis of acquired syphilis, and Charcot joints in tabes dorsalis. The large joints are the most frequently involved, such as the knees followed by the elbow and less frequently by the ankle, shoulder, and wrist joints (Wile and Senear, 1916).

In late congenital syphilis, joint involvement is usually the result of infection of the synovial membrane to produce Clutton joints. The lesions are almost always bilateral and commonly affect the knees. The inflammation is mainly confined to the soft tissues of the joint, but may affect the bone in chronic cases resembling pyogenic arthritis.

In acquired syphilis the arthritic changes more often result from direct extension of gummatous destruction from the metaphysis to the epiphysis. This occurs in the large diarthrodial joints with capsules extending to the metaphyseal area, because gummata seldom erode through the articular cartilage. The joint involvement is unilateral and rarely affects more than one joint. Inflammation of the joint causes production of granulation tissue and pannus formation. The articular cartilage is then destroyed and typical osteoarthritic bone changes develop such as osteophyte formation, marginal lipping, and eburnation (see Chapter IX). Definite areas of destruction may be present in the articular surfaces as the result of gummatous involvement and ankylosis may occur.

The Charcot joint is not a true syphilitic lesion, but it may be found in late syphilis. It is a neuropathic arthrosis resulting from tabes dorsalis. Tabes dorsalis in turn occurs in many cases of advanced tertiary syphilis and is a degeneration of dorsal root ganglia producing loss of joint sense. Similar neuropathic joint changes may be found in leprosy, diabetes mellitus, and severe nutritional deficiency (Eichenholz, 1966).

The knee is most frequently involved followed by the ankle, hip, and spine. The gross morphology of a Charcot joint justifiably resembles osteoarthritis since it is a progressive degeneration of an insensitive joint. Marginal hyperplasia is excessive and fractures through the osteophytes and articular facets are common producing even more new bone formation. Porotic atrophy of both old and new bone occurs through disuse of the limb in severe cases.

A Differential Diagnosis of Syphilis

The lesions of syphilis often imitate other infections, hereditary disorders, and neoplastic processes and a thorough knowledge of a differential diagnosis is particularly important. Thickening of the table and widening of the diploic spaces may form in the calvarium as in meningioma and the hereditary anemias. However, the presence of syphilitic infection in postcranial bones usually makes a differential diagnosis possible. Sometimes a multiple myeloma or metastatic carcinoma may be confused with the destructive syphilitic perforations of the skull. However, these tumors appear as small, not especially necrotic, defects in the cranium accompanied by little if any bone regeneration. The tumorous lesions are widely scattered and rarely coalesce unlike the large areas of necrosis in cranial syphilis.

Skeletal tuberculosis often proves difficult to distinguish in a differential diagnosis of osseous syphilis. In the early stages, tuberculosis tends to be progressive with predominant bone destruction. Even in later stages there is very little new bone production. The hypertrophic "swelling" of the long bone diaphyses rarely occurs in tuberculosis compared to syphilis. Tubercular erosion of the cranium is infrequent but usually begins in the inner table instead of the frequent involvement of the outer table by syphilis.

The confusion of syphilis with primary osteosarcoma has already been noted in the discussion of syphilitic osteoperiostitis. Osteosarcomas affect a much younger age group and usually affect only one bone while syphilitic lesions often occur bilaterally in the long bones and are characterized by multiple lesions in these bones.

Pyogenic osteomyelitis may closely resemble the osseous lesions of syphilis in the initial stages of periostitis and osteitis. However, more bones are involved in advanced cases of syphilis and there is a notable lack of suppurative cloaca and sequestra formation in postcranial syphilis. Pyogenic osteomyelitis rarely involves the cranium except as a secondary infection of syphilitic gummata in this area.

Paget's disease may have postcranial lesions very similar to syphilis. However, the massive thickening of the skull vault is very different from cranial syphilis. Furthermore, histologic examination of macerated bone affected by Paget's disease reveals a mosaic architecture not found in syphilis.

NONVENEREAL SYPHILIS (ENDEMIC SYPHILIS, BEJEL)

Nonvenereal syphilis, also known as endemic syphilis or bejel, is a form of treponematosis with clinical manifestations intermediate between those of venereal syphilis and yaws. Cardiovascular lesions occur with some frequency, but involvement of the central nervous system is rare. Like yaws, nonvenereal syphilis is usually acquired in childhood under unhygienic conditions through body contact or use of common eating and drinking utensils. Nonvenereal syphilis is found today only in underdeveloped countries with warm, arid climates such as parts of north and south Africa, a large part of the Middle East, Yugoslavia, Central Asia and the Australian interior (see Hackett, 1963).

With the influx of civilization and better living conditions, nonvenereal syphilis is rapidly disappearing from these areas. It is therefore not surprising that nonvenereal syphilis was more widespread centuries ago when large areas of the world were inhabited by people living under primitive and unhygienic conditions. Historically, one finds nonvenereal syphilis as the sibbens of Scotland, button scurvy of Ireland, radesyge of Norway, saltfluss of Sweden, spirocolon of Greece and Russia, and frenga of Bosnia (Grin, 1935; Willcox, 1960; Morton, 1967).

Frequency of Bone Involvement

Bone lesions are uncommon in the early stage of nonvenereal syphilis and the few that do occur usually heal spontaneously. Even in the late stage, bone lesions are relatively uncommon. Among the Bedouins and villagers of Syria, about 3.6 percent of 9,052 cases of nonvenereal syphilis had demonstrable bone lesions (computed from Hudson, 1958a). In a study of nonvenereal syphilis in Bechuanaland, 1.1 percent of 9,656 cases exhibited bone lesions. When only active cases were considered, bone involvement was present in 22.2 percent of the cases (Murray et al., 1956). Nasal-palatal destruction and osteoperiostitis were found in about 5 percent of 3,507 cases of nonvenereal syphilis in Iraq (Csonka, 1953). Thus, a skeletal series from an area where nonvenereal syphilis is endemic might be expected to reveal bone lesions due to the disease in roughly 1 to 5 percent of the entire series.

Location of Bone Lesions

Unlike venereal syphilis, the cranial vault is rarely affected by nonvenereal syphilis, comprising only 4 percent of ninety-nine cases involving bone (Murray et al., 1956). Even when involved, a localized osteitis without extensive destruction is the only result. On the other hand, destruction of the nasal-palatal region is a prominent and frequent feature of nonvenereal syphilis (Csonka, 1953; Hudson, 1958a). Of the long bones, the tibia and fibula are by far the most frequently affected followed in rapidly decreasing order by the ulna and radius, clavicle, phalanges, and calcaneus (Murray et al., 1956). The diaphyses and metaphyses are the favored sites of bone changes and joint lesions are rare.

Gross Bone Pathology

The bone lesions of nonvenereal syphilis are identical to those found in venereal syphilis, differing only in the relative frequency of various types of lesions. In this latter respect, nonvenereal syphilis closely resembles yaws which also primarily infects children. The bone changes of yaws are fully discussed in the following section so a brief description of the bone lesions in nonvenereal syphilis is presented here.

Nonvenereal syphilis closely resembles late congenital syphilis. Osteochondritis and dental stigmata are absent because these occur only when the treponemes invade the fetus *in utero*. The earliest manifestation is usually a localized subperiosteal apposition of bone. The periosteal reaction is mainly in the middle portion of the shaft, creating a spindle-shaped bone. The tibia is often involved resulting in the typical saber-shin tibia. The layers of new bone are usually parallel to the shaft, but may sometimes project at right angles from the shaft (Rost, 1942; Merriweather, 1953).

The early lesions may heal spontaneously or progress to the stage where the medullary cavity is narrowed by endosteal bone.

In addition, gummatous destruction of cortical tissue ensues creating circumscribed areas of rarefaction as revealed by the roentgenogram (see Figs. 56 & 57). The shaft of the long bone at this stage appears greatly distorted by the periosteal proliferation and localized areas of cortical destruction.

The nasal-palatal destruction is identical to the "gangosa" of tertiary yaws and is almost as frequent as the saber-shin tibia (Murray et al., 1956). All degrees of destruction may be found, ranging from perforation of the hard palate or destruction of the nasal bones to horrible destruction of all bony structure in the region (see Jones, 1953).

Paleopathology

Studies of nonvenereal syphilis in Syria, Iraq, and southern Africa state that the natives regard the disease as of very ancient

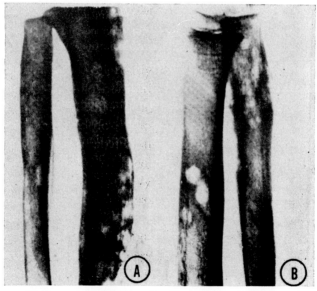

Figure 56. Nonvenereal syphilis (bejel). (A.) Lower leg of a six-year-old boy who had had nonvenereal syphilis one year previously. The medullary cavity is narrowed and the bones are irregularly thickened by subperiosteal bone. Scattered areas of destruction are evident. (B.) Distal end of the forearm in the same individual showing localized areas of bone destruction. (Courtesy of Dr. E. H. Hudson, 1958.)

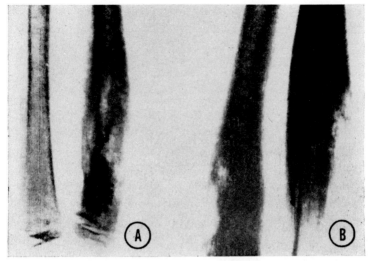

Figure 57. Nonvenereal syphilis (bejel). (A.) Enlargement of the ulna in a child who had acquired bejel several years earlier. The location and appearance of the lesions are similar to that found in syphilis. (B.) Swelling in both ulna and radius produced by subperiosteal bone apposition. (Courtesy of Dr. E. H. Hudson, 1958.)

origin. Csonka briefly examined "a few crates of skeletons and individual bones estimated to be 3000 to 4000 years old" but did not find any evidence of osteoperiostitis attributable to any form of treponematosis. A solitary skull from Iraq and dated around 500 A.D. has lesions suggestive of nonvenereal syphilis or yaws (Kail and de Froe, 1953; see also Guthe and Willcox, 1954). Two crater-like depressions are evident in the frontal bone of this specimen. The largest depression is about 2 cm in diameter and extends to the inner table. The other depression is much smaller and not as deep. A slight sclerosis of bone is found around both lesions. The appearance of these lesions is very similar to that found in cranial yaws, but it should be noted that Iraq is currently an area where nonvenereal syphilis is common and it is outside the tropical belt where yaws is endemic.

From the arid and semiarid regions of Australia have come many possible precontact and early historic specimens of nonvenereal syphilis (McKay, 1938). A Tasmanian skull of an aged male living in the early 1800's exhibits extensive osteitis caused by venereal or perhaps nonvenereal syphilis (Møller-Christensen and Inkster, 1965; Hackett, 1974).

YAWS (FRAMBOESIA, PIAN, BUBAS)

According to the Unitarian theory, yaws is one of the four syndromes of treponematosis caused by *T. pallidum* which includes: venereal syphilis, nonvenereal syphilis (bejel), yaws, and pinta. Others insist that yaws is a separate disease caused by *T. pertenue*. Endemic yaws is a disease of the hot, humid tropics and rarely found outside the 20th parallels north and south of the equator. Prominent localities are the West Indies, tropical South America, an immense portion of tropical Africa, India, Southeast Asia, various Polynesian and Melanesian islands, New Guinea and northern Australia (Guthe, 1969; Williams, 1935).

Yaws is primarily a disease of childhood, but the condition may last for many years into adulthood, often in latent form. The cutaneous lesions are highly infectious and transmission is predominantly by body contact between scantily clad children and by contaminated insects such as flies. In endemic areas as much as 10 to 25 percent of the population will exhibit some form of active yaws and a much larger portion will show evidence of past infection (Taneja, 1968; Guthe, 1969).

Frequency of Bone Involvement

Studies of yaws give widely differing frequencies of bone involvement. In several reports the complaint of bone pain constituted bone involvement and very high frequencies resulted (see Harley, 1933; Taneja, 1968). Radiographic studies have shown that many cases of bone pain reveal no evidence of change in the bone architecture. The bone lesions of secondary yaws often heal spontaneously, returning to normal bone architecture within several months. The more significant bone lesions mainly occur in tertiary yaws. They are chronic and cause permanent changes in the bone. Such bone changes were found in 15.3 percent of 699 cases of tertiary yaws (Van Nitsen, 1920). Approximately 10 percent of all yaws infections progress to the tertiary stage. Thus, a skeletal series from an area of endemic yaws might be expected to reveal bone lesions due to yaws in roughly 1 to 5 percent of the entire series. This estimate is supported by a study of 1,423 consecutive cases of yaws in which 4.7 percent had demonstrable bone lesions (Wilson and Mathis, 1930).

Location of Bone Lesions

The distribution of bone lesions in yaws is illustrated in Figure 58. The tibia is by far the most common bone involved followed by the fibula, medial portion of the clavicle, femur, ulna, and bones of the hands and feet (Maul, 1919; Goldman and Smith, 1943; Helfet, 1944; and Oosthuizen, 1949). The skull is infrequently involved although the nasal area may exhibit prominent destruction. The diaphyses and metaphyses of the long bones are the most common sites of bone changes. The joints are rarely involved with significant bone changes.

Gross Bone Pathology

The early literature on the bone lesions of yaws was thoroughly reviewed by Cecil J. Hackett in 1946. Except for his classic monograph, little has been added to our knowledge of yaws affecting the bone (Hackett, 1951; see also Delahaye et al., 1970). Hacket and many others have repeatedly emphasized the close similarity of the bone lesions in both yaws and syphilis (see Lunn, 1961; Wilson, 1973). Indeed, except for the osteochondritis and dental stigmata of congenital syphilis, no lesion is found in one disease that may not be observed in the other. The differences are merely quantitative, and it is this aspect that will be stressed in this section.

DACTYLITIS. Prominent bone changes in the hand are often found in younger individuals with yaws in contrast to the infrequency of dactylitis in syphilis. The metacarpals, metatarsals, and phalanges are the major bones involved. The affected bones appear "swollen" by subperiosteal bone apposition parallel to the cortex and resorption of the original cortex. Many trabeculae are resorbed and the remaining ones appear coarsened in the

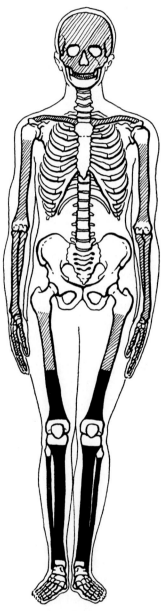

Figure 58. Skeletal distribution of yaws. The solid black areas indicate the most frequent sites and the diagonal lines mark occasional sites.

roentgenogram. The dactylitis often affects several bones simultaneously except in adult cases where only one bone is usually affected (see Fig. 59).

SABER-SHIN TIBIA. The saber-shin tibia found in late congenital syphilis and acquired syphilis is a prominent feature of yaws as well. About 40 percent of all cases of osseous yaws affected the tibia (Maul, 1919; Goldman and Smith, 1943). As in late congenital syphilis, there may be true forward bowing of the tibia in addition to the marked subperiosteal apposition of bone along the anterior aspect of the tibia. This true bowing explains the term "boomerang leg" describing the legs of many Australian aborigines (Hackett, 1936; see also Garner et al., 1970).

CRANIAL YAWS. In contrast to the frequency and severity of cranial syphilis, the involvement of the calvarium by yaws is less common and less destructive (Williams, 1935). In 119 cases of osseous yaws, the cranial vault was affected twenty times (16.8 percent), and in only one instance did the bone destruction approach that found in syphilis (Butler, 1951). Instead, the bone changes of yaws consist of a few crater-like depressions located mainly in the frontal bone. The depressions do not perforate the inner table and are surrounded by a slight bony thickening (Hackett, 1951).

GANGOSA (RHINOPHARYNGITIS MUTILANS). The word "gangosa" is Spanish for "nasal voice" and describes the extensive destruction of the nasal region in tertiary yaws. Fortunately for its victims, the condition is very uncommon in yaws. It occurs in about 1.0 percent of all cases of yaws or roughly 7.5 percent of all cases of osseous yaws (Taneja, 1968). Perforation of the hard palate alone occurs in another 5 percent of cases. Nasal involvement is common in syphilis, but the destruction is seldom as severe (see Church, 1939). The ulcerating process begins in the soft tissues of the nose and rapidly spreads to the bone (Bittner, 1926). In severe cases, the process creates a single space of the orbital cavities, nose, and pharynx with destruction of the nasal bones, hard palate, and parts of the maxilla (Ash and Spitz, 1945). Bone sclerosis occurs at the periphery of the destruction but the regeneration is not prominent.

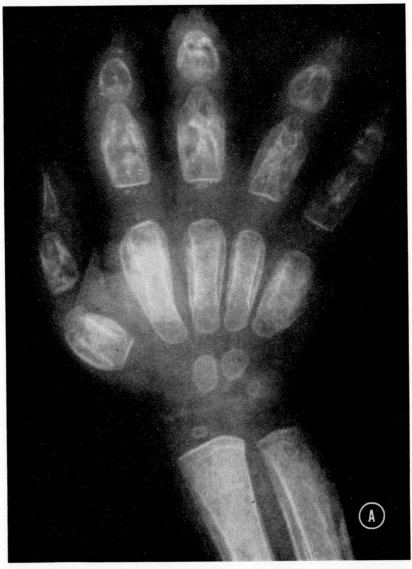

Figure 59. Dactylitis in yaws. (A.) Right hand of a two-year-old boy who had yaws for three months. There is prominent subperiosteal apposition of bone in the hand and forearm bones. In several of the phalanges, the original cortical outlines are lost so that the bones appear swollen. Many trabeculae are resorbed and the remainder are coarsened. (Courtesy of Dr. Cecil J. Hackett, 1951.)

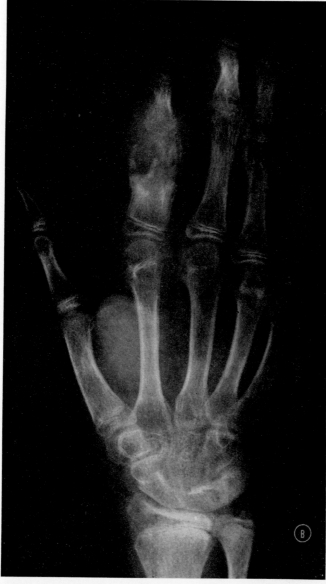

Figure 59B. Right hand of a fourteen-year-old girl with yaws. There is expansion and destruction of the second proximal phalanx. Other bones in the hand appear osteoporotic due to disuse of the hand for several months. (Courtesy of Dr. Cecil J. Hackett, 1951.)

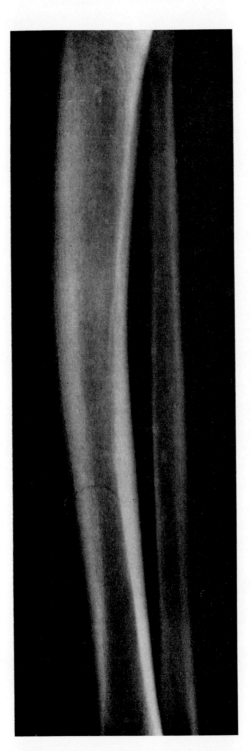

Figure 60. Lateral view of the right leg of a seven-year-old boy with yaws. There is slight tibial bowing and thickening of the cortex anteriorly. (Courtesy of Dr. Cecil J. Hackett, 1951.)

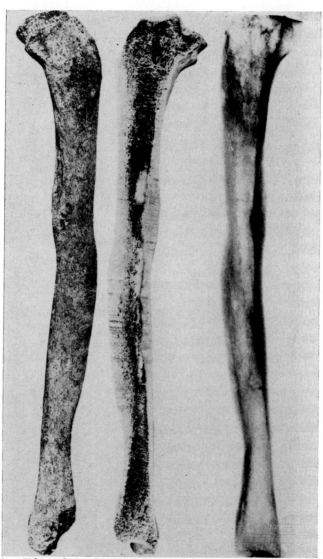

Figure 61. Tibia of an aborigine from Northern Australia. The sectioned specimen and roentgenogram reveal nodular areas of subperiosteal bone apposition due to yaws. (Courtesy of Dr. Cecil J. Hackett, 1936.)

150   *Paleopathological Diagnosis and Interpretation*

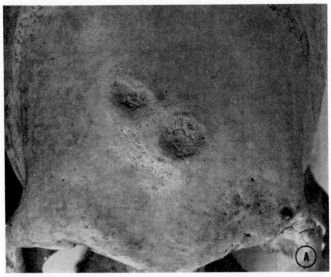

Figure 62. Cranial yaws. (A.) Crater-like lesions on the frontal bone of an Australian aborigine with yaws.

Figure 62B. Possible cranial yaws in an adolescent from the Mariana Islands and dating around 850 A.D. (Courtesy of Dr. T. Dale Stewart from Stewart and Spoehr, 1952.)

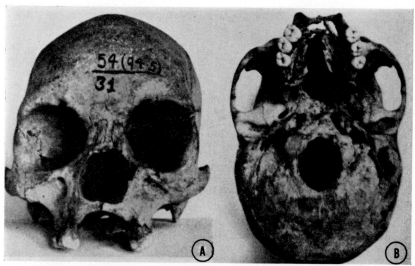

Figure 63. Gangosa in an aborigine from Victoria, Australia and probably predating European contact. (A.) Frontal view of the skull showing extensive destruction of the central maxilloalveolar region. The nasal bones and nasal processes of the maxillae are thickened and spongy. (B.) Basal view of the skull showing destruction of most of the hard palate. The alveolar process is represented only as a narrow bar of bone below the nasal orifice. (Courtesy of Dr. Cecil J. Hackett, 1936.)

Skeletal evidence of gangosa without associated postcranial lesions is not pathognomonic of yaws because several other diseases can present this picture. The three forms of Leishmaniasis have an almost identical geographical distribution and each can cause destruction of the nasopharynx. *American leishmaniasis*, also known as *Uta* or *Espundia*, is particularly apt to cause extensive destruction of this area. The term "tropical ulcer" refers mainly to the cutaneous lesions of leishmaniasis and treponematosis, but also includes chronic ulcers of the legs caused by mixed bacteria in persons suffering from starvation and neglect. The bone changes in the distal tibia and foot closely resemble those seen in yaws (Brown and Middlemiss, 1956). Finally, leprosy may produce similar destruction of the nasal-palatal bones, but the periosteal reaction in the long bones is not as common.

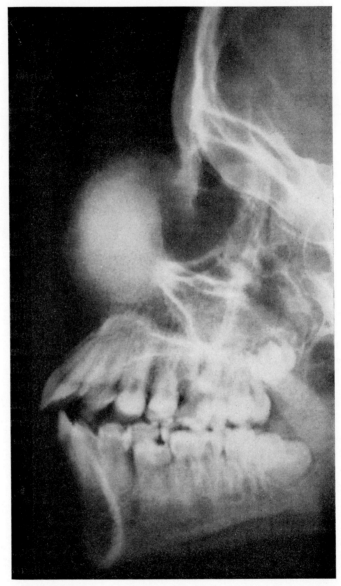

Figure 64. Goundou in a young child from Liberia. The clinical roentgenogram reveals the tumor-like expansion of the superior maxillae bilaterally. (Courtesy of Dr. E. B. D. Neuhauser.)

GOUNDOU. Goundou is an African term meaning "large nose" and describes a tumor-like expansion of the superior maxilla, often bilaterally (Cornet and Vilasco, 1971). The condition is even less common than gangosa and found almost exclusively in children. The fluid-containing cysts are surrounded by a thin rim of bone and may become so enlarged that vision is obstructed (see Fig. 64).

GUMMATOUS OSTEOMYELITIS OF THE LONG BONES. Localized cortical destruction produced by gummata is a prominent feature of tertiary yaws. The lesions are often irregularly oval with the long axis parallel to the shaft of the bone. They range in size from a few millimeters to 3 cm (Maul, 1919).

The periosteal reaction in syphilitic osteomyelitis is usually much more pronounced than in yaws (Goldman and Smith, 1943). Bone reaction to the gummatous destruction in yaws may be minimal giving the bones a "punched out" appearance (Oosthuizen, 1949). However, the frequency of saber-shin tibia in yaws must be emphasized, and in most cases the periosteal reaction is localized and nodal. Osteoperiostitis, with spicules of bone projecting at right angles from the shaft, is much more common in syphilis than in yaws (Helfet, 1949).

Paleopathology

Strong paleopathological evidence of yaws comes from Tinian, one of the Mariana Islands (Stewart and Spoehr, 1952). The material consists of a skull, three associated long bones, and a portion of a tibia from another individual (see Fig. 68). Radiocarbon dating places the remains around 850 A.D. The skull and long bones are from an individual thirteen to fourteen years old and the tibia is also from a subadult. As Figure 62 illustrates, a crater-like lesion is present in the frontal bone and similar lesions are evident on both parietals. New bone production is minimal. Circumscribed areas of cortical destruction are evident in the three long bones, again with minimal bone response. The fragmentary tibia exhibits marked subperiosteal apposition of bone

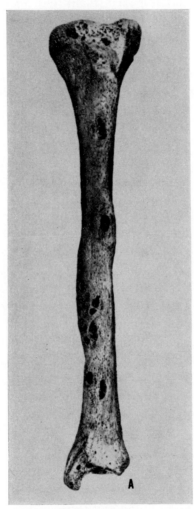

Figure 65. Gummatous osteomyelitis of yaws. (A.) Left tibia of a native from eastern New Guinea showing nodular swelling and many necrotic perforations of the cortex.

Treponema Infection: Syphilis, Bejel, and Yaws

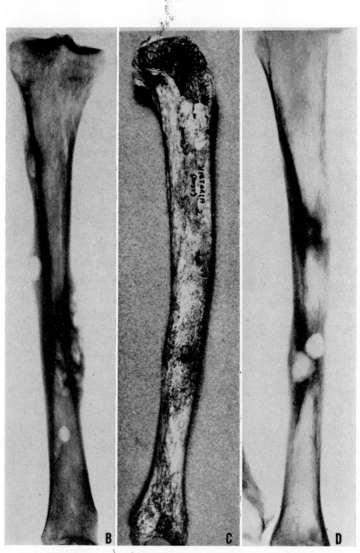

Figure 65. (B.) Roentgenogram of the same specimen showing the localized areas of cortical destruction. Many are located in regions of subperiosteal new bone formation. (C.) Right tibia of an aborigine from Victoria, Australia showing necrotic perforations in a region of nodular swelling. (D.) Roentgenogram of the same specimen showing the punched-out appearance of the lesions. (Courtesy of Dr. Cecil J. Hackett, 1936.)

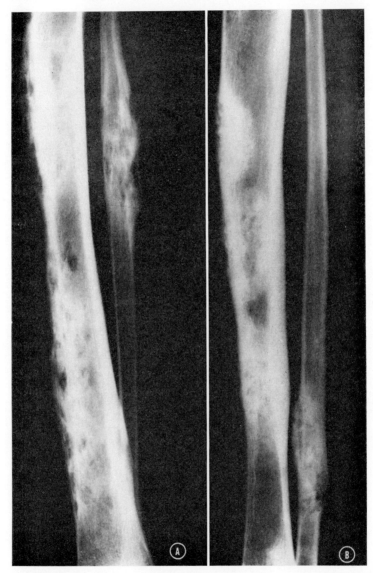

Figure 66. Gummatous osteomyelitis of yaws. (A.) Lateral view of the right leg of a six-year-old boy who had complained of painful swelling for the past six months. Subperiosteal apposition of bone is prominent anteriorly together with focal areas of destruction. Fusiform swelling of the fibula is similar to that occurring in syphilis. (B.) Lateral view of the right leg of a twenty-five-year-old woman who had complained of painful swelling for the past two years. The localized cortical swellings interspersed with numerous small rarefied foci are identical to osseous lesions of syphilis. (Courtesy of Dr. Cecil J. Hackett, 1951.)

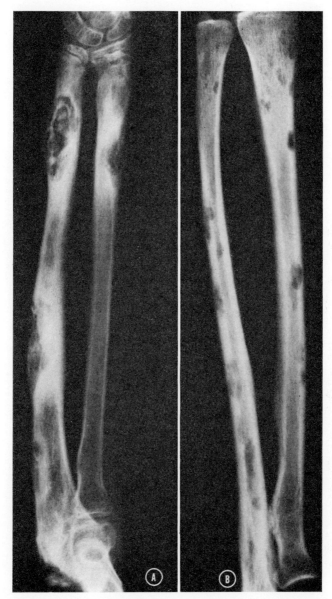

Figure 67. Gummatous osteomyelitis of yaws. (A.) Lateral view of the forearm of a ten-year-old girl showing irregular areas of bone destruction showing localized cortical swelling and perforations of the cortex. (B.) Lateral view of the forearm of a twenty-eight-year-old woman showing many irregularly oval areas of cortical destruction. (Courtesy of Dr. Cecil J. Hackett, 1951.)

along the anterior aspect associated with several areas of cortical destruction.

Australian aboriginal remains perhaps predating European contact show undoubted lesions of yaws (Hackett, 1936; McKay,

Figure 68. Possible evidence of yaws from the Mariana Islands. (*left.*) Three fragmentary long bones from a thirteen- to fourteen-year-old person showing localized cortical swellings and performations of the cortex. (*right.*) Tibia from another individual of similar age showing prominent thickening of the shaft anteriorly. (Courtesy of Dr. T. Dale Stewart from Stewart and Spoehr, 1952.)

1938). Several tibiae have the typical saber-shin appearance and others exhibit more localized areas of osteomyelitis and cortical destruction. One skull shows the typical bone changes of gangosa and two skulls reveal severe osteitis caused by yaws or venereal syphilis. Many of the specimens described by McKay from more arid regions may be the result of nonvenereal syphilis.

Possible evidence of yaws has been reported on the tropical Pacific island of Tonga (Pietrusewsky, 1971). About ten adults from a skeletal series of ninety-nine precontact burials exhibit bone lesions in the skull or postcranial bones entirely consistent with osseous yaws. Two skulls display multiple foci of cratering and new bone formation. Among the other bones affected are several tibiae, fibulae, ribs, a humerus and ulna.

An adult male skull from Easter Island around 1100-1850 A.D. may have bone changes due to yaws (Murrill, 1968). A small triangular depression surrounded by a rim of new bone is located about 2.5 cm above the left orbit. A large series of precontact and early historic burials from Oahu, Hawaii contains very little evidence suggestive of yaws. A thirty-year-old male with a perforated palate and six isolated bones with osteoperiostitis from other individuals comprise the only possible evidence from among 864 burials (Bowers, 1966; Snow, 1974).

Five skulls from Puerto Rico have been described showing periostitis and osteitis possibly caused by syphilis (Gejvall and Henschen, 1971). Radiocarbon dating of two specimens gives estimates of 1630 and 1480 A.D. $\pm$ 100 years. It is therefore hard to say whether these specimens are pre-Columbian. Although the extensive osteitis in two of the specimens is more suggestive of syphilis, yaws must also be considered. This is of course an academic point if one regards both diseases as syndromes of treponematosis. Nevertheless, determination of the particular syndrome endemic to a population can provide important information regarding the living conditions of that population.

From caves in the Candellaria Mountains of Mexico have come numerous skeletons dated between the 6th and 16th centuries A.D. Out of this material, twenty skulls show evidence of treponematosis: eighteen skulls with generalized osteitis and two skulls with more localized lesions attributed to yaws (Goff, 1963; 1967). It is more likely that all of these specimens are the result of venereal syphilis.

TREPONEMAL INFECTIONS: VENEREAL SYPHILIS, NON VENEREAL SYPHILIS (BEJEL), AND YAWS BIBLIOGRAPHY

Acsádi, G. and J. Nemeskéri: *History of Human Life Span and Mortality,* Akadémai Kiadó, Budapest, 1970.
Akrawi, F.: "Is bejel syphilis?" *Brit. J. Vener. Dis.,* 25:115-123, 1949.
Andersen, J. G.: "Studies in the mediaeval diagnosis of leprosy in Denmark," *Danish Med. Bull.,* 16 Suppl.:1-142, 1969.
Anderson, J. E.: "Human skeletons of Tehuacan," *Science,* 148:496-497, 1965.
Ash, J. E. and S. Spitz: *Pathology of Tropical Diseases,* American Registry of Pathology, Washington, D.C., 1945.
Ashmead, A. S.: "Pre-Columbian syphilis in Yucatan," *Amer. Anthrop.,* 9:106-109, 1896.
Barlow, T.: "Specimens of disease of skull in congenital syphilis," *Trans. Path. Soc. London,* 30:333-339; 350-353, 1879.
Bittner, L. H.: "Some observations on the tertiary lesions of framboesia tropica, or yaws," *Amer. J. Tropical Med.,* 6:123-130, 1926.
Bowers, W. F.: "Pathological and functional changes found in 864 pre-Captain Cook contact Polynesian burials from the sand dunes at Mokapu, Oahu, Hawaii." *Int. Surg.,* 45:206-217, 1966.
Bradlaw, R. V.: "The dental stigmata of prenatal syphilis," *Oral Surg.,* 6:147-158, 1953.
Brothwell, D. R.: "The real history of syphilis," *Science J.,* 6:27-32, 1970.
Brown, J. S. and J. H. Middlemiss: "Bone changes in tropical ulcer," *Brit. J. Radiol.,* 29:213-217, 1956.
Brues, A. M.: "Skeletal material from the Horton site," *Bull. Okla. Anthrop. Soc.,* 6:27-32, 1958.
———: "Skeletal material from the Morris site," *Bull. Okla. Anthrop. Soc.,* 7:63-70, 1959.
———: Discussant in S. Jarcho (ed.): *Human Palaeopathology,* pp. 107-112, Yale Univ. Press, New Haven, 1966.
Brühl, G.: "On the pre-Columbian existence of syphilis in the Western Hemisphere," *Cincinnati Lancet Clinic,* May, 8:487-493, 1890.
Bruusgaard, E.: "Ueber das schicksal der nicht spezifisch behandelten luetiker," *Arch. f. Dermat. u. Syph.,* 157:309-332, 1929.
Buchman, J. and H. S. Lieberman: "Prevalence of syphilis of the bones and joints," *Arch. Derm. Syph.,* 44:1-12, 1941.

Bullen, A. K. :"Paleoepidemiology and distribution of prehistoric treponemiasis (syphilis) in Florida," *Florida Anthrop.*, 25:133-174, 1972.
Buret, F.: *Syphilis Today and Among the Ancients*, F. A. Davis, Philadelphia, 1891.
Butler, C. S.: *Syphilis sive Morbus Humanus*, Science Press, Lancaster, Pa., 1936.
Caffey, J.: "Syphilis of the skeleton in early infancy: the non-specificity of many of the roentgenographic changes," *Amer. J. Roentgenol.*, 42:637-655, 1939.
Campell, W. C.: "An analysis of bone and joint lesions of known syphilitic origin," *Radiology*, 5:122-131, 1925.
Cannefax, G. T., L. C. Norins, and E. J. Gillespie: "Immunology of syphilis," *Ann. Rev. Med.*, 18:471-482, 1967.
Castiglioni, A.: *History of Medicine*, Knopf, New York, 1941.
Church, F. H.: "Syphilis of the center of the face," *Bull. Hist. Med.*, 7:705-718, 1939.
Cockburn, T. A.: "The origin of the treponematoses," *Bull. World Health Org.*, 24:221-228, 1961.
——————: *The Evolution and Eradication of Infectious Diseases*, Johns Hopkins Press, Baltimore, 1963.
Cole, H. N.: Congenital and prenatal syphilis," *J. Amer. Med. Assoc.*, 109:580-585, 1937.
Cole, H. N., J. C. Harkin, B. S. Kraus, and A. R. Moritz: "Pre-Columbian osseous syphilis," *Arch. Derm. Syph.*, 71:231-238, 1955.
Cornet, L., and J. Vilasco: "Pian osseux et goundou," *J. Chirurg.*, 102:101-104, 1971.
Coulson, W. J.: *A Treatise on Syphilis*, Churchill, London, 1869.
Crosby, W. A.: "The early history of syphilis: a reappraisal," *Amer. Anthrop.*, 71:218-227, 1969.
——————: *The Columbian Exchange*, Greenwood, Westport, Conn., 1972.
Csonka, G. W.: "Clinical aspects of bejel," *Brit. J. Vener. Dis.*, 29:95-103, 1953.
Delahaye, R. P., P. Many, R. Misson, P. Boursiquot, and A. Combes: "Les lésions osseouses observées lors des tréponématoses sérologiques," *Semaine des Hopitaux*, 46:189-194, 1970.
Denninger, H. S.: "Prehistoric syphilitic lesions. Example from North America," *Southwestern Med.*, 19:202-204, 1935.
——————: "Syphilis of Pueblo skull before 1350," *Arch. Path.*, 26:724-727, 1938.
Derums, V. Y.: "Some paleopathologic data on Baltic coast inhabitants," *Fed. Proc. (Trans. Suppl.)*, 24:225-230, 1965.
Eaton, G. F.: "The collection of osteological material from Machu Picchu," *Mem. Conn. Acad. Arts and Sciences*, 5:1-96, 1916.

Eichenholtz, S. N.: *Charcot Joints,* Thomas, Springfield, 1966.
Evans, W. A.: "Syphilis of the bones in infancy. Some possible errors in the roentgen diagnosis," *J. Amer. Med. Assoc., 115*:197-200, 1940.
Fournier, A.: *Traité de la Syphilis,* Rueff, Paris, 1899.
Gann, T.: "Recent discoveries in Central America proving the pre-Columbian existence of syphilis in the New World," *Lancet,* 968-970, 1901.
Garner, M. F., J. L. Blackhouse, and G. J. Tibbs: "Yaws in an isolated Australian aboriginal population," *Bull. World Health Org., 43*:603-606, 1970.
Gejvall, N-G.: *Westerhus: Medieval Population and Church in the Light of Skeletal Remains,* Boktryckeri, Lund, 1960.
Gejvall, N-G., and F. Henschen: "Anatomical evidence of pre-Columbian syphilis in the West Indian islands," *Beitr. Path., 144*:138-157, 1971.
Goff, C. W.: "New evidence of pre-Columbian bone syphilis in Guatemala," in R. B. Woodbury and A. S. Trik (ed.): *The Ruins of Zaculeu, Guatemala,* pp. 312-319, Wm. Byrd Press, Richmond, 1953.
———: "New evidence of syphilis (?), yaws (?) from Cueva de la Candelaria," *Amer. J. Phys. Anthrop., 21*:412, 1963.
———: "Syphilis," in D. R. Brothwell and A. T. Sandison (ed.): *Diseases in Antiquity,* Thomas, Springfield, 1967.
Goldman, C. H. and S. J. Smith: "X-ray appearances of bone in yaws," *Brit. J. Radiol., 16*:234-238, 1943.
Goldstein, M. S.: "The skeletal pathology of early Indians in Texas," *Amer. J. Phys. Anthrop., 15*:299-312, 1957.
Gramberg, K. P. C. A.: "Leprosy and the Bible," *Trop. Geograph. Med., 11*:127-139, 1959.
Grin, E. I.: "Endemic syphilis in Bosnia and Herzegovia," *Urol. Cutan. Rev., 39*:482-487, 1935.
———: "Endemic syphilis in Bosnia: clinical and epidemiological observations on a successful mass-treatment campaign," *Bull. World Health Org., 7*:1-74, 1952. Also Monograph #11 of World Health Org. Geneva, 1953.
Guthe, T.: "Clinical, serological, and epidemiological features of framboesia tropica (yaws) and its control in rural communities," *Acta Derm. Vener., 49*:343-368, 1969.
Guthe, T. and R. R. Willcox: "Treponematoses: a world problem," *Chron. World Health Org., 8*:37-113, 1954.
Hackett, C. J.: *Boomerang leg and Yaws in Australian Aborigines,* Monograph 1, Roy. Soc. Trop. Med. Hyg., 1936.
———: "A review of references to the bone lesions of yaws," *Trop. Dis. Bull., 43*:1091-1104, 1946.
———: *Bone Lesions of Yaws in Uganda,* Blackwell, Oxford, 1951.

———: "On the origin of the human treponematoses," *Bull. World Health Org.*, 29:7-41, 1963.
———: "Some aspects of treponematoses in past populations," *J. Anthrop. Soc. South. Australia*, 6(9):5-14, 1968.
———: "Possible treponemal changes in a Tasmanian skull," *Man*, 9:436-443, 1974.
Haltom, W. L. and A. R. Shands: "Evidence of syphilis in mound builder's bones," *Arch. Path.*, 25:228-242, 1938.
Hamlin, H.: "The geography of treponematosis," *Yale J. Biol. Med.*, 12:29-50, 1939.
Harley, G. W.: "The symptomatology of yaws in Liberia," *J. Trop. Med. Hyg.*, 36:217-223, 1933.
Harrison, L. W.: "The origin of syphilis," *Brit. J. Vener. Dis.*, 35:1-7, 1959.
Hazen, H. H.: *Syphilis*, C. V. Mosby, St. Louis, 1921.
Helfet, A. J.: "Acute manifestations of yaws of bone and joint," *J. Bone Joint Surg.*, 26B:672-681, 1944.
Holcomb, R. C.: *Who Gave the World Syphilis? The Haitian Myth*, Froben Press, New York, 1930.
———: "Christopher Columbus and the American origin of syphilis," *U.S. Nav. Med. Bull.*, 32:401-430, 1934.
———: "The antiquity of syphilis," *Med. Life*, 42:275-325, 1935.
———: "Syphilis of the skull among Aleuts and the Asian and North American Eskimo about the Bering and Arctic Seas," *U.S. Nav. Med. Bull.*, 38:177-192, 1940.
Hooton, E. A.: *The Indians of Pecos Pueblo*, Yale Univ. Press, New Haven, 1930.
Hoyme, L. E. and W. M. Bass: "Human skeletal remains from the Tollifero (Ha6) and Clarksville (Mc14) sites, John H. Kerr Reservoir Basin, Virginia," *Bur. Amer. Ethnol. Bull.*, 182:329-400, 1962.
Hrdlička, A.: "The anthropology of Florida," *Publ. Florida State Hist. Soc.*, 1922.
———: "Disease, medicine and surgery among the American aborigines," *J. Amer. Med. Assoc.*, 99:1661-1662, 1932.
Hudson, E. H.: "Syphilis in the Euphrates Arab," *Amer. J. Syph.*, 16-17: 447-469, 10-14, 1932-1933.
———: *Non-venereal Syphilis: A Sociological and Medical Study of Bejel*, Livingstone, London, 1958a.
———: "The treponematoses—or treponematosis?" *Brit. J. Vener. Dis.*, 34:22-23, 1958b.
———: "Endemic syphilis—heir of the syphiloids," *Arch. Intern. Med.*, 108:1-4, 1961a.
———: "Historical approach to the terminology of syphilis," *Arch. Derm.*, 84:545-562, 1961b.

———: "Treponematosis and anthropology," *Ann. Intern. Med.*, 58: 1037-1048, 1963a.
———: "Treponematosis and pilgrimage," *Amer. J. Med. Sci.*, 246:645-656, 1963b.
———: "Treponematosis and African slavery," *Brit. J. Vener. Dis.*, 40:43-52, 1964.
———: "Treponematosis and man's social evolution," *Amer. Anthrop.*, 67:885-901, 1965a.
———: "Treponematosis in perspective," *Bull. World Health Org.*, 32:735-748, 1965b.
———: "Christopher Columbus and the history of syphilis," *Acta Tropica*, 25:1-16, 1968.
———: "Diagnosing a case of venereal disease in 15th century Scotland," *Brit. J. Vener. Dis.*, 48:146-153, 1972.
Hussein, M. K.: "Quelques specimens de pathologie osseouses chez les anciens egyptiens," *Bull. Inst. Egypte (Cairo)*, 32:11-17, 1949.
Hutchinson, J.: *A Clinical Memoir on Certain Diseases of the Eye and Ear Consequent on Inherited Syphilis*, J. Churchill, London, 1863.
Hyde, J. N.: "A contribution to the study of pre-Columbian syphilis in America," *Amer. J. Med. Sci.*, 102:117-131, 1891.
Jaffe, H. L.: *Metabolic, Degenerative, and Inflammatory Diseases of Bones and Joints*, Lea and Febiger, Philadelphia, 1972.
Jarcho, S.: "Some observations on diseases in prehistoric America," *Bull. Hist. Med.*, 38:1-19, 1964.
Jeans, P. C. and J. V. Cooke: *Prepubescent Syphilis*, Appleton, New York, 1930.
Joklik, W. K. and D. T. Smith: *Zinsser Microbiology*, Appleton, New York, 1972.
Jones, J.: "Exploration of the aboriginal remains of Tennessee," *Smithson. Contrib. Knowledge*, Vol. 22, Washington, 1876.
Jones, L. G. G.: "Mutilating bejel," *Brit. J. Vener. Dis.*, 29:104-105, 1953.
Kail, F. and A. de Froe: "Der schädel von Zakho. Ein frühchristlicher schädel aus nordirak mit auffälligen pathologischen (syphillitischen?) veränderungen," *Hautarzt*, 4:82, 1953.
Kampmeier, R. H.: "The late manifestations of syphilis; skeletal, visceral, and cardiovascular," *Med. Clin. N. Amer.*, 48:667-697, 1964.
Karnosh, L. J.: "Histopathology of syphilitic hypoplasia of the teeth," *Arch. Derm. Syph.*, 13:25-42, 1926.
Keyes, E. L.: *Syphilis*, Appleton, New York, 1908.
King, A. J. and R. D. Catterall: "Syphilis of bones," *Brit. J. Vener. Dis.*, 35:116-127, 1959.
Krumbhaar, E. B.: "A pre-Columbian Peruvian tibia exhibiting syphilitic (?) periostitis," *Ann. Med. Hist.*, 8:232-235, 1936.

Lamb, D. S.: "Pre-Columbian syphilis," *Proc. Assoc. Amer. Anat.*, *10*:63-69, 1898.

Lunn, H. F.: "Treponematosis of bone; a case for discussion," *J. Trop. Med. Hyg.*, *64*:166-168, 1961.

MacCurdy, G. G.: "Human skeletal remains from the highlands of Peru," *Amer. J. Phys. Anthrop.*, *6*:218-329, 1923.

MacFarlane, W. V. and I. Rannie: "Gummatous osteitis of the skull," *Brit. J. Vener. Dis.*, *34*:153-159, 1958.

Magnuson, H. J., E. W. Thomas, S. Olansky, B. I. Kaplan, L. Mello, and J. C. Cutler: "Inoculation syphilis in human volunteers," *Medicine*, *35*:33-82, 1956.

Maul, H. G.: "Bone and joint lesions in yaws with x-ray findings in 20 cases," *Amer. J. Roentgen.*, *6*:423-433, 1919.

McDonagh, J. E. R.: *The Biology and Treatment of Venereal Diseases*, Lea and Febiger, Philadelphia, 1916.

McKay, C. V.: "Some pathological changes in Australian aboriginal bones," *Med. J. Aust.*, *2*:537-555, 1938.

McLean, S.: "Osseous lesions of congenital syphilis: summary and conclusions in 102 cases," *Amer. J. Dis. Child.*, 130-152; 363-395; 607-675; 887-922; 1411-1418, 1931.

Means, H. J.: " A roentgenological study of the skeletal remains of the prehistoric mound builder Indians of Ohio," *Amer. J. Roentgen.*, *13*: 359-367, 1925.

Merriweather, A. M.: "A case illustrating the bone lesions seen in cases of extra-venereal treponematosis in the Bechuanaland protectorate," *Trans. Roy. Soc. Trop. Med. Hyg.*, *47*:242-245, 1953.

Møller-Christensen, V.: *Bogen om Aebelholt Kloster* (with English summary), Danish Science Press, Copenhagen, 1958.

————: "The history of syphilis and leprosy—an osteoarchaeological approach," *Abbottempo*, *1*:20-25, 1969.

Møller-Christensen, V. and R. G. Inkster: "Cases of leprosy and syphilis in the osteological collection of the department of anatomy at the University of Edinburgh, with a note on the skull of King Robert the Bruce," *Danish Med. Bull.*, *12*:11-18, 1965.

Møller-Christensen, V. and W. H. Jopling: "An examination of the skulls in the Catacombs of Paris," *Med. Hist.*, *8*:187-188, 1964.

Morant, G. M. and M. F. Hoadley: "A study of the recently excavated Spitalsfields crania," *Biometrika*, *23*:222, 1931.

Morgan, E. L.: "Pre-Columbian syphilis," *Virginia Med. Semi-Month.*, *21*:1042, 1894.

Morse, D.: "Two cases of possible treponema infection in prehistoric America," in W. D. Wade (ed.): *Miscellaneous Papers in Paleopathology*, *Mus. North. Ariz. Tech. Ser.*, *7*:48-61, 1967.

————: "Pathology and abnormalities of the Hampson skeletal collection," in D. F. Morse (ed.): *Nodena,* Arkansas Arch. Survey Publ. in Archeology Res. Series #4, 1973.

Morton, R. S.: "The sibbens of Scotland," *Med. Hist., 11*:374-380, 1967.

Murray, J. F., A. M. Merriweather, and M. L. Freedman: "Endemic syphilis in the Bakwena reserve of the Bechuanaland protectorate," *Bull. Wrld. Health Org.. 15*:975-1039, 1956.

Murrill, R. I.: *Cranial and Postcranial Skeletal Remains from Easter Island,* Univ. Minn. Press, Minneapolis, 1968.

Ong, H. A. and M. A. Selinger: "Some observations on the symptoms and treatment of congenital syphilis in the special out-patient clinic," *Internat. Clin., 2*:230-238, 1922.

Oosthuizen, S. F.: "Yaws," *Brit. J. Radiol., 22*:276-279, 1949.

Orton, S. T.: "A study of the pathological changes in some Mound Builder's bones from the Ohio Valley, with especial reference to syphilis," *Univ. Penn. Med. Bull. Phila., 18*:36-44, 1905.

Pales, L.: *Paléopathologie et Pathologie Comparative,* Masson et Cie, Paris, 1930.

Parmalee, A. H. and L. J. Halpern: "The diagnosis of congenital syphilis," *J. Amer. Med. Assoc., 105*:563-566, 1935.

Parramore, T. C.: "Non-venereal treponematosis in colonial North America," *Bull. Hist. Med., 44*:571-581, 1970.

Parrot, J.: "The osseous lesions of hereditary syphilis," *Lancet, 1*:696-698, 1879.

————: Same article in French: "Les lésions osseuses de la syphilis héréditaire," *Trans. Path. Soc. London, 30*:339-350, 1879.

Pendergrass, E. P. and R. S. Bromer: "Congenital bone syphilis," *Amer. J. Roentgen., 22*:1-21, 1929.

Pendergrass, E. P., R. L. Gilman, and K. B. Castleton: "Bone lesions in tardive heredosyphilis," *Amer. J. Roentgen., 24*:234-257, 1930.

Pietrusewsky, M.: "An osteological study of cranial and infracranial remains from Tonga," *Rec. Auckland Inst. and Mus., 6*:287-402, 1971.

Pusey, W. A.: *The History and Epidemiology of Syphilis,* Thomas, Springfield, 1933.

Putkonen, T.: "Dental changes in congenital syphilis," *Acta Derm.-Vener., 42*:44-62, 1962.

Rabkin, S.: "Dental conditions among prehistoric Indians of Northern Alabama," *J. Dent. Res., 21*:211-222, 1942.

Rat, J. N.: *Frambesia (Yaws),* Waterlow and Sons, London, 1891.

Rokhlin, D. G. and A. E. Rubasheva: "New data on the age of syphilis," *Vestnik Rentgenol., 21*:183-1938.

Roney, J. G.: "Palaeopathology of a California archaeological site," *Bull. Hist. Med., 33*:97-109, 1959.

———: "Paleoepidemiology: an example from California," in S. Jarcho (ed.): *Human Palaeopathology*, Yale Univ. Press, New Haven, 1966.
Rosahn, P. D.: "Studies in syphilis VII: the end results of untreated syphilis," *J. Vener. Dis. Inform.*, 27:293-301, 1946.
Rost, G. S.: "Roentgen manifestations of bejel ("endemic syphilis") as observed in the Euphrates River Valley," *Radiology*, 38:320-325, 1942.
Saul, F. P.: "The human skeletal remains of Altar de Sacrificios: an osteobiographic analysis," *Papers of the Peabody Museum of Archaeology and Ethnology*, Vol. 63 #2, Harvard Univ., Cambridge, 1972.
Shattuck, G. C.: "Lesions of syphilis in American Indians," *Amer. J. Trop. Med.*, 18:577-586, 1938.
Singer, C. and E. A. Underwood: *A Short History of Medicine*, Oxford Univ. Press, New York, 1962.
Smith, E. G. and W. Jones: "Report on the human remains," *Archaeological Survey of Nubia, Report of 1907-1908*, Cairo, Ministry of Finance, 1910.
Snow, C. E.: "Physical anthropology of Adena," *Univ. Ky. Rep. Anthrop. Archeol.*, 6:247-309, 1945.
———: "Indian Knoll skeletons of site Oh 2, Ohio County, Kentucky," *Univ. Ky. Rep. Anthrop. Archeol.*, 4:371-554, 1948.
———: "Indian burials from St. Petersburg, Florida," *Contrib. Flor. State Mus. (Soc. Sci)*, #8, 1962.
———: *Early Hawaiians, an Initial Study of Skeletal Remains from Mokapu, Oahu*, Univ. Ky. Press, Lexington, 1974.
Speed, J. S. and H. B. Boyd: "Bone syphilis," *South. Med. J.*, 29:371-377, 1936.
Steindler, A.: "Congenital syphilis of bones and joints," *Urol. Cutan. Rev.*, 49:568-575, 1945.
Stewart, D. M.: "Roentgenological manifestations in bone syphilis," *Amer. J. Roentgen.*, 40:215-223, 1938.
Stewart, T. D.: "Some historical implications of physical anthropology in North America," *Smithson. Misc. Coll.*, 100:15-50, 1940.
———: "Skeletal remains from Xochicalco, Morelos," *Estudios Anthrop. Mexico, D.F.*, pp. 131-156, 1956.
———: *The People of America*, Scribners, New York, 1973.
Stewart, T. D. and A. Spoehr: "Evidence on the paleopathology of yaws," *Bull. Hist. Med.*, 26:538-553, 1952.
———: Also in D. R. Brothwell and A. T. Sandison (ed.): *Diseases in Antiquity*, Thomas, Springfield, 1967.
Stokes, J. H.: *Modern Clinical Syphilology*, Saunders, Philadelphia, 2nd. ed., 1934.
Stokes, J. H., H. Beerman, and N. R. Ingraham: *Modern Clinical Syphilology*, Saunders, Philadelphia, 3rd. ed., 1944.

Sudhoff, K.: *The Earliest Printed Literature on Syphilis,* (1495-1498), Florence, 1925.
―――――: *Essays in the History of Medicine,* Medical Life Press, New York, 1926.
Symmers, D.: "Anatomic lesions in late acquired syphilis," *J. Amer. Med. Assoc.,* 66:1457-1462, 1916.
Taneja, B. L.: "Yaws: clinical manifestations and criteria for diagnosis," *Indian J. Med. Res.,* 56:100-113, 1968.
Taylor, R. W.: *Syphilitic Lesions of the Osseous System in Infants and Young Children,* William Wood and Co., New York, 1875.
Tello, J. C. and H. U. Williams: "An ancient syphilitic skull from Paracas in Peru," *Ann. Med. Hist.,* 2:515-529, 1930.
Thoma, K. H.: *Oral Pathology,* Mosby, St. Louis, 3rd. ed., 1950.
Turner, T. B. and D. H. Hollander: *Biology of the Treponematoses,* World Health Organization, Monograph 35, Geneva, 1957.
Van Nitsen, R.: "Les manifestations tertiaires du pian dans la région du Tanganika-Moëro," *Ann. Soc. Belge de Med. Trop.,* 1:39-54, 1920.
Virchow, R.: "Ueber die natur der constitutionell-syphilitischen Affectionenem," *Virchow's Arch. Path. Anat. Physiol.,* 15:217-336, 1858.
―――――: "Beiträge zur geschichte der lues," *Dermat. Z.,* 3:1-9, 1896.
Von Zeissl, H.: *Outlines of the Pathology and Treatment of Syphilis,* Appleton, New York, 1886.
Wakefield, E. G., S. C. Dellinger, and J. D. Camp: "A study of the osseous remains of the 'mound builders' of eastern Arkansas," *Amer. J. Med. Sci.,* 193:488-495, 1937.
Whitney, J. L.: "A statistical study of syphilis," *J. Amer. Med. Assoc.,* 65:1986-1989, 1915.
Whitney, J. L. and W. I. Baldwin: "Syphilis of the spine," *J. Amer. Med. Assoc.,* 65:1989-1994, 1915.
Whitney, W. F.: "On the existence of syphilis in America before the discovery by Columbus," *Bost. Med. Surg. J.,* 108:365-366, 1883.
Wile, U. J. and F. E. Senear: "A study of the involvement of the bones and joints in early syphilis," *Amer. J. Med. Sci.,* 152:689-693, 1916.
Willcox, R. R.: "Endemic syphilis in Africa: the njovera of southern Rhodesia," *S. Afr. Med. J.,* 25:501-504, 1951.
Willey, G. R.: "New World prehistory," *Science,* 131:73-85, 1960.
Williams, H. U.: "American origin of syphilis, with citations from early Spanish authors collected by Dr. Montejo y Robledo," *Arch. Derm. Syph.,* 16:683-696, 1927.
―――――: "The origin and antiquity of syphilis: the evidence from diseased bones," *Arch. Path.,* 13:779-814, 931-983, 1932.
―――――: "Pathology of yaws," *Arch. Path.,* 20:596-630, 1935.
―――――: "The origin of syphilis: evidence from diseased bones, a supplementary report," *Arch. Derm. Syph.,* 33:783-787, 1936.

Wilson, J.: "Syphilis and yaws: diagnostic difficulties and case report," *New Zealand Med. J.*, 78:18-21, 1973.

Wilson, P. W. and M. S. Mathis: "Epidemiology and pathology of yaws. Based on study of 1423 consecutive cases in Haiti," *J. Amer. Med. Assoc.*, 94:1289-1292, 1930.

Wright, D. J. M.: "Syphilis and Neanderthal man," *Nature*, 229:409, 1971.

Zimmermann, E. L.: "The early pathology of syphilis especially as revealed by accounts of autopsies on syphilitic corpses (1497-1553)," *Janus*, 38:1-24; 37-69, 1934.

——————: "The pathology of syphilis as revealed by autopsies performed between 1563 and 1761," *Bull. Hist. Med.*, 3:355-399, 1935.

Zinsser, H.: *Rats, Lice and History*, Little, Brown Co., Boston, 1935.

Chapter V

TUBERCULOSIS

TUBERCULOSIS IN MAN is caused by *Mycobacterium tuberculosis,* a nonmotile, acid-fast bacillus. In all probability tuberculosis is older than the human race with other species in the genus Mycobacterium causing the disease in mammals, birds, reptiles, and fish (Breed et al., 1957). Also included in the genus are commensals and saprophytes very similar to the tubercle bacillus and capable of producing tubercular lesions in experimental animals and man (Rich, 1944; Joklik and Smith, 1972). Tuberculosis in man and other animals, both warm-blooded and cold-blooded, may thus have arisen from a common Mycobacterial ancestor many millions of years ago. It is possible that the genus evolved much more recently and spread rapidly to all these animals and habitats, but the range of distribution is so great that it appears highly unlikely (Cockburn, 1963).

Other evidence suggests that tuberculosis in man (at least in endemic form) may be of much more recent origin. *Mycobacterium tuberculosis* occurs in two forms, the human and bovine. The former can infect any organ and tissue, but the pulmonary form of the disease is by far the most common. The bovine variety can cause pulmonary tuberculosis in man, but is more likely to infect lymph nodes, the abdomen, and the skeleton. Bovine tuberculosis does not infect wild animals, but only those that have been domesticated. This is probably because animal-to-animal transmission is facilitated by close contact in herds and even more so when stalled in barns. Bovine tuberculosis has never established itself entirely as a human parasite. It is quite possible that the human variety is a mutant of the bovine form (Hare, 1967). Since cattle were first domesticated during the Neolithic period (6000-7000 B.C.) in the Old World (Clark, 1962), it appears that the tubercle bacillus may have first become an endemic human disease during this period.

PALEOPATHOLOGICAL EVIDENCE

Egypt

Tuberculosis has existed for thousands of years in Egypt as shown by art forms and many examples of Pott's disease or tuberculosis of the spine in ancient Egyptian skeletal material (Smith and Jones, 1910; Derry, 1938; Rowling, 1960). In a recent review of this material, sixteen previously reported specimens were reexamined and fifteen new cases described dating from 3700-1000 B.C. (Morse et al., 1964). Perhaps the most convincing case of skeletal tuberculosis comes from the 21st dynasty or about 1000 B.C. It is a mummy of a twenty-five to thirty-year-old Egyptian priest of Amun near Thebes. The spinal column exhibits destruction of the lower thoracic and upper lumbar vertebrae creating an angular kyphosis. In addition, there is a huge abscess cavity in the right psoas muscle (Smith and Ruffer, 1910; Ruffer, 1921; Cave, 1939).

It is interesting to note that four possible cases of spinal tuberculosis were excavated from two neighboring graves (Derry, 1938). One grave contained the remains of a man and a woman while the other contained the skeleton of two adult males and a nine-year-old boy. Some workers have regarded this cluster of cases as evidence of a cemetery for a tuberculosis sanitarium. However, it is more likely that the individuals are related and acquired the disease from a common source of infection within the family.

The Egyptian medical papyri do not contain any clear descriptions of tuberculosis. The earliest literary references to the disease come from the Vedic Hymns of India around 2000 B.C. and possible descriptions in the Bible around 700 B.C. The Greek and Roman literature from 900 B.C. on contains many descriptions and dissertations on tuberculosis (Meinecke, 1927).

Europe

The earliest evidence of tuberculosis in Europe comes from a Neolithic grave near Heidelberg dated about 5000 B.C. (Bartels, 1907). The skeleton of a young man exhibits destruction and fusion of the 3rd and 4th thoracic vertebrae resulting in kyphosis. Another recently excavated case from Neolithic times is that of a twenty-two to thirty-year-old woman from Denmark around 2500-1500 B.C. (Sager et al., 1972). Here again there is kyphosis produced by destruction and fusion of the 3rd and 4th thoracic vertebrae.

Three cases of spinal tuberculosis have been reported from Russia, one from the late Bronze Age or 1500-500 B.C. (Rokhlin and Maikova-Stroganova, 1938). The other two cases are dated approximately 200-100 B.C. From the Baltic coast of Lithuania come the remains of a child with destruction and fusion of three midthoracic vertebrae. This possible case of spinal tuberculosis is dated 200-400 A.D. (Derums, 1965). A Romano-British skeleton dating 330-350 A.D. exhibits lesions of the right foot and left knee perhaps caused by tuberculosis (Wells, 1964). Two cases of tuberculosis affecting the spine have been described in Saxon burial grounds in Britain dated 425-850 A.D. (Brothwell, 1961).

Beginning with the Middle Ages (after 1100 A.D.), at least fifteen possible cases of skeletal tuberculosis have been reported in northern Europe (Tordeman et al., 1939; Isager, 1941; Sjövall, 1942; Møller-Christensen, 1958; Gejvall, 1960; Derums, 1965; Weiss and Møller-Christensen, 1971). One of the most interesting and unusual of these cases occurred in an adult male with leprous bone changes (Weiss and Møller-Christensen, 1971). During the careful excavation of this individual from a medieval leprosaria cemetery in Denmark, three sets of plate-like calcifications were found within the thoracic cavity near the 8th, 9th and 10th ribs posterolaterally. The spinal column revealed partial destruction and fusion of the 2nd through 4th thoracic vertebrae and there was marked bony proliferation of the inferior margins of the ribs near their articulations with the spine. The recovery of the pleural calcifications provided important additional evidence of tuberculosis in this individual.

To summarize the antiquity of tuberculosis in the Old World, there is no evidence of the disease prior to Neolithic times. Possible cases of spinal tuberculosis have been described in Europe and Egypt around 5000 B.C. and 3700 B.C. respectively. From medieval times to the present, an increasing number of cases have been reported. The paleopathological evidence obtained thus far neither proves nor refutes the origin of endemic tuberculosis sometime during the Neolithic.

New World

Until quite recently the evidence for pre-Columbian tuberculosis in the New World has been questionable because of inaccurate dating and ambiguous skeletal pathology (Morse, 1961; 1967). Possible specimens of spinal tuberculosis have been reported in definitely pre-Columbian remains from California, Tennessee, Illinois, and New York (Ritchie, 1952; Lichtor and Lichtor, 1957; O'Bannon, 1957; Roney, 1959; Morse, 1961). Most of these cases are dated 1100-1600 A.D. although the specimen from California could be as early as 500 B.C. Less likely cases or cases from postcontact sites have been reported in New Mexico, Tennessee, Louisiana, Venezuela, and Peru (Whitney, 1886; Hrdlička, 1913; Moodie, 1927; Hooton, 1930; Requena, 1945).

Peruvian mummies have provided us with the most information about the antiquity of tuberculosis. A definitely pre-Columbian mummy exhibited destruction and fusion of the 7th thoracic through 1st lumbar and histologic examination of the rehydrated lungs showed a large amount of fibrous tissue in the right apex (Garcia-Frias, 1940). Conclusive evidence of pre-Columbian tuberculosis is found in a young boy about eight years old from the Nazca Culture of southern Peru and radiocarbon dated around 700 A.D. (Allison et al., 1973). This young boy had kyphosis resulting from destruction of the first three lumbar vertebrae. A psoas abscess is found adjacent to this vertebral destruction. Rehydration and staining of the organs reveals acid-fast bacilli and tubercles in the lung, liver, pleura, and right kidney. All of this evidence points toward tuberculosis as the only possible cause and establishes its pre-Columbian presence in the New World.

In all likelihood the occurrence of tuberculosis among the American Indians was very infrequent as the following may indicate. During the migrations of America's first inhabitants across the Bering Strait, only the strongest and healthiest could have endured the rigors of the elements. Furthermore, the Far North acted as a screen preventing the flow of many pathogenic microorganisms along with their hosts. Thus, the arctic trek may have eliminated many of the diseases of the American Indians as they moved southward (Stewart, 1960). The high susceptibility of the Indians to many of the diseases brought over by the first white settlers supports this possibility (see Dubos, 1966).

The European and Asiatic races, through long contact with the disease, have a distinct natural resistance to tuberculosis consisting of an ability to develop an immune response to the primary infection. American Indians and Eskimos have much less ability to develop an effective immune response to the primary infection which then becomes rapidly progressive (Stead, 1974). For example, in 1890 tuberculosis hit the Saskatchewan Indian Reservation in epidemic proportions. The *annual* tuberculosis death rate soon reached the high figure of almost 10 percent of the total population (Ferguson, 1934). Similar evidence of a low resistance by American Indians to tuberculosis comes from several reports reviewed by Hrdlička, 1909.

One of the major reasons for the decrease in tuberculosis today has been the eradication of bovine tuberculosis by milk pasteurization and slaughter of infected cattle. In fact, 75 to 85 percent of all cases of tuberculosis were due to the bovine variety of tubercle bacillus (Knaggs, 1926; Girdlestone and Somerville, 1940). Since the American Indian did not have domesticated cattle, this major source of tuberculous infection was not even present.

Finally, before the implementation of agriculture, most American Indians lived in very small groups making the existence of endemic human tuberculosis in these groups difficult if not impossible (see Cockburn, 1963). It is interesting to note that the case of tuberculosis reported by Allison comes from the Nazca culture which is characterized by well-developed agriculture and urbanization (Kidder, 1956; Bushnell, 1966).

DIAGNOSIS OF TUBERCULOSIS

Age

Due to an advanced standard of living and better hygiene, the age incidence of tuberculosis is higher today than in the past. In earlier studies when tuberculosis was much more common and living conditions poor, many were exposed to massive infection in the first few years of life. In a group of 5461 patients treated for skeletal tuberculosis, 87 percent were less than fourteen years of age (Whitman, 1927). In a recent study of 1000 cases of skeletal tuberculosis in Hong Kong (with squalid living conditions and high incidence of tuberculosis), 70 percent of cases were under the age of ten years (Hodgson et al., 1969). Since tuberculosis often lasts for years in a chronic form, the skeletal remains will often be of adults who had or have had the disease since childhood.

Frequency of Bone Involvement

By far the major forms of tuberculosis are *pulmonary* and *lymphatic*. With rare exception *skeletal* tuberculosis is a secondary infection from either the lungs or the lymph nodes. Early studies giving the frequency of skeletal involvement in tuberculosis of the lungs and lymph nodes provide important information for the paleopathologist. In 5161 tuberculosis patients, 4.7 percent had discernible bone lesions (Jacono, 1937). 5.2 percent of 6682 adult cases of tuberculosis had skeletal involvement (LaFond, 1958). Out of 2997 patients with tuberculosis, 6.3 percent had skeletal lesions (Cheyne, 1911). In 1628 cases of pulmonary and lymphatic tuberculosis among the American Indians, 7.0 percent exhibited lesions of the bone (Hrdlička, 1909). From the close agreement of the data above, one may conclude that approximately 5 to 7 percent of all cases of tuberculosis will involve the bones and joints.

Location of Skeletal Lesions

The great majority of statistical analyses indicate that the three major sites of skeletal tuberculosis are the vertebral column, the hip, and the knee (see Table VIII). Approximately 25 to 50 percent of all cases of skeletal tuberculosis involve the vertebral column which makes this site of vital importance to the paleopathologist. The hip joint is affected in 15 to 30 percent of all cases, and the knee is the site of involvement in 10 to 20 percent of all cases. Other major sites of tubercular destruction are the ankle joint, bones of the fingers and toes, and the sacroiliac joint. Because of its hematogenous spread however, tuberculosis may involve any bone of the skeleton. In general, the joints of the lower extremities are affected much more often than are the joints of the upper extremities. Moreover, when lesions of the arm and shoulder do develop, they are usually found in adults.

Multiple Lesions

The frequency of multiple bone lesions in tuberculosis differs among workers according to their definition of what qualify as multiple foci of infection. When tubercular lesions in several fingers or toes (spina ventosa) are counted as separate lesions, the frequency may be as high as 21 percent (Johansson, 1926). When only those lesions that appeared in entirely distinct localizations are counted, the frequency drops to 10.5 percent (Sanchis-Olmos, 1948). In general, tubercular involvement of two or more different skeletal areas is rather uncommon. When it does occur, the spine, hip, and knee are often affected simultaneously.

TUBERCULOSIS OF THE VERTEBRAL COLUMN

(Pott's Disease)

Since 25 to 50 percent of all cases of skeletal tuberculosis involve the vertebral column, it is not surprising that virtually all paleopathological reports of tuberculosis have been based on vertebral lesions found in archaeological material. However, very similar lesions may be produced by various diseases and conditions. Compression fractures of one or more vertebrae, pyogenic osteomyelitis, and blastomycosis produce vertebral destruction

TABLE VIII
LOCATION OF BONE LESIONS IN TUBERCULOSIS

Bone	%	Bone	%	Bone	%	Bone	%
spine	23.2	spine	33.0	spine	40.8	spine	50.1	spine	47.5
knee	16.5	hip	14.8	hip	28.9	hip	36.7	knee	10.0
hip	14.6	long bones, fingers, & toes	10.8	knee	9.6	ankle	7.9	sacroiliac	7.9
tarsus & fingers, & toes	14.4	knee	8.4	foot	5.8	knee	2.7	hip	4.8
ankle	6.3	sacroiliac	7.5	fingers & toes	4.6	wrist	1.0	foot	3.1
elbow	6.0	sternum	3.9	skull & face	2.7	shoulder	0.8	ankle	2.8
wrist & hand	6.0	elbow	3.9	shoulder	2.4	elbow	0.8	shoulder	2.8
sternum, clavicle, & ribs	5.2	ribs	3.4	elbow	1.5	TOTAL	100.0	elbow	2.4
skull & face	5.0	shoulder	3.4	long bones	1.3			ribs	3.0
pelvis	3.5	ankle	2.9	hand & wrist	1.1			hand	2.2
femur, tibia, & fibula	3.5	wrist	2.0	sacroiliac	0.9			ilium	1.3
shoulder	1.5	tarsus	2.0	bones of thorax	0.6			wrist	1.0
scapula, ulna, & radius	1.0	ilium, sacrum, & mandible	1.5	TOTAL	100.0			humerus	1.0
humerus	0.8	other	2.5					tibia	1.0
patella	0.1	TOTAL	100.0					os pubis	0.6
other	1.4							clavicle	0.6
TOTAL	100.0							other	8.0
								TOTAL	100.0

(compiled from well over 1000 cases by Cheyne, 1891)

(data from 160 cases in Rosencrantz et al., 1941)

(data from 408 cases of skeletal tuberculosis in children in Sanchis-Olmos, 1948)

(data from 3820 cases of skeletal tuberculosis in children. From Foote, 1927a)

(data from 230 cases of skeletal tuberculosis in adults. From LaFond, 1958)

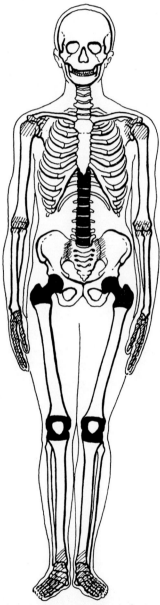

Figure 69. Skeletal distribution of tuberculosis. The solid black areas indicate the most frequent sites and diagonal lines mark occasional sites.

and collapse very similar to spinal tuberculosis. Other conditions which may be confused with tuberculosis of the spine include actinomycosis, coccidiodomycosis, Paget's disease, typhoid spine, and neoplasms such as hemangioma, aneurysmal bone cysts, and metastatic carcinoma. Examination of the rest of the skeleton provides a valuable means of eliminating several of the disorders listed above from consideration.

Spinal tuberculosis is found mainly in the lower thoracic and upper lumbar vertebrae. The cervical vertebrae and sacrum are infrequently involved. When a barograph is constructed using 587 cases of spinal tuberculosis, a binomial distribution is found ascending in regular order to a peak at the first lumbar vertebra (Hodgson et al., 1969). The evidence suggests that the initial site of spinal tuberculous infection is in the first lumbar vertebra from the adjacent left and right kidneys with subsequent spread through the paravertebral venous plexus or lymph nodes (see Batson, 1940).

Tuberculosis of the spine usually involves two to four vertebrae. In 192 cases of spinal tuberculosis, 82.0 percent affected four vertebrae or less. Over 50 percent affected only two vertebrae (Hallock and Jones, 1954). Occasionally one or more normal vertebrae may be found between two foci of infection in the vertebrae.

The gross appearance of spinal tuberculosis may vary considerably depending on the virulence of the organism and the resistance of the host to the infection. Thus, there may be marked bone destruction with little new bone formation or a chronic infection with bone regeneration. The central and anterior portions of the vertebral bodies are the most common sites of tubercular destruction. The infection reaches the center of the vertebral body through the venous plexus and the anterior portions often become involved by extension of a paravertebral abscess to the bone. In the latter case, three or more adjacent vertebrae are often affected. Tuberculous infection rarely involves the transverse processes, pedicles, lamina, or spinous processes of the vertebrae.

Tubercle formation begins in the marrow of the vertebral centra and causes resorption of the surrounding trabeculae. The focus enlarges with the formation of new tubercles and onset of caseation. The osseous destruction is apparently caused by proteolytic enzymes from leukocytes. Such destruction and necrosis often extends to and perforates the cortex. Complete destruction of one or more vertebral bodies and associated intervertebral disks often occurs. Bone regeneration is usually very minimal; the trabeculae surrounding the tubercular abscess may become sclerotic. Even after healing the abscess itself is rarely filled in with new bone, but remains filled with caseous matter.

As a result of the extensive trabecular and cortical destruction without new bone formation, collapse of the involved vertebral

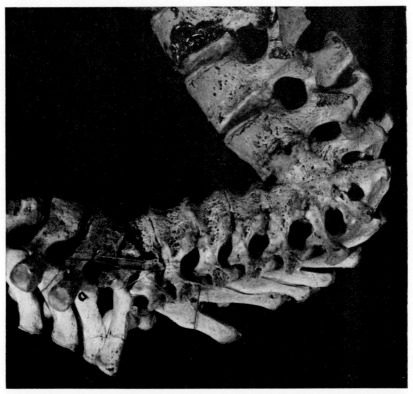

Figure 70. Spinal tuberculosis (Pott's disease). The bodies of the 9th thoracic through 1st lumbar vertebrae are almost completely destroyed and have collapsed into an irregular mass of bone. The vertebral arches are not involved and the spinal canal is unaffected. Bone regeneration is minimal. (WAM 5117)

bodies almost invariably occurs. The anterior portion of the vertebral body is usually most seriously involved and unequal collapse results in an angular posterior deformity or kyphosis (see Figs. 70 & 71). The degree of angulation is very marked

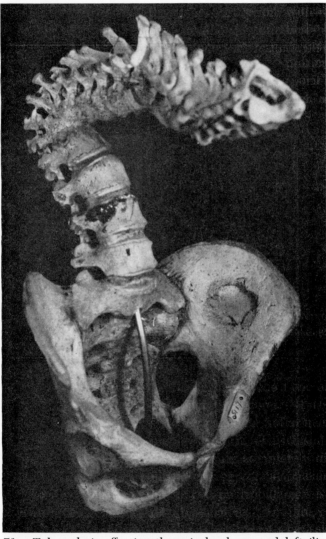

Figure 71. Tuberculosis affecting the spinal column and left ilium. The tubercular abscess in the ilium has almost completely perforated it. There is very little evidence of new bone formation around the lesion. (WAM 5117)

in the midthoracic region and less so when the collapse occurs in the cervical or lumbar regions. A lateral curvature or scoliosis of the spine often occurs, particularly in the thoraco-lumbar region. In spite of such vertebral collapse and deformity, the diameter of the spinal canal is infrequently narrowed. However, paralysis may result when excessive granulation tissue exerts pressure on the spinal cord.

Body bridging or fusion of the collapsed vertebral bodies is not very common. Among 100 cases of spinal tuberculosis, 10 percent exhibited body bridging (Cofield, 1922). All of these ten cases had certain characteristics in common. All of the patients were adults and the lesions were in the thoraco-lumbar region—a highly mobile part of the spine. The bony bridging is apparently due to ossification of the paravertebral ligaments (Hodgson et al., 1969).

TUBERCULOSIS OF THE JOINTS AND METAPHYSES

The tubercle bacillus is carried by the bloodstream and sets up a tubercular inflammation in the bone or synovial membrane of a joint, particularly the hip or knee. The bacillus usually lodges in the spongiosa of the metaphysis or epiphysis, probably due to the looped arrangement of the end arteries. The tuberculous destruction advances to and includes the articular cartilages thus affecting the joint and spreading to the synovial membrane. In other cases the synovial membrane is the initial site of infection, and the reverse process occurs although the final result is the same. The destruction is produced by the vascular and phagocytic action of the granulation tissue and leukocytes in the joint fluid.

The erosion of the articular surfaces may be widespread in the joint or localized to a small circular area. A characteristic feature of tuberculous arthritis is that it tends to produce an identical lesion on the two opposing surfaces of the joint. In advanced cases with articular destruction and edema, subluxation may occur producing further arthritic deformity. Bony ankylosis may take place, particularly in the knee. However, the natural

Figure 72. Spinal column sectioned to show traumatic fracture and collapse of the 12th thoracic vertebrae. A compression fracture can closely resemble tuberculosis although bone repair is often more prominent. (WAM 4629)

healing of a tuberculous joint by bone formation is slow and imperfect. The initial resorptive process in the spongiosa leaves a permanent mark in the form of a coarse cancellous network affecting the ends of both bones in the involved joint.

Besides extending from the metaphysis into the joint space, tuberculous destruction may also spread to the diaphysis. In some cases the joint is not affected while the metaphysis and adjacent portion of the diaphysis are extensively involved. The destruction does not always break through the cortex making it necessary to radiograph the bones which otherwise may appear normal. The radiological changes are typical of a granuloma.

184 *Paleopathological Diagnosis and Interpretation*

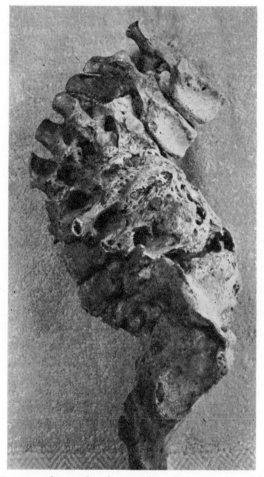

Figure 73. Portion of spinal column of a thirty-six-year-old male Indian from the Mississippian Culture of Illinois and dated around 1200-1300 A.D. All 5 lumbar and lower 3 thoracic vertebrae are involved by destruction and subsequent new bone formation. The large number of vertebrae affected and marked bone regeneration are atypical of tuberculosis. (USNM 381,852)

The abscess originates in the metaphyseal cancellous tissue and gradually enlarges with the resorption of the trabeculae. The trabeculae bordering the abscess may become slightly sclerotic but complete encapsulation of the abscess is rare. This differs from the marked sclerosis surrounding a Brodie's abscess.

Tuberculosis

If a large part of the cancellous tissue is invaded by granulation tissue, a sequestrum forms. Sequestration is less common than in pyogenic osteomyelitis, and the sequestra are usually much smaller. Moreover, they are irregular and ill-defined in the roentgenogram compared to the sharply demarcated sequestra in pyogenic osteomyelitis (Shanks and Kerley, 1950). Occasionally, conical shaped tuberculous sequestra are located in the articular ends of the bones on opposite sides of the joint and have been described as "kissing sequestra" (Pomeranz, 1933). Such sequestra are more common in adults than in children and are found at the points of maximum weight-bearing in the joint (Phemister and Hatcher, 1933).

Periostitis is present in about 30 percent of cases and appears to depend on the proximity of the medullary destruction to the cortex. If the cortex is osteoporotic and the pressure within

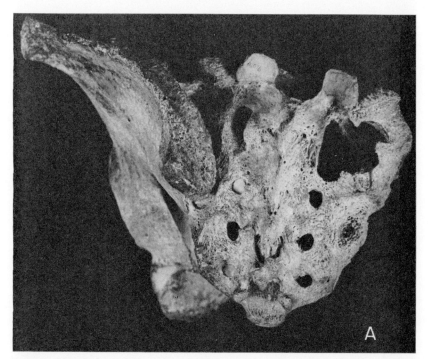

Figure 74. Tuberculosis in a forty-year-old male Indian from Yucatan. (A.) Posterior view of sacrum showing large abscess at site of articulation with the ilium.

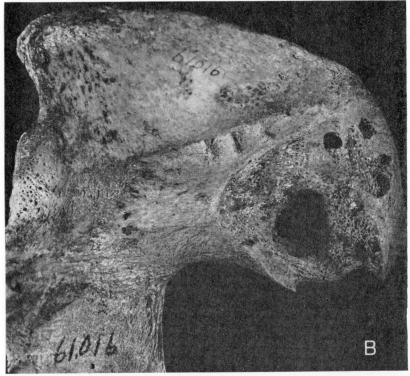

Figure 74B. Anterior view of the ilium showing similar destruction by the abscess with little evidence of bone regeneration. (PM N/61016)

the bone is raised by excessive granulation tissue, a local expansion tends to occur. Further expansion occurs with subperiosteal bone apposition although this is not as marked as in pyogenic osteomyelitis (Knaggs, 1926).

DIFFERENTIAL DIAGNOSIS

Since many conditions closely resemble vertebral tuberculosis, it is important to radiograph the rest of the skeleton for evidence of tuberculous abscess formation in the cancellous tissue of the metaphyses. A tuberculous abscess closely resembles a Brodie's abscess but lacks the marked sclerosis walling off the lesion. Larger abscesses extending into the diaphyses may simulate

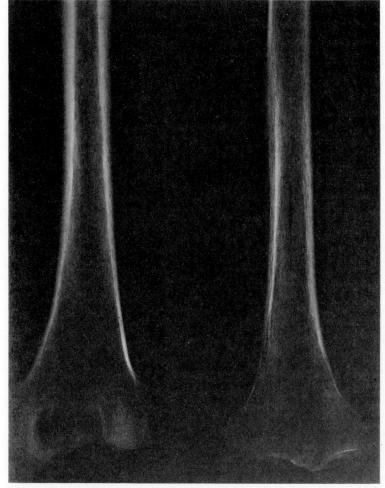

Figure 75. Roentgenogram of the left femur and tibia from the same individual as in the previous figure. The distal epiphysis of the femur reveals lytic destruction by a tubercular abscess. Both bones are extremely light and brittle. (PM N/61016)

pyogenic osteomyelitis but differ in certain respects as discussed above. Furthermore, the epiphyseal cartilage acts as a very effective barrier to the spread of sepsis from the shaft, but offers no resistance to tuberculosis. In syphilis, the lesions are usually multiple and involve a larger area of the shaft. The trabeculation in the medullary cavity is extensively altered with areas of sclerosis and rarefaction. The cortex is uneven with areas of destruction and other regions with marked subperiosteal bone apposition. These characteristics make syphilis easily differentiated from tuberculosis. The arthritic changes in tuberculosis may suggest a severe osteoarthritis except the other joints will appear normal.

TUBERCULOSIS BIBLIOGRAPHY

Allison, M. J., J. Mendoza, and A. Pezzia: "Documentation of a case of tuberculosis in pre-Columbian America," *Amer. Rev. Resp. Dis., 107*: 985-991, 1973.

Auerbach, O., and M. G. Stemmerman: "Roentgen interpretation of pathology in Pott's disease," *Amer. J. Roentgen., 52*:57-63, 1944.

Bartels, P.: "Tuberkulose in der jüngeren Steinzeit," *Arch. Anthrop., 6*: 243-250, 1907.

Batson, D. V.: "The function of the vertebral veins and their role in the spread of metastases," *Ann. Surg., 112*:138-149, 1940.

Bosworth, D. M. and J. Levine: "Tuberculosis of the spine," *J. Bone Joint Surg., 31A*:267-274, 1949.

Bourke, J. B.: "The palaeopathology of the vertebral column in ancient Egypt and Nubia," *Med. Hist., 15*:363-375, 1971.

Breed, R. S., E. G. D. Murray, and N. R. Smith (ed.): *Bergey's Manual of Determinative Bacteriology*, Williams and Wilkins, Baltimore, 1957.

Brothwell, D. R.: "The palaeopathology of early British man," *J. Roy. Anthrop. Inst., 91*:318-344, 1961.

Bushnell, G. H. S.: *Peru*, F. A. Praeger, New York, 1966.

Cave, A. J. E.: "The evidence for the incidence of tuberculosis in ancient Egypt," *Brit. J. Tuberc., 33*:142-152, 1939.

Cheyne, W. W.: "Lectures on the pathology of tuberculous disease of bone and joints," *Brit. Med. J.*, pp. 896-901, 1891.

——————: *Tuberculous Diseases of Bones and Joints, Their Pathology, Symptoms, and Treatment*, Henry Frowde, London, 1911.

Clark, G.: *World Prehistory*, Cambridge Univ. Press, Cambridge, 1962.

Cockburn, A.: *The Evolution and Eradication of Infectious Diseases*, The Johns Hopkins Press, Baltimore, 1963.

Cofield, R. B.: "Bony bridging in tuberculosis of the spine," *J. Amer. Med. Assoc.*, pp. 1-9, 1934.

Derry, D. E.: "Pott's disease in ancient Egypt," *Med. Press, 197*:196-199, 1938.

Derums, V. Y.: "Some paleopathologic data on Baltic Coast inhabitants," *Fed. Proc. (Tran.s Suppl.), 24*:225-230, 1965.

Dubos, R.: *Man Adapting*, Yale Univ. Press, New Haven, 1966.
Ferguson, R. G.: "A study of tuberculosis among Indians," *Trans. Amer. Clin. Climat. Assoc.*, pp. 1-9, 1934.
Garcia-Frias, J. E.: "La tuberculosis en los antiguos Peruanos," *Actualidad Med. Peruana*, 5:274-291, 1940.
Gejvall, N-G.: *Westerhus: Medieval Population and Church in the Light of Skeletal Remains*, Boktryckeri, Lund, 1960.
Girdlestone, G. R. and E. W. Somerville: *Tuberculosis of Bone and Joint*, Oxford Med. Pub., London, 1940.
Hallock, H. and J. B. Jones: "Tuberculosis of the spine," *J. Bone Joint Surg.*, 36A:219-240, 1954.
Hare, R.: "The antiquity of diseases caused by bacteria and viruses, a review of the problem from a bacteriologist's point of view," in D. R. Brothwell and A. T. Sandison (ed.): *Diseases in Antiquity*, Thomas, Springfield, 1967.
Hodgson, A. R., W. Wong, and A. Yau: *X-ray Appearances of Tuberculosis of the Spine*, Thomas, Springfield, 1969.
Hooton, E. A.: *Indians of Pecos Pueblo*, Yale Univ. Press, New Haven, 1930.
Hrdlička, A.: "Tuberculosis among certain Indian tribes," *Bureau of Amer. Ethnol. Bull.*, 42:1-48, 1909.
——————: "A report on a collection of crania and bones from Sorrel Bayou, Iberville Parish. Louisiana," *J. Phila. Acad. Nat. Sci.*, 16:97, 1913.
Isager, K.: *Krankenfürsorge des Dänischen Zisterzienserkloster Øm. cara Insulina 1172-1560*, Munksgaard, Copenhagen, 1941.
Jacono, G.: "Rapportie e correlazione evolutive fra tubercolosi pulmonare e tubercolosi extrapulmonare," *Arch. Med. Chir.*, 6:3-25, 1937.
Johansson, S.: *Knochen und Gelenktuberkulose in Kindesalter*, Fischer, Jena, 1926.
Joklik, W. K. and D. T. Smith (ed.): *Zinsser's Microbiology*, Meredith Corp., New York, 1972.
Kidder, A.: "Settlement patterns—Peru," in G. Willey (ed.): *Prehistoric Settlement Patterns in the New World*, Wenner-Gren Foundation, New York, 1956.
Knaggs, R. L.: *Diseases of Bone*, W. Wood, New York, 1926.
LaFond, E. M.: "An analysis of adult skeletal tuberculosis," *J. Bone Joint Surg.*, 40A:346-364, 1958.
Lichtor, J. and A. Lichtor: "Paleopathological evidence suggesting pre-Columbian tuberculosis of the spine," *J. Bone Joint Surg.*, 39A:1398-1399, 1957.
Meinecke, B.: "Consumption (tuberculosis) in classical antiquity," *Ann. Med. Hist.*, 9:379-402, 1927.

Møller-Christensen, V.: *Bogen om Aebelholt Kloster,* Dansk Videnskabs Forlag, Copenhagen, 1958.

Moodie, R. L.: "Pott's disease in early historic Peru," *Ann. Med. Hist.,* 9:356, 1927.

Morse, D.: "Prehistoric tuberculosis in America," *Amer. Rev. Resp. Dis.,* 83:489-504, 1961.

————: "Tuberculosis," in D. R. Brothwell and A. T. Sandison (ed.): *Diseases in Antiquity,* Thomas, Springfield, 1967.

Morse, D., D. R. Brothwell, and P. J. Ucko: "Tuberculosis in ancient Egypt," *Amer. Rev. Resp. Dis.,* 90:524-541, 1964.

Nathanson, L. and W. Cohen: "Statistical and roentgenological analysis of 200 cases of bone and joint tuberculosis," *Radiology,* 36:550-567, 1941.

O'Bannon, L. G.: "Evidence of tuberculosis of the spine from a Mississippi stone box burial," *Tenn. Archeologist,* 13:75-80, 1957.

Phemister, D. B. and C. H. Hatcher: "Correlation of pathological and roentgenological findings in the diagnosis of tuberculous arthritis," *Amer. J. Roentgen,* 29:736-752, 1933.

Pomeranz, M. M.: "Roentgen diagnosis of bone and joint tuberculosis," *Amer. J. Roentgen.,* 29:753-762, 1933.

Poppel, M. H., L. R. Lawrence, H. G. Jacobson, and J. Stein: "Skeletal tuberculosis: roentgenographic survey with reconsideration of diagnostic criteria," *Amer. J. Roentgen.,* 70:936-963, 1953.

Requena, A.: "Evidencia de tuberculosis en la America pre-Columbia," *Acta Venezolana,* 1:1-20, 1945.

Rich, A. R.: *The Pathogenesis of Tuberculosis,* Thomas, Springfield, 1944.

Ritchie, W. A.: "Pathological evidence suggesting pre-Columbian tuberculosis in New York State," *Amer. J. Phys. Anthrop.,* 10:305-317, 1952.

Rokhlin, D. G. and V. S. Maikova-Stroganova: "Tuberculous injury of the spinal column in paleopathologic material," *Vestnik Rentgenol.,* 19:182, 1938.

Roney, J. G.: "Palaeopathology of a California archaeological site," *Bull. Hist. Med.,* 33:97-109, 1959.

Rosencrantz. E., A. Piscitelli, and F. C. Bost: "An analytical study of bone and joint lesions in relation to chronic pulmonary tuberculosis," *J. Bone Joint. Surg.,* 23A:628-638, 1941.

Rowling, J. T.: "Disease in ancient Egypt: evidence from pathological lesions found in mummies," M.D. Thesis, Univ. of Cambridge, 1960.

Ruffer, M. A.: "Pott's disease in an Egyptian mummay from the 21st Dynasty (1000 B.C.)," in R. L. Moodie (ed.): *Studies in the Palaeopathology of Egypt,* Univ. of Chicago Press, Chicago, 1921.

Sager, P., M. Schalimtzek and V. Møller-Christensen: "A case of spondylitis tuberculosa in the Danish Neolithic age," *Danish Med. Bull.,* 19:176-180, 1972.

Sanchis-Olmos, V.: *Skeletal Tuberculosis*, Williams and Wilkins, Baltimore, 1948.
Senn, N.: *Tuberculosis of Bones and Joints*, F. A. Davis, Philadelphia, 1892.
Shanks, S. C. and P. Kerley (ed.): *A Text-Book of X-ray Diagnosis*, vol. 4, H. K. Lewis, London, 1950.
Sjövall, E.: "Undersökning av gamla skelettdelar i kulturhistoriens tjänst," *Naturens Verden*, 26:297, 1942.
Smith, G. E. and M. A. Ruffer: "Pott'sche Krankheit an einer agyptischen Mumie aus der Zeit der 21 Dynastie (um 1000 V. Chr.)," in Sudhoff's *Zur historischen Biologie der Krankheitserreger*, Leipzig, pp. 9-16, 1910.
Stead, W. W.: "Mycobacterial diseases," in Wintrobe (ed.): *Harrison's Principles of Internal Medicine*, McGraw-Hill, New York, 1974.
Stewart, T. D.: "A physical anthropologist's view of the peopling of the New World," *Southwestern J. Anthrop.*, 16:259-273, 1960.
Tordeman, B., P. Nørlund, and B. E. Ingelmark: *Armour from the Battle of Wisby*, Stockholm, 1939.
Weiss, D. L. and V. Møller-Christensen: "An unusual case of tuberculosis in a Medieval leper," *Danish Med. Bull.*, 18:11-14, 1971.
Wells, C.: "An early case of birth injury. Multiple abnormalities in a Romano-British skeleton," *Develop. Med. Child. Neurol.*, 6:397-402, 1964.
Whitman, R.: *A Treatise on Orthopaedic Surgery*, Lea and Febiger, Philadelphia, 1927.
Whitney, W. F.: "Notes on the anomalies, injuries, and diseases of the bones of the native races of North America," *Report of the Peabody Museum*, 3:433-448, 1886.

Chapter VI

LEPROSY

HISTORY

MORE THAN ANY other disease with the possible exception of syphilis, leprosy is cloaked with centuries of misconceptions and legends regarding its origin, history, and epidemiology. The earliest evidence of leprosy is found in the ancient medical literature of India and China. Reasonably good descriptions of the clinical features and treatment of this disease have been dated at approximately 600 B.C. in these two countries (Cochrane, 1964). It has been suggested that leprosy was brought to the Mediterranean area by the soldiers of Alexander the Great returning from the Indian campaign of 327-326 B.C. (Andersen, 1969). This is supported by the lack of any references to leprosy prior to 300 B.C., including the writings of Hippocrates (460-377 B.C.). The first descriptions of the disease in this region were those of the Graeco-Roman authors such as Celsus (25 B.C.-37 A.D.), Plinius 23-79 A.D.), Aretaios (200 A.D.), and Galen 130-201 A.D.).

These authors used the word *elephantiasis* or *elephas* to denote true leprosy. Much confusion has arisen by the translation of *tsara'ath* from the Hebrew Bible into the latin word *lepra* by Jewish scholars around 300-200 B.C. From the writings of authors such as Hippocrates, Plinius, Aribasios (326-403 A.D.), and Paulos of Aegina (about 600 A.D.), it is clear that the word *lepra* was not used to describe true leprosy. It was not until the 8th century A.D. that *lepra* was used to describe true leprosy through the writings of the Arab physician John of Damascus (Hulse, 1972).

From the Mediterranean area, leprosy spread to all parts of Europe and Africa. By the Middle Ages numerous leper hospitals had been established all over Europe, including Scandinavia and Britain (see Richards, 1960). As discussed earlier, it is believed that many of the "lepers" were actually victims of syphilis or other diseases, but careful and extensive archaeological excavations have failed to confirm this belief (Møller-Christensen, 1969; Andersen, 1969). The prevalence of leprosy in Europe reached a peak around the time of the Crusades (1096-1221). There is evidence that leprosaria actually helped disseminate leprosy at that time by the inclusion of noninfected cripples in such institutions (Jarcho, 1971). In any case, leprosy rapidly declined in Europe around the time of the Renaissance, perhaps due to improved living conditions.

Leprosy was probably introduced to the New World through the slave trade from highly endemic areas in West Africa (Cochrane, 1964). Pre-Columbian clay figurines showing mutilations of the face have been cited as evidence of leprosy, but are more likely the result of syphilis or leishmaniasis (Ashmead, 1901, 1902; Pesce, 1955).

PALEOPATHOLOGICAL EVIDENCE

Through the pioneering work of Dr. Vilhelm Møller-Christensen and his associates, we have extensive skeletal evidence of leprosy in earlier human populations. Dr. Møller-Christensen, a Danish physician, located and meticulously excavated several Medieval cemeteries associated with leprosy hospitals in Denmark. He has described several hundred skeletons with bone lesions definitely caused by leprosy (Møller-Christensen, 1953, 1958, 1961; see also Andersen, 1969). Dr. Møller-Christensen has recently examined large skeletal series from many parts of the world for evidence of leprosy and his results together with a brief review of the literature on paleopathological cases of leprosy are presented here.

Geographical Distribution of Leprosy

Egypt

Although the earliest literary references to leprosy came from India and China, very little effort has been made to examine the somewhat scanty skeletal material from these areas for leprotic bone changes. The earliest known skeletal evidence of leprosy comes from Egypt around 500 A.D. Out of the many thousands of skeletons excavated from this area, only two individuals exhibit definite bone changes of leprosy—and both come from the same Coptic cemetery near Aswan, Nubia. One of the individuals is a mummified adult male and the leprotic changes in the hands and feet were first described by Smith and Derry (1910) and confirmed by Møller-Christensen (1967). The skull of an adult female from the same cemetery was originally reported as a case of chronic rhinitis (Smith and Jones, 1910). Reexamination of this specimen in the light of new knowledge of *facies leprosa* makes this a likely case of leprosy (Møller-Christensen and Hughes, 1966).

Africa

Sub-Saharan Africa today is a vast area with a high prevalence of leprosy. Very little effort has been made to examine skeletal material from this area for evidence of leprosy. One questionable case of leprosy was noted among 122 19th century skeletons from Africa (Møller-Christensen, 1967). Similarly, leprosy is currently endemic to the Middle East but examination of 695 skeletons from Palestine (700-600 B.C.) reveals no evidence of leprosy (Møller-Christensen, 1967). This latter finding supports the possible introduction of leprosy into the area at a later date as suggested by Andersen (1969).

Europe

Leprosy was probably brought to Britain during the Roman invasions and the earliest skeletal evidence of the disease comes from five Saxon skeletons around 600 A.D. (Brothwell, 1958, 1961; Møller-Christensen and Hughes, 1962; Wells, 1962). Four other cases from Medieval times have been reported (Møller-Christensen, 1967).

By far the greatest number of leprotic skeletons have come from Denmark through the careful excavations of Dr. Møller-Christensen. These skeletons date from the Middle Ages or 1200-1500 A.D. Over 300 skeletons with bone changes characteristic of leprosy have been described (Møller-Christensen and Faber, 1952; Møller-Christensen, 1953, 1958, 1961; Møller-Christensen and Weiss, 1971).

New World

There is no definite pre-Columbian evidence of leprosy in the New World, which supports the belief that the disease was introduced by the Spaniards or African slaves. Only one possible case of pre-Columbian leprosy has been reported, but the bone lesions were more appropriately attributed to frostbite (Post and Donner, 1972). The remains are those of a mummified adult male from the highlands of northern Chile around the 1st century A.D. Both well-preserved feet exhibit complete loss of the distal phalanges and the right foot lacks the middle and greater part of the proximal phalanges as well. The complete lack of metatarsal absorption in this individual and the absence of leprotic changes in the numerous other mummies examined makes this an unlikely case of leprosy. The bone destruction in this mummy closely resembles clinical descriptions of frostbite (see Vinson and Schatzki, 1954).

Only one case of post-Columbian leprosy has been reported in American Indian skeletal material. The specimen is a skull of an adult Comox Indian from Vancouver Island during the 19th century (Møller-Christensen and Inkster, 1965). Leprosy is currently endemic in large parts of South America and scattered pockets of the disease may be found in Central America and the Gulf states, particularly Louisiana. Examination of skeletal remains from these areas may reveal further evidence of leprosy in the New World.

BIOLOGY AND EPIDEMIOLOGY

Leprosy is a chronic, infectious disease caused by *Mycobacterium leprae,* an acid-fast bacillus first described by Hansen in 1874. Although *M. leprae* was the first pathogenic microorganism described and associated with a specific disease, it has never fulfilled Koch's postulates. The bacteria are very numerous in lepromatous lesions but can not be grown in artificial media or tissue cultures. Only recently have certain animals been successfully inoculated.

Transmission

The usual mode of transmission is by direct contact with an infected person or by inhaling bacilli released from the nasal mucosa of an infected individual (see Davey, 1974). The leprosy bacilli penetrate the skin and spread along peripheral nerve fibers. In most cases exposure must be prolonged and explains why the disease is frequently a family infection. Even in close family contacts with a case of lepromatous leprosy, the attack rate is only 5 to 10 percent—disproving the popular misconception that leprosy is a highly contagious disease. Leprosy can be contracted at any age but children are more susceptible because of growth and change in the skin during childhood and the special liability of the cutaneous nerve plexuses to damage (Khanolkar, 1951). The incubation period for leprosy is extremely long, ranging on the average from three to six years, and cases have been reported extending the period to several decades (Doull, 1962).

Distribution

History has shown that climate is not an important factor in the geographical distribution of leprosy, although it is predominantly a tropical disease today. The disease was prevalent in northern Europe during the Middle Ages and declined through improved living conditions. It is interesting to note that leprosy remained prevalent along the coast of Norway well into the 18th and 19th centuries because of poor housing conditions on shore and close living quarters of the fishermen at sea. In general, the poor living conditions and frequent overcrowding explains the geographical distribution of leprosy today in tropical and subtropical regions.

Prevalence

As shown by its low attack rate and long incubation period, leprosy is not an easily communicable disease. Even in areas where the disease is endemic, the prevalence is rarely more than 1 percent of the population. Leprosy is a sporadic disease and does not affect an entire area uniformly. Two villages less than 100 yards apart may have prevalences of 5 and 0 percent respectively. Sociological factors are important in determining this pattern such as the caste system or restriction of exposure to persons outside of the extended family (Cochrane, 1964).

Classification

The clinical presentation of leprosy is a broad spectrum depending on the immune response of the host (Fasal, 1971). With complete resistance, infection will not occur. If resistance is not complete, an indeterminate form of leprosy develops which either spontaneously heals or develops into one of two forms of the disease: *tuberculoid* or *lepromatous*. In tuberculoid leprosy, the host-immune response is effective in limiting the disease to the skin and nerves. The skin lesions are sharply demarcated and contain few bacilli. In lepromatous leprosy, there is very little resistance to the disease which is extensive, diffuse, and bilaterally symmetrical. Almost any organ can be affected and the bacilli are numerous in these lesions.

The skeletal and historical evidence suggest that lepromatous leprosy, also known as low-resistance leprosy, was by far the most common form of the disease in all areas of the world until relatively recent times (Andersen, 1969). Today the tuberculoid form of the disease predominates in areas such as India and Africa where leprosy has long been endemic, and this shift may reflect the length of endemicity.

Improved living conditions and hygiene may explain the decline of leprosy in Europe at the beginning of the Renaissance, but it is quite possible that an immunologic factor was involved as well. There is evidence that tuberculosis (also caused by a species of Mycobacteria) provides partial cross-immunity with leprosy, and the decline of leprosy may be correlated with a rise of the more infectious tuberculosis around the time of the Renaissance (Chaussinand, 1948). Tuberculosis is often a fatal complication of lepromatous leprosy because there is low host resistance to both diseases. (For an interesting paleopathological case of tuberculosis in a Medieval leper see Weiss and Møller-Christensen, 1971a). On the other hand, tuberculosis is a rare complication of tuberculoid leprosy.

FREQUENCY OF BONE CHANGES

Most studies of bone changes in leprosy report a very high incidence of bone involvement because they use institution populations with severe and protracted cases of leprosy. For example, Table IX shows that studies of leprosaria patients report frequencies of bone changes as high as 50 to 68 percent of all cases. Perhaps the best estimate of the incidence of osseous leprosy is the evidence of bone changes in 15 percent of 150 randomly selected leprous patients (Chamberlain et al., 1931).

With a prevalence of leprosy in endemic areas of roughly 1 percent and 15 percent frequency of bone lesions in these cases, one would expect to find a very low incidence of the disease in a representative skeletal series. In such instances several hundred or even thousands of skeletons must be examined to find evidence of leprosy (see Møller-Christensen, 1967). However, the isolation of lepers in leprosaria, a practice which has been maintained for well over 1000 years, provides a skeletal population with a

TABLE IX

FREQUENCY OF BONE LESIONS IN CASES OF LEPROSY

Percent with Bone Changes	Number of Cases	Location	Study
15	150	Hawaii	Chamberlain et al., 1931
29	505	Louisiana	Faget and Mayoral, 1944
50	140	Hawaii	Murdock and Hutter, 1932
54	894	Hong Kong	Paterson, 1961
68	483	Columbia	Esguerra-Gomez and Acosta, 1948
77	202	Denmark	Andersen, 1969*

* paleopathologic study

very high incidence of the disease. For example, 77 percent of 202 skeletons excavated from a Medieval Danish leprosarium exhibit evidence of leprosy (Andersen, 1969). This high incidence is similar to that reported in modern leprosaria and includes many cases of leprous changes in the rhinomaxillary region not previously appreciated by modern clinicians (see Møller-Christensen, 1965). Information from these paleopathological studies will be utilized along with clinical studies to describe the bone changes of leprosy in the following sections.

Classification of Bone Lesions

The bone changes in leprosy are of three main types: specific bone destructive changes, nonspecific bone absorptive changes, and osteoporosis. Specific bone changes are those produced by the direct action of leprosy bacilli on bone. Formerly such changes were thought to be very rare, but the description of *facies leprosa* and its common occurrence reported by Møller-Christensen has provided an important means of diagnosing leprosy both paleopathologically and clinically. Other less common specific bone changes include enlarged nutrient foramina and small areas of localized cortical destruction in the small bones of the hands and feet due to lepromatous granulomas.

Nonspecific bone absorptive changes are by far the most common bone lesions found in leprosy. The bones of the hands and feet are almost exclusively involved with the absorptive or neurotrophic changes. In severe cases the weakened bones are liable to fracture and subluxation of the joints may occur resulting in arthritic changes and ankylosis. Secondary pyogenic infections of bone are common and must be carefully differentiated from the primary process. Osteoporosis as a result of disuse is the third type of bone change and is present only in severe cases.

Location

The skeletal distribution of specific and nonspecific leprous bone changes is illustrated in Figure 76. The bones of the nasal region and phalanges, metacarpals, and metatarsals of the hands and feet are almost exclusively involved. Less commonly, the tibia and fibula may exhibit nonspecific periosteal thickening.

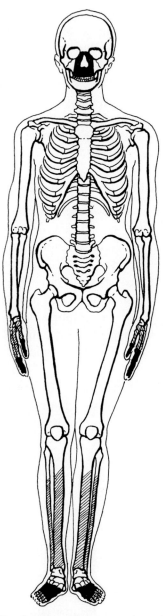

Figure 76. Skeletal distribution of leprosy. The solid black areas indicate the most frequent sites and diagonal lines mark occasional sites.

In forty-one complete skeletons exhibiting leprous bone changes, the rhinomaxillary bones were involved in virtually all cases and associated lesions in the hands and feet were present in about 66 percent of these skeletons. The presence of leprotic changes in the hands and feet without involvement of the rhinomaxillary bones is very infrequent (Møller-Christensen, 1961; Andersen, 1969).

GROSS BONE PATHOLOGY

Skull

Excavated crania from leprosaria cemeteries exhibited a high incidence of well-defined bony changes referred to as *facies leprosa* (Møller-Christensen and Faber, 1952). The changes characterizing this condition are:

1) Atrophy of the anterior nasal spine.
2) Atrophy and recession of the maxillary alveolar margin with loosening and possible antemortem loss of the incisors.
3) Inflammatory changes on the superior surface of the hard palate. The thinning, pitting, or perforation of the hard palate must be present for a diagnosis of leprosy. The first two features listed above are not always present.

It is interesting to note that Dr. Møller-Christensen's descriptions of *facies leprosa* in archaeological material provided important impetus to the clinical study of leprosy (see Møller-Christensen et al., 1952). In a radiographic study of ninety-six lepers from the Congo, the characteristic bone changes of *facies leprosa* were present in 60 percent (Lechat and Chardome, 1955). In a similar sampling of fifty patients from India, 82 percent exhibited rhinomaxillary changes (Andersen, 1969). It should again be emphasized that the high incidences reported are the result of examining only severe cases of lepromatous leprosy. A better perspective is provided by a general survey of 488 cases of lepromatous leprosy revealing perforation of the nasal septum in only 9.6 percent. The more extensive bone changes were probably even less frequent (Job et al., 1966).

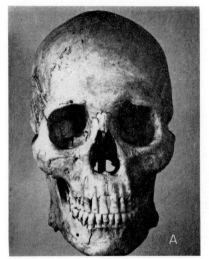
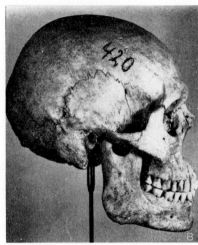

Figure 77. Successive stages of *facies leprosa* as illustrated by Medieval lepers from Denmark.
Figure 77A and B. Smoothing of the nasal aperture and slight erosion of the upper alveolar margin.

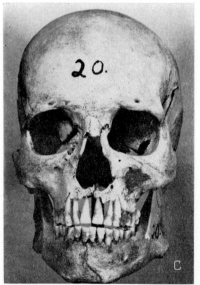
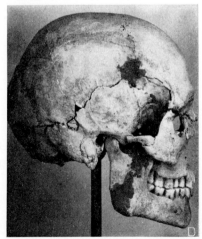

Figure 77C and D. Extensive absorption of the nasal aperture and alveolar margin with loosening of the upper incisors.

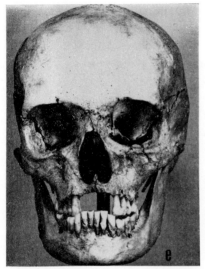 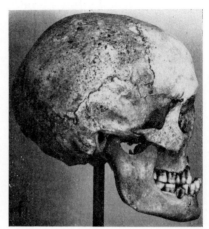

Figure 77E and F. Antemortem loss of the upper incisors. (Courtesy of Dr. Vilhelm Møller-Christensen.)

It is now well known that the nose contains large numbers of leprosy bacilli in nearly all cases of lepromatous leprosy. The inferior turbinate is initially infected leading to ulceration of the nasal septum by the lepromatous granulation tissue. Eventually the bone changes of *facies leprosa* may result (see Barton, 1974; Davey, 1974).

Hands and Feet

Bone changes in the hands and feet are prominent features in severe cases of both lepromatous and tuberculoid leprosy. The relatively uncommon specific leprous changes such as pseudocysts (localized areas of bone destruction) are hard to distinguish in archaeological material because of confusion with postmortem changes in the small bones. Furthermore, the antemortem destruction of these already fragile bones greatly decreases the possibility of preservation. Enlarged nutrient foramina are present in a significant number of leprosy cases (Murdock and Hutter, 1932). However, they are not reliable evidence of the disease because enlarged foramina may be found in normal individuals.

The nonspecific absorptive changes are by far the most common lesions found in the hands and feet. These lesions are well described in the literature so that only a few important points need be mentioned here (see Paterson, 1955, 1961; Paterson and Job, 1964; Lechat, 1962). The pathogenesis of the bone absorption with very little reactive bone formation remains unclear, but certainly the impairment of sensory and motor innervation and neurovascular changes are important factors.

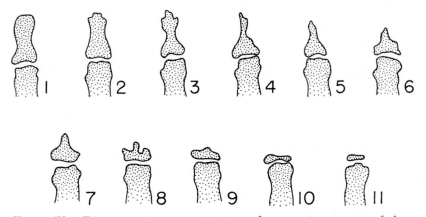

Figure 78. Diagrammatic representation of successive stages of bone absorption in the distal phalanges of the hand. (Courtesy of Dr. Michel Lechat, 1962.)

In the hands, the bone absorption or acroosteolysis begins in the distal phalanges. It is often bilateral but is rarely truly symmetrical. The initial manifestation is a notching of the distal phalangeal tuft. This distal absorption may continue until the distal phalanx completely disappears and the process continues in the more proximal phalanx. In some cases severe destruction of the proximal phalanges is evident without involvement of the distal phalanges. In rare cases even the metacarpals become involved, but the process seldom extends further. Reactive bone formation is not a prominent feature. New bone is laid down on the medullary side of the cortex while the absorption begins at the periosteal surface creating the "melted bone" appearance.

Similar bone changes occur in the feet except the phalangeal absorption frequently starts in the metatarsophalangeal joints leaving the distal phalanges intact. This difference may be due to the increased trauma and pressure to the metatarsal bones in normal use similar to the increased trauma to the distal phalanges in the fingers (Cooney and Crosby, 1944). In any case, the bone absorption beginning at the m-p joint progresses in both the proximal phalanx and metatarsal resulting in "pencilling" or "point" formation (see Fig. 80). The two pointed bones face each other. Bone repair plays an important role in this point

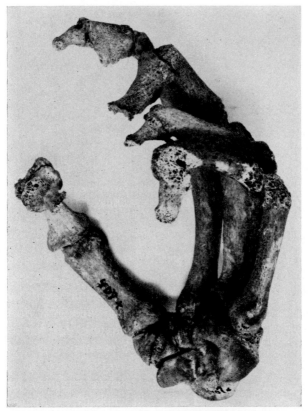

Figure 79. Hand of a Medieval leper showing the absorptive changes in the distal phalanges. (Courtesy of Dr. Vilhelm Møller-Christensen.)

formation (see Gondos, 1968). In severe cases, the bone destruction may extend to both the distal phalanges and even tarsal bones.

Arthritic changes are common in both hands and particularly feet as a result of muscle contractures, abnormal gait and functional use of hands following motor and sensory nerve impairment. Ankylosis is not common but may occur in any of the interphalangeal joints and tarsal joints.

Osteitis is rare in the bones of the hands and feet and is probably due to secondary pyogenic infection in soft tissue ulcers. This is almost certainly the case if sequestrum formation, sinus formation, or hyperostosis has occurred.

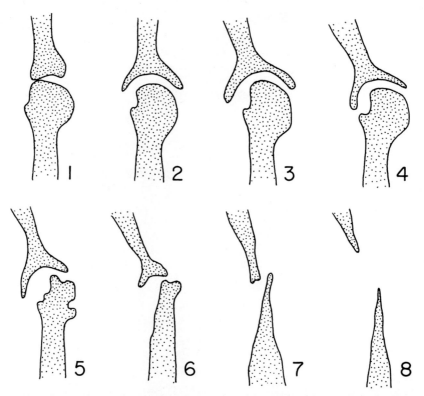

Figure 80. Diagrammatic representation of successive stages of bone absorption originating in the metatarsophalangeal joints of the feet. (Courtesy of Dr. Michel Lechat, 1962.)

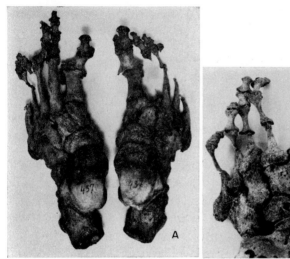

Figure 81. Absorptive changes in the feet of Medieval lepers. (A.) Note the pencilling of the metatarsals and absence of involvement in many of the phalanges. (B.) Absorption particularly evident in the metatarsals. (Courtesy of Dr. Vilhelm Møller-Christensen.)

Tibia and Fibula

Irregular subperiosteal bone deposits in the long bones at the ankle and wrist have been infrequently noted in clinical studies (Chamberlain et al., 1931; Murdock and Hutter, 1932). However, Møller-Christensen found subperiosteal deposits in the tibia and fibula in 78 percent of ninety-five Medieval leper skeletons. Andersen has confirmed this finding in a clinical study revealing similar bone changes in 36 percent of fifty-six leprosy patients.

The irregular subperiosteal deposits may occur anywhere along the length of the tibia or fibula, but are usually located in the distal third. They are almost invariably on adjacent surfaces, i.e. lateral aspect of the tibia and medial aspect of the fibula. The tibia is usually more severely involved than the fibula. In many cases, prominent transverse striations cross the subperiosteal deposits and may be vascular grooves.

DIFFERENTIAL DIAGNOSIS

Various infectious, vascular, neurological, and neoplastic conditions may produce bone changes at least partially resembling those caused by leprosy. Fortunately, very few produce lesions in both the skull and extremities and several of the conditions listed below are exceedingly rare.

Pyogenic osteomyelitis may affect the bones of the hands and feet following exposure of the bones through trauma or ulceration. Reactive bone formation is a prominent feature of this disease which is not present in leprosy. In addition sequestrum formation and sinus formation are not features of leprosy. Since pyogenic osteomyelitis may occur secondarily in leprotic ulceration, the bones must be carefully examined to ascertain if features of both diseases are present. Pyogenic osteomyelitis does not produce the bone lesions of *facies leprosa*.

Syphilis and yaws may involve both the facial bones and bones of the hands and feet. Perforation of the palate may occur and severe cases may also extend the destruction to the maxilla. The nasal bones are commonly involved in contrast to the absence of leprotic destruction of these bones. The dactylitis of syphilis and yaws rarely resembles the absorptive changes of leprosy. There is often phalangeal "expansion" through subperiosteal apposition rather than concentric absorption with endosteal bone formation. Some cases of yaws have been reported with absorption of the phalangeal shafts and subsequent shortening (Riseborough et al., 1961). The bone absorption occurs bilaterally but it rarely involves the epiphyses of the phalanges in contrast to leprosy. Neurosyphilis may result in mixed hypertrophic and atrophic changes of the digits but the large joints are usually involved (Charcot joints). In the majority of such cases, typical syphilitic bone lesions will be present in the skull vault and tibia.

In tuberculosis, the phalanges of the hand and metatarsals of the foot may reveal multiple destructive foci with marked periostitis resulting in cortical "expansion." These lesions are infrequently bilateral and should be easily differentiated from leprosy (Feldman et al., 1971).

Fungal infections such as coccidioidomycosis, actinomycosis, and torulosis may present as solitary bone lesions in the hand or foot, but they often affect the bone eminences such as the olecranon, malleolus, and patella (Rosen and Jacobson, 1966).

Several conditions may produce absorptive changes very similar to leprosy because of similar vascular and neurotrophic mechanisms. Frostbite may affect both the hands and feet and almost always bilaterally. Enlarged nutrient foramina are a common finding in this condition as in leprosy. Unlike leprosy, the absorptive changes in the feet begin in the distal phalanges rather than m-p joints (Vinson and Schatzki, 1954). Diabetes mellitus may also produce absorptive bone changes in the hands and feet as a result of neurotrophic and vascular factors (Friedman and Rakow, 1971; Gondos, 1968; Plessis, 1970). However, it is unlikely that such individuals with severe diabetes could have survived long enough without modern treatment to exhibit such bone lesions. Other less common conditions producing similar bone lesions include syringomyelia, scleroderma, Raynaud's disease, and congenital indifference to pain.

Neoplastic conditions such as multiple enchondromata, eosinophilic granuloma, or fibrous dysplasia may produce focal areas of destruction resembling the pseudocysts of leprosy. Cortical expansion often occurs in such cases and other bones are frequently affected.

LEPROSY BIBLIOGRAPHY

Andersen, J. G.: "Studies in the mediaeval diagnosis of leprosy in Denmark," *Danish Med. Bull.*, 16 suppl.:1-142, 1969.

Ashmead, A. S.: "Deformations on American (Incan) pottery not evidence of pre-Columbian leprosy," *St. Louis Med. Surg. J.*, 80:177-192, 1901.

―――: "Introduction of leprosy into America from Spain—that disease was not pre-Columbian in the Western Hemisphere, but syphilis was," *St. Louis Med. Surg. J.*, 83:65-82, 1902.

Barnetson, J.: "Osseous changes in neural leprosy," *Acta Radiol.*, 34:47-56, 1950.

Barton, R. P.: "A clinical study of the nose in lepromatous leprosy," *Leprosy Rev.*, 45:135-144, 1974.

Brothwell, D. R.: "Evidence of leprosy in British archaeological material," *Med. Hist.*, 2:287-291, 1958.

―――: "The palaeopathology of early British man," *J. Roy. Anthrop. Inst.*, 91:318-344, 1961.

Browne, S. G.: "How old is leprosy," *Brit. Med. J.*, 3:640-641, 1970.

Chamberlain, W. E., N. E. Wayson, and L. H. Garland: "Bone and joint changes of leprosy. A roentgenological study," *Radiology*, 17:930-939, 1931.

Chaussinand, R.: "Tuberculose et lèpre, maladies antagoniques. Eviction de la lèpre per la tuberculose," *Int. J. Leprosy*, 16:431-438, 1948.

Cochrane, R. G.: "The history of leprosy and its spread throughout the world," in R. G. Cochrane and T. F. Davey (ed.): *Leprosy in Theory and Practice*, J. Wright, Bristol, 2nd ed., 1964.

Cochrane, R. G. and T. F. Davey (ed.): *Leprosy in Theory and Practice*, J. Wright, Bristol, 2nd ed., 1964.

Cooney, J. P. and E. H. Crosby: "Absorptive bone changes in leprosy," *Radiology*, 42:14-19, 1944.

Davey, T. F.: "The nose in leprosy: steps to a better understanding," *Leprosy Rev.*, 45:97-103, 1974.

Doull, J. A.: "The epidemiology of leprosy: present status and problems," *Int. J. Leprosy*, 30:48-66, 1962.

Esguerra-Gómez, G. and E. Acosta: "Bone and joint lesions in leprosy," *Radiology*, 50:619-631, 1948.

Faget, G. H. and A. Mayoral: "Bone changes in leprosy: A clinical and roentgenologic study of 505 cases," *Radiology*, 42:1-13, 1944.

Fasal, P.: "A primer in leprosy," *Cutis*, 7:525-542, 1971.

Feldman, F., R. Auerbach, and A. Johnston: "Tuberculous dactylitis in the adult," *Amer. J. Roentgen.*, 112:460-479, 1971.

Friedman, S. A. and R. B. Rakow: "Osseous lesions of the foot in diabetic neuropathy," *Diabetes*, 20:302-307, 1971.

Gondos, B.: "Roentgen observations in diabetic osteopathy," *Radiology*, 91:6-13, 1968.

Honeij, J. A.: "Bone changes in leprosy," *Amer. J. Roentgen.*, 4:494-511, 1917.

Hopkins, R.: "Bone changes in leprosy," *Radiology*, 11:470-473, 1928.

Hulse, E. V.: "Leprosy and ancient Egypt," *Lancet*, 2:1024, 1972.

Jarcho, S.: "Lazar houses and the dissemination of leprosy," *Med. Hist.*, 15:401, 1971.

Job, C. K.: "Pathology of leprous osteomyelitis," *Int. J. Leprosy*, 31:26-33, 1963.

Job, C. K., A. B. Karat, and S. Karat: "The histopathological appearance of leprous rhinitis and pathogenesis of septal perforation in leprosy," *J. Laryngol. Otol.*, 80:718-732, 1966.

Khanolkar, V. R.: "Studies in the histology of early lesions in leprosy," *Indian Coun. Med. Res., Special Report Series* #19, 1951.

Lechat, M. F.: "Bone lesions in leprosy," *Int. J. Leprosy*, 30:125-137, 1962.

Lechat, M. F. and J. Chardome: "Altérations radiologiques des os de la face chez le lépreux congolais," *Ann. Soc. Belge Méd. Trop.*, 35:603-612, 1955.

Lurie, M. B.: "A pathogenic relationship between tuberculosis and leprosy: the common denominators in the tissue response to Mycobacteria," pp. 340-343 in *Ciba Foundation Symposium on Experimental Tuberculosis*, Little, Brown and Co., Boston, 1955.

Møller-Christensen, V.: *Ten Lepers from Naevsted in Denmark*, Danish Science Press, Copenhagen, 1953.

―――― : *Bogen om Aebelholt Kloster*, Dansk Videnskabs Forlag, Copenhagen, 1958.

―――― : *Bone Changes in Leprosy*, Munksgaard, Copenhagen, 1961.

―――― : "New knowledge of leprosy through paleopathology," *Int. J. Leprosy*, 33:603-610, 1965.

―――― : "The history of syphilis and leprosy—an osteoarchaeological approach," *Abbottempo*, 1:20-25, 1969.

Møller-Christensen, V., S. N. Bakke, R. S. Melsom, and A. E. Waaler: "Changes in the anterior nasal spine and the alveolar process of the maxillary bone in leprosy," *Int. J. Leprosy*, 20:335-340, 1952.

Møller-Christensen, V. and B. Faber: "Leprous changes in a material of mediaeval skeletons from the St. George's Court, Naevsted," *Acta Radiol.*, 37:308-317, 1952.

Møller-Christensen, V. and D. R. Hughes: "Two early cases of leprosy in Great Britain," *Man*, 62:177-179, 1962.

―――― : "An early case of leprosy from Nubia," *Man*, 66:242-245, 1966.

Møller-Christensen, V. and R. G. Inkster: "Cases of leprosy and syphilis in the osteological collections of the Department of Anatomy, University of Edinburgh," *Danish Med. Bull.*, 12:11-18, 1965.

Møller-Christensen, V. and D. L. Weiss: "One of the oldest datable skeletons with leprous bone-changes from the Naevsted leprosy hospital churchyard in Denmark," *Int. J. Leprosy*, 39:172-182, 1971.

Murdock, J. R. and H. J. Hutter: "Leprosy: a roentgenological survey," *Amer. J. Roentgen.*, 28:598-621, 1932.

Pesce, H.: "Lepra en el Perú precolombino," *Anales Fac. Med. Univ. Nac. Mayor San Marcos de Lima*, 38:48-64, 1955.

Paterson, D. E.: "Bone changes in leprosy: their incidence, progress, prevention, and arrest," *Int. J. Leprosy*, 29:393-422, 1961.

Paterson, D. E. and C. K. Job: "Bone changes and absorption in leprosy," in R. G. Cochrane and T. F. Davey (ed.): *Leprosy in Theory and Practice*, J. Wright, Bristol, 2nd ed., 1964.

Plessis, J. D.: "Lesions of the feet in patients with diabetes mellitus," *S. Afr. J. Surg.*, 8:29-46, 1970.

Post, P. W. and D. D. Donner: "Frostbite in a pre-Columbian mummy," *Amer. J. Phys. Anthrop.*, 37:187-191, 1972.

Richards, P.: "Leprosy in Scandinavia," *Centaurus*, 7:101-133, 1960.

Riordan, D. C.: "The hand in leprosy," *J. Bone Joint Surg.*, 42A:661-690, 1960.

Riseborough, A. W., R. A. Joske, and B. F. Vaughan: "Hand deformities in aborigines due to yaws," *Clin. Radiol.*, 12:109-113, 1961.

Rosen, R. S. and G. Jacobson: "Fungus disease of bone," *Seminars in Roentgen.*, 1:370-391, 1966.

Smith, G. E. and D. E. Derry: "Anatomical report," *Archaeological Survey of Nubia*, Bulletin 5-6, 11-25, 9-30, 1910.

Smith, G. E. and F. W. Jones: "Report on the human remains," *Archaeological Survey of Nubia Report for 1907-1908*, Vol. 2.

Vinson, H. A. and R. Schatzki: "Roentgenological bone changes encountered in frostbite, Korea 1950-1951," *Radiology*, 63:685-695, 1954.

Vogelsang, T. M.: "Leprosy in Norway," *Med. Hist.*, 9:29-35, 1965.

Weiss, D. L. and V. Møller-Christensen: "An unusual case of tuberculosis in a mediaeval leper," *Danish Med. Bull.*, 18:11-14, 1971a.

――――――: "Leprosy, echinococcosis and amulets: a study of a mediaeval Danish inhumation," *Med. Hist.*, 15:260-267, 1971b.

Wells, C.: "A possible case of leprosy from a Saxon cemetery at Beckford," *Med. Hist.*, 6:383-386, 1962.

Chapter VII

HEMATOLOGIC DISORDERS—
THE ANEMIAS

THE PROBLEM OF OSTEOPOROSIS SYMMETRICA
(SPONGY HYPEROSTOSIS)

IN 1913 THE GREAT American physical anthropologist, Aleš Hrdlička, discovered a large number of pre-Columbian Peruvian Indians with peculiar lesions on their skull vaults. These lesions were characterized by their highly symmetrical distribution on both sides of the parietals and on the occipital above the crest (see Fig. 82). The facial bones were not involved except for a vascularized pitting of the orbits. The lesions rarely involved the suture lines of the vault. Hrdlička coined the term *symmetrical osteoporosis* to describe the sieve-like porosity involving the outer surface of the skull (Hrdlička, 1914).

Similar lesions affecting other parts of the skull have been described. *Cribra orbitalia,* also known as usura orbitae, is a bilateral pitting of the orbital portion of the frontal bone and was noted by Hrdlička in the Peruvian crania. *Cribra cranii* refers to cribriform changes or porosities of the internal table of the skull, particularly in the frontal region (Henschen, 1961). The exact relationship of these two lesions with symmetrical osteoporosis is not clear and will be discussed in a later section.

Unfortunately, the term symmetrical osteoporosis has caused continuing confusion between the primarily hyperostotic or bony overgrowth present in these lesions and the bony tissue loss manifestations referred to as osteoporosis. A better term for the

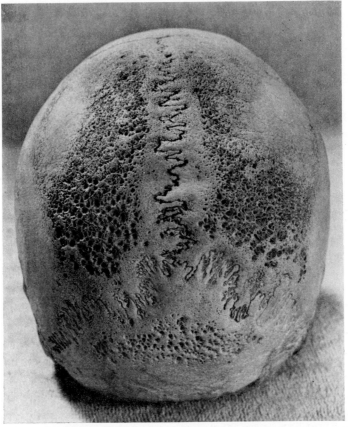

Figure 82. Posterior view of an adult Peruvian skull from Valley Chicama. This skull was collected by Dr. Aleš Hrdlička and clearly shows why he termed these lesions "symmetrical osteoporosis." A more appropriate term for the condition is spongy hyperostosis. (USNM 264,543)

lesions of symmetrical osteoporosis is *spongy hyperostosis* (Hamperl and Weiss, 1955) or *porotic hyperostosis* (Angel, 1966, 1967). The only true osteoporosis which occurs in spongy hyperostosis is thinning and often complete destruction of the outer table of the cranial vault. This is caused by pressure atrophy produced by hypertrophy of the hematopoietic diploë between the inner and outer tables. The sieve-like or coral appearance of the affected area is due to the complete destruction of the outer table and exposure of the hypertrophied cancellous bone or diploë. The skull vault in the affected area is thicker than normal due to the hypertrophied bone which protrudes to a slight or moderate degree over the normal external contour of the skull.

Earlier workers suggested several possibilities regarding the etiology and pathogenesis of spongy hyperostosis. Similar symmetrical lesions of the skull had been noted in cases of rickets (Williams, 1929). A monkey with calcium deficiency also exhibited such lesions (Todd cited by Williams, 1929). Toxic disorders or endocrine disturbances were mentioned as possible

Figure 83. Superior view of the cranial vault of a two-year-old Indian child from Canaveral, Florida. There is complete loss of the outer table and exposure of the hypertrophied diploic tissue in this active case of spongy hyperostosis. (USNM 377,454)

causes of the lesions (Hrdlička, 1914). Still others ascribed the lesions to congenital syphilis (Parrot cited by Williams, 1929) or tuberculosis (Virchow cited by Henschen, 1961). It was even suggested that chronic irritation and pressure produced by carrying water pots on the head (Smith and Jones, 1908) or by strapping an infant's head to a cradle board (Williams, 1929) might have produced the lesions.

The first real step toward determining the etiology of spongy hyperostosis came in 1930 with Hooton's report on the Indians of Pecos Pueblo (see also Moore, 1929). Hooton found the condition to be common in the skulls of young children from the Sacred Cenote of Chichen Itza in Yucatan (900-1200 A.D.). In twenty-one well-preserved skulls of children between six and twelve years from that site, marked spongy hyperostosis was noted in fourteen of them or 66 percent. Similar lesions in various phases of healing were found in older individuals from the Sacred Cenote as well as from Pecos Pueblo (800 A.D.-1800 A.D.) in New Mexico. Hooton compared radiograms of these skulls with

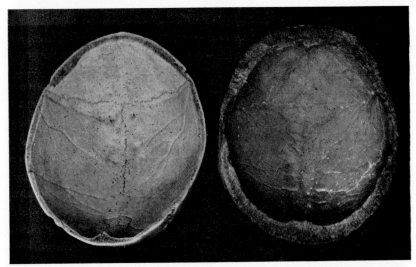

Figure 84. Sectioned calvarium with normal on the left for comparison. Hypertrophy of the hematopoietic tissue has caused the vault thickening. The outer table is thinned by the expanding tissue beneath. This is an autopsy specimen of thalassemia.

radiograms of living children with thalassemia major (Cooley's anemia) and noted their close similarity. Both skulls revealed overgrowth of the marrow in the skull resulting in widening of the diploic space. The coarsened trabeculae were arranged perpendicular to the inner table, creating a "hair-on-end" appearance in the roentgenograms which is characteristic of thalassemia (see Fig. 85). An early comparison was also made between spongy hyperostotic lesions in a Mayan skull and sickle-cell anemia (Moore, 1929).

Very little has been added to the solution of this interesting problem since then—leading many to believe that these prehistoric and early historic people suffered from a congenital hemolytic anemia such as thalassemia or sickle-cell anemia. This misconception has recently given rise to misleading statements concerning the prehistoric distribution of abnormal hemoglobins (Zaino, 1964, 1967).

Recent advances in hematology and radiology have shown

Figure 85. Autopsy specimen of thalassemia. (A.) Roentgenogram showing the "hair-on-end" appearance. Lateral view.

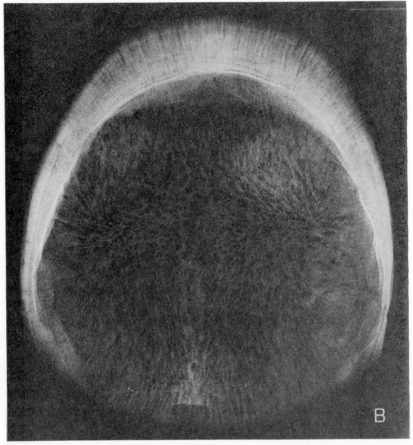

Figure 85B. Roentgenogram showing the symmetrical pattern produced by the hypertrophied diploë on both sides of the sagittal suture. Superior view.

that several hematologic disorders are capable of producing a "hair-on-end" pattern of bone change in the skull. Thus, in addition to thalassemia major and sickle-cell anemia, the lesions of spongy hyperostosis may be found in hereditary spherocytosis, thalassemia minor, hereditary elliptocytosis and other less common hematologic disorders. Noninheritable conditions, particularly iron deficiency anemia, have been found to produce similar bone changes (Moseley, 1965). The following table is a partial list of conditions which must be considered by the paleopathologist as possible producers of spongy hyperostosis.

The pathologic process common to all the above disorders is hypertrophy of the bone marrow in infancy or childhood in order to produce more red blood cells to compensate for the high turnover rate of the abnormal red blood cells. The hyperblastic cellular marrow between the tables of the skull widens the diploic space and thins the outer table. The new bone formed is arranged parallel to the marrow blood vessels which are perpendicular to the inner table. Radiologically, this is the "hair-on-end" appearance of the skull.

The appearance of the calvarium alone, however, does not provide any sound criteria for differential diagnosis by the paleopathologist since the pattern of bone change is the same for all the disorders listed in Table X. However, certain differences do exist in the reactions or lack of them of other parts

TABLE X

PARTIAL LIST OF CONDITIONS CREATING A HAIR-ON-END RADIOGRAPHIC PATTERN IN THE CRANIUM*

I. CONGENITAL HEMOLYTIC ANEMIAS
  A. Thalassemias
    Thalassemia Major (Mediterranean Disease, Cooley's Anemia, Erythroblastic Anemia)
    Thalassemia Intermedia—severe heterozygous
    Thalassemia Minor—mild heterozygous
  B. Sickle Cell Disease
    Sickle Cell Anemia (Hemoglobin S homozygous)
    Hemoglobin C—homozygous
    Hemoglobin E—homozygous
    Hemoglobin S—C
    Hemoglobin S—Thalassemia
    Other less common abnormal hemoglobins
  C. Hereditary Nonspherocytic Hemolytic Anemia
    Glucose-6-phosphate Dehydrogenase Deficiency
    Pyruvate Kinase Deficiency
    Probably other deficiencies
  D. Hereditary Spherocytosis (Spherocytic Anemia, Congenital Hemolytic Jaundice).
  E. Hereditary Elliptocytosis (rare)
II. IRON DEFICIENCY ANEMIA
III. CYANOTIC CONGENITAL HEART DISEASE (rare)
IV. POLYCYTHEMIA VERA IN CHILDHOOD (rare)

* Modified from Moseley, 1965.

of the skeleton to hematologic processes. Therefore, let us look at each of the main hemolytic disorders and iron deficiency anemia more closely.

THALASSEMIA MAJOR AND THALASSEMIA MINOR

Thalassemia is a hereditary disorder of hemoglobin synthesis which has a very widespread distribution. There are several different types of thalassemia, but pathologically all may be grouped under one heading. Genetically, thalassemia may result from one of several mutant genes which lead to amino acid substitution in either the alpha or beta chain of the hemoglobin molecule (Cavalli-Sforza and Bodmer, 1971).

Clinically, there are two main forms of thalassemia. Thalassemia minor occurs in individuals who are heterozygous for the autosomal dominant trait. Thalassemia major is much more severe in its manifestations and is found in individuals who are homozygous for the mutant gene. Both the homozygous and the heterozygous states however, show considerable variation in their degree of severity.

The Skull

Classical thalassemia major (Mediterranean Disease, Cooley's Anemia, Erythroblastic Anemia) usually becomes evident within the first year or two of life. Skull changes are more common and tend to be more striking in thalassemia major than in the other congenital hemolytic anemias. The earliest changes in the skull occur most often in the frontal bone (Caffey, 1937). There is usually no involvement of the occipital squamosa inferior to the internal occipital protuberance. Characteristic changes occur in the maxillary, sphenoidal, and temporal bones. Overgrowth of marrow in these bones impedes pneumatization of the paranasal sinuses and mastoids and in some cases completely suppresses it. Inhibition of pneumatization is most evident in the maxillary and sphenoid sinuses. Overgrowth of marrow in the upper maxilla may result in swelling of the bone and in older patients may produce serious malocclusion of the jaws (Moseley, 1963).

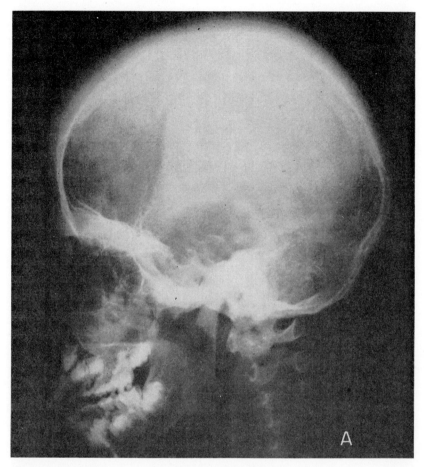

Figure 86. Clinical roentgenogram of thalassemia major. (A.) Young child showing moderate thickening of the diploë, particularly in the frontal region.

Changes in the paranasal sinuses, mastoids, and facial bones are seldom found in sickle-cell or the other congenital anemias. Although experience with skull changes in iron deficiency anemia is still limited, they have not been found in this condition either (Aksoy et al., 1966). Thus, retardation of pneumatization in the bones mentioned above may be an important point in differential diagnosis by the paleopathologist.

Postcranial

The presence or absence of several pathologic processes are notable in the postcranial skeleton of patients with thalassemia major. There is a generalized osteoporosis of the vertebral column, but the compression of the vertebrae is seldom seen in thalassemia major. This is of diagnostic value because vertebral compression or cupping is common in sickle-cell anemia (Carroll, 1957).

Early Changes

The earliest skeletal changes are found in the smaller bones, particularly the metacarpals and metatarsals (see Fig. 87). In thalassemia major, these changes are more pronounced than in other congenital anemias. The long bones show marrow hypertrophy with loss of the normal bone contour. The medullary cavity is widened and the cortices are thinned. There is a mottled osteoporosis, particularly in the metacarpals and metatarsals. The trabeculae are coarsened in these bones to form a honeycomb pattern in the swollen bones. As the figure shows, the metacarpals have lost their normal external contours and tend to be rectangular. Coarsened trabeculae are also present in the distal and proximal ends of the long bones as Figure 87 demonstrates. Several of the ribs may be involved with cortical thinning and medullary expansion, particularly in the posterior portions.

Later Changes

Few children with thalassemia major survive beyond adolescence. However, in those surviving infancy and early childhood there is a marked regression of the changes in the peripheral skeleton as age advances. With progressing age, active marrow recedes from the distal portions of the skeleton toward the trunk and the earlier changes mentioned above in the metacarpals, metatarsals, and long bones tend to disappear (Moseley, 1963). In the central segments of the skeleton however, the bone changes may become more pronounced. Thus, in later childhood and adulthood the skull, spine, and pelvis may show advanced radio-

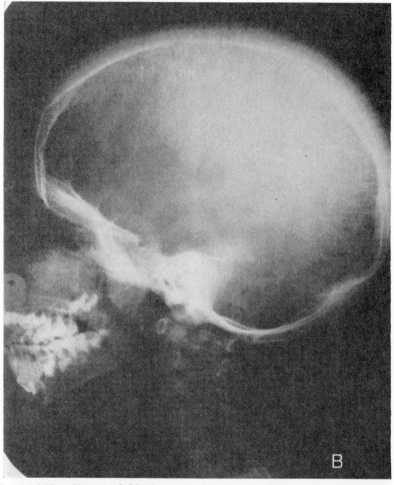

Figure 86B. Young child showing the classic "hair-on-end" appearance of severe thalassemia. (Courtesy of Dr. E. B. D. Neuhauser.)

graphic bone changes while the long bones appear normal. The hands, which are very important for diagnosis in infants and young children, are the least diagnostic parts in older children and young adults. This changing pattern of skeletal involvement with increasing age has been well demonstrated in longitudinal studies of patients from early childhood through adulthood (Caffey, 1951). Of further note to paleopathological diagnosis

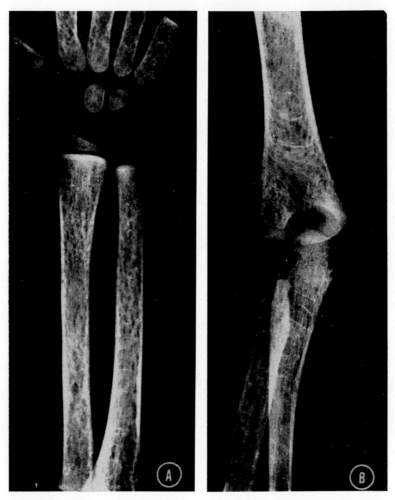

Figure 87. Clinical roentgenograms of thalassemia major. (A.) The radius, ulna, and metacarpals of a three-year-old child show marrow hypertrophy with loss of the normal bone contours. The medullary cavity is widened and the cortices are thinned. The coarse trabeculae in the metacarpals form a "honeycomb" pattern. (B.) The bones of the elbow show similar widening of the medullary cavities and cortical thinning. The trabeculae are considerably coarsened in the metaphyses. (By permission of J. E. Moseley: *Bone Changes in the Hematologic Disorders*, 1963, Grune and Stratton.)

is the premature fusion of the epiphyses in the long bones of many patients with homozygous thalassemia. This occurs after the age of ten years, and the most common sites are the proximal ends of the humeri and the distal ends of the femora. The fusion may involve only a segment of the epiphysis, and the arrested growth at this site often causes skeletal dwarfism with possible deformity (Currarino and Erlandson, 1964).

Thalassemia minor is much less severe than thalassemia major and is much harder to diagnose because it closely resembles the bone changes in iron deficiency anemia. However, if the skull changes were due to thalassemia minor, then skeletons showing changes consistent with thalassemia major should be found among the archaeological remains as well.

SICKLE-CELL ANEMIA

Bone changes in sickle-cell anemia are found in many individuals who are homozygous for the mutant hemoglobin S gene. The author also includes under this heading individuals who are homozygous for the abnormal hemoglobin C and E genes which are now recognized polymorphisms (Cavalli-Sforza and Bodmer, 1971). All of these genetic polymorphisms, which have widespread distribution, will be discussed in a later section with relation to malaria. The sickling gene (hemoglobin S) is found almost exclusively in Negroes, whereas the other mutant genes may be found in various racial groups. Clinically, all have similar pathological manifestations, but the racial differences may be of some use to the paleopathologist in making a differential diagnosis of spongy hyperostosis among racially distinct archaeological remains. A large proportion of the individuals who are homozygous for the abnormal hemoglobins S, C, and E will die during the first year of life. Many of those living on into childhood will die before the age of ten.

Bone changes in the skull and postcranial bones are due to two fundamental processes. The first is that of marrow hyperplasia in an effort to produce more red blood cells. The second process is that of thrombosis or blood clotting which occurs most often in the long bones. When the abnormal red blood cells assume an elongated, rigid shape in response to trauma, they have difficulty in getting through capillaries and therefore blood stasis occurs. Clot-activating substances are liberated and this initiates the coagulation of blood. Thrombosis then prevents nutrients from reaching areas of the bone (as well as other tissues and organs) and aseptic necrosis may follow. The combination of these two pathologic processes results in a pattern of bone change strongly diagnostic of sickle-cell anemia.

Skull

Bone changes in the skull are mainly the result of marrow hyperplasia which results in a slight radial arrangement of the trabeculae. The "hair-on-end" pattern is rarely as pronounced as that of thalassemia major. In fact, the "hair-on-end" appearance may be present in only 5 percent of cases (Loiacano and Reeder, 1969). As in thalassemia, the frontal bone is usually the first bone affected. However, unlike thalassemia, there is no significant facial bone and sinus involvement in sickle-cell anemia and this is of some importance in a differential diagnosis.

Bone changes as a result of thrombosis may occur in the skull but not as frequently as in the long bones. The lesions caused by thrombosis resemble the punched-out lesions of multiple myeloma in the skull and occur in 7.3 percent of patients with sickle-cell anemia (Carroll, 1957). A differential diagnosis is possible here because multiple myeloma occurs almost exclusively in people over fifty years of age, whereas the lesions of sickle-cell anemia are found in young children.

Postcranial

Marrow hyperplasia also occurs in the postcranial skeleton, but this is mainly limited to the vertebrae and pelvis. Although vertebral compression is virtually absent in thalassemia, it is present in 30 percent of those with sickle-cell anemia (Carroll, 1957). Such a high frequency makes this feature of great importance to the paleopathologist when confronted with a

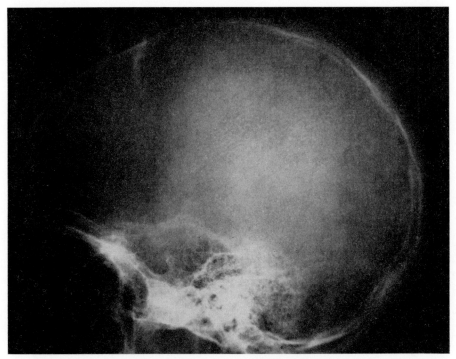

Figure 88. Clinical roentgenogram of homozygous sickle-cell anemia in a fourteen-year-old girl. The diploic trabeculae are arranged radially with thinning of the outer table. (Courtesy of Dr. E. B. D. Neuhauser.)

case of spongy hyperostosis. The morphology of vertebral cupping is different from that present in other disorders such as senile osteoporosis. In the latter disorders, protrusion of normal intervertebral tissue into the upper and lower surfaces of the weakened vertebral body produces rounded depressions which extend all the way to the periphery of the centrum. This is often called the "fish vertebra" sign. In sickle-cell anemia the defect does not involve the entire end-plate with a rounded depression. Instead, the depression is confined to the central zone so that the periphery retains its normal flat surface. The base of the depression is formed by a flat base of dense bone (Fig. 89). Circulatory stasis and ischemia rather than osteoporotic thinning is the primary cause of this distinctive lesion in sickle-cell anemia (Reynolds, 1966). Unlike the porosis involving the peripheral

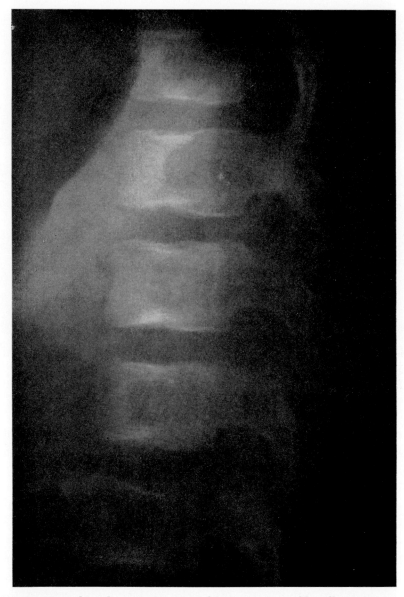

Figure 89. Clinical roentgenogram of homozygous sickle-cell anemia in a ten-year-old boy. The lower thoracic and lumbar vertebrae have central depressions with flat bases of dense bone. (Courtesy of Dr. E. B. D. Neuhauser.)

portions of the skeleton which tend to disappear during adolescence, the changes in the spine persist into adult life. It is frequently present when no other osteoporosis is evident and when seen in combination with the thrombotic changes of the long bones it is highly diagnostic of sickle-cell anemia.

The "hand-foot syndrome" of sickle-cell anemia is very similar to that found in thalassemia and is therefore of little use in differential diagnosis. Of much greater importance is the process of thrombosis which is a major characteristic of sickle-cell anemia in the long bones. Definite changes due to thromboses in bone are found in approximately 50 percent of all individuals with sickle-cell anemia. The general reaction of a long bone to thrombosis is thickening of the cortex and narrowing of medullary cavity. This is exactly opposite to the cortical thinning which occurs in thalassemia (see Fig. 90).

The thrombosis and resultant aseptic necrosis frequently involve an entire ossification center such as an epiphysis. Aseptic necrosis of the femoral head occurs in 7.3 percent of all cases. When other epiphyses are included, this percentage rises to over 16 percent. Arthritic changes may also occur in the joints as a reaction to the necrosis of an adjacent epiphysis (Carroll, 1957).

Individuals with sickle-cell anemia are more susceptible to osteomyelitis and this should be noted by the paleopathologist. The reason for this increased incidence is probably due to thrombosis in the wall of the gut allowing enteric organisms such as salmonella to enter the bloodstream. The organisms may also enter through soft tissue ulcers in the leg, particularly in the ankle region. The resultant hematogenous osteomyelitic destruction, usually beginning at the site of bone infarct, should not be confused with aseptic necrosis although they may grossly resemble each other. However, the osteomyelitic infection will spread throughout the shaft of the bone whereas the aseptic necrosis mainly involves the epiphyses at the ends of the bones.

Thus, when a paleopathologist finds a case of spongy hyperostosis cranii, it is very important to examine the postcranial remains radiologically to differentiate between thalassemia and sickle-cell anemia. Of even greater importance is a differential

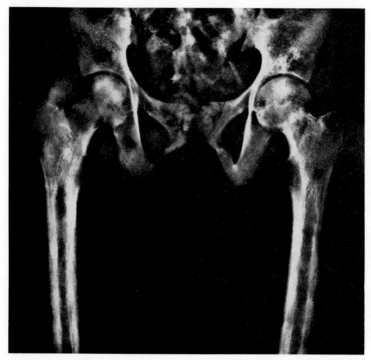

Figure 90. Clinical roentgenogram of homozygous sickle-cell anemia. Aseptic necrosis and secondary degenerative arthritic bone changes are apparent in the femoral heads. There is thickening of the trabeculae in the neck and upper shaft. The medullary cavity is considerably narrowed by thickening of the cortex. (By permission of J. E. Moseley: *Bone Changes in the Hematologic Disorders,* 1963, Grune and Stratton.)

diagnosis of iron deficiency anemia which is not a genetic disorder but a nutritional one and therefore closely related to the environment. The bone changes of iron deficiency have just recently been observed (Eng, 1958) and this anthropologically important disorder is described in the following section.

IRON DEFICIENCY ANEMIA

Iron deficiency anemia is by far the most common cause of anemia in every part of the world at the present time (Davidson and Passmore, 1969). However, bone changes are only occasionally encountered and often do not reflect the severity of the anemia. Why certain people exhibit bone lesions whereas others with more profound iron deficiency anemia do not, is not completely understood at present.

Iron deficiency anemia resembles thalassemia in many respects except etiology. As the name implies, the anemia is caused by inadequate dietary iron content, inability to absorb iron from the intestine, excessive losses of iron from the body, or disturbances of iron metabolism by infection or other mechanisms. Other dietary factors may be involved such as chronic deficiencies of amino acids and minerals necessary for bone formation.

Skull

Early workers described only the skull changes in iron deficiency anemia and either ignored the postcranial changes or found no gross lesions evident (Moseley, 1963). More recent workers however, have found postcranial lesions in iron deficiency anemia closely resembling thalassemia. Bone changes induced by iron deficiency anemia may occur in early infancy due to inadequate diet or gastrointestinal diseases. Such changes may also occur in adulthood as a result of diet or chronic blood loss through parasitic infection (Eng, 1958). Skull changes have been found in 50 percent of severely iron-deficient patients with main involvement in the occipital region instead of the frontal (Aksoy et al., 1966). The spongy hyperostosis of the cranium is usually not as severe as that found in thalassemia and there is no noticeable retardation of the pneumatization of the maxillary sinuses. This feature is of some importance in a differential diagnosis, but one must emphasize the variability in degree of bone change and its location in thalassemia.

Postcranial

Iron deficiency anemia does not often involve the postcranial skeleton. The most common bone changes are in the bones of the elbow. There is marked osteoporosis and coarse trabecular striation in the distal end of the humerus and proximal portions of the radius and ulna. The hands and feet may also be involved

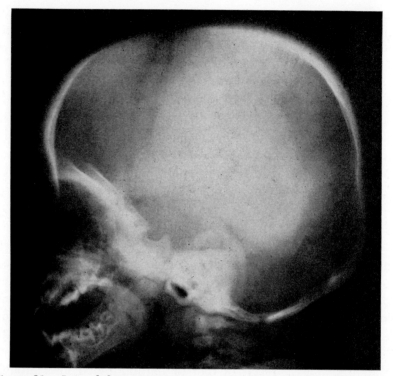

Figure 91. Iron deficiency anemia in a nine-month-old infant. This lateral view shows the radial pattern of the newly formed trabeculae with a maximum vault thickness greater than one centimeter. Radiographic examination of the postcranial skeleton was normal. (Courtesy of Dr. E. B. D. Neuhauser.)

but not to the extent found in thalassemia major. There is a significant lack of thrombosis which differentiates iron deficiency anemia from sickle-cell anemia. Of further importance is the minimal osteoporosis of the vertebral bodies in iron deficiency anemia and almost complete absence of vertebral compression (Aksoy et al., 1966).

It should be emphasized that malnutrition is rarely selective for only one vital dietary component. Malnutrition (including malabsorption and excessive loss of nutrients) is almost always multiple, resulting in deficiency of several or many nutrients to varying degrees. For example, in a study of fifteen children

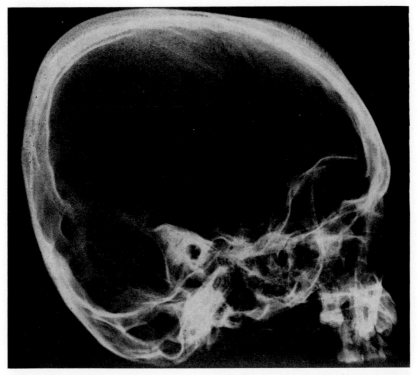

Figure 92. Spongy hyperostosis in an ancient Peruvian child. The frontal bone is particularly affected as in many cases of iron deficiency anemia.

aged fourteen months to three years with skeletal manifestations of iron deficiency anemia, six of the children also had bone changes due to rickets (Lanzkowsky, 1968).

The cranial bossing evident when rickets is present in addition to anemia, differs from that due to iron deficiency alone. In the former case, the forehead bulges anteriorly much more than in the latter. This is probably due to the presence of excessive osteoid tissue outside the outer table in rickets whereas only the diploic spaces expand in anemia. The diploic space is uniformly widened from the frontal to occipital regions in most patients with iron deficiency anemia. When rachitic changes are also present, the frontal and parietal bossing coupled with delayed fontanelle closure results in nonuniform widening of the diploic space (Lanzkowsky, 1968).

Cases of cranial rickets unassociated with iron deficiency anemia may be confused with the latter unless the different locations of osseous hypertrophy described above are kept in mind. In the healed state one might confuse the two conditions if only the skull vault is available for examination (Angel, 1967). When postcranial remains are preserved, the flaring at the ends of the shafts, short diaphyseal length, and often bowing deformity present provide clear means of distinguishing between rickets and the less marked changes of iron deficiency anemia (see also Chapter VIII).

More work must be done using larger series of afflicted individuals to determine the characteristic radiographic patterns of bone change in iron deficiency anemia. Critical examination of the postcranial skeleton is imperative. Only then can the paleopathologist successfully attach a specific etiology to the case of spongy hyperostosis under study.

MALARIA AND THE HEREDITARY ANEMIAS

The mutant genes causing thalassemia, sickle-cell anemia, and other hereditary hemolytic anemias are balanced polymorphisms with relatively stable gene frequencies and widespread geographic distribution. How can such lethal genes maintain a stable gene frequency within a population? This question puzzled scientists for decades until it was found that people who are heterozygous for thalassemia or the sickling trait have a higher resistance to malaria (Allison, 1954, 1961). This resistance is mainly due to the short lifespans of the red blood cells in the heterozygous individuals which prevent the malaria parasites from establishing an infection in the bloodstream.

Distributional maps reveal the close correlation between the geographic distribution of malaria and that of the hemolytic anemias previously described. This correlation cannot be expected to be extremely close for several reasons. There may be fluctuations in the intensities of local infection. People with the polymorphism may move away from infected areas to nonmalarious regions. And finally, there may be genetic adaptations other than the abnormal hemoglobins and thalassemia which help resist malaria and thereby decrease the correlation. Yet in spite of these possibilities for deviation, the correlation as shown in such distributional maps is indeed striking (see Cavalli-Sforza and Bodmer, 1971).

Thus, lesions of bone diagnostic of the hereditary anemias may also provide indirect evidence of the antiquity and distribution of malaria. In early Neolithic (around 6500 B.C.) and Bronze Age Greek skeletal material, bone changes in both children and adults resembling thalassemia strongly suggest the presence of endemic malaria in that area (Angel, 1964; 1971a; 1971b). It is noteworthy that both malaria and thalassemia are still common in Greece. Still earlier evidence suggestive of thalassemia comes from the Upper Paleolithic of Italy (Zaino, 1964). It has been suggested that the thalassemia mutation originated in Upper Paleolithic times in the Mediterranean area (Angel, 1971a).

In Africa, spongy hyperostosis is rarely found in Late Paleolithic remains from Nubia but is more prevalent in Mesolithic Nubians and Predynastic Egyptians (Anderson, 1968). Iron deficiency anemia is probably the major factor in many of these cases (see Carlson et al., 1974). Two Egyptian mummies dating 332-30 B.C. and 0-200 A.D. respectively have evidence of bone infarction in several long bones (Gray, 1968). Several conditions besides sickle-cell anemia can produce bone infarction, but these cases illustrate the importance of radiographing skeletal material. Two femora from Nigeria around 1200 A.D. have localized endosteal thickening and a possible juxtacortical infarct suggestive of sickle-cell anemia (Bohrer and Connah, 1971).

In Europe the earliest reported case of spongy hyperostosis comes from the Bronze Age of Wales (1650-1550 B.C.). The fragmentary skull of a six-year-old child exhibits bossing on both sides of the frontal bone suggestive of iron deficiency anemia or rickets (Cule and Evans, 1968). The skull of a two-year-old infant from Medieval Sweden has marked spongy hyperostosis (Gejvall, 1960).

Although there is strong evidence supporting Angel's belief

that the spongy hyperostosis found in several early Greeks is due to thalassemia, one should not casually diagnose every such case as evidence of thalassemia. Unfortunately, cases of spongy hyperostosis from all parts of the world have been attributed to the affects of thalassemia. Such specimens should be reexamined, particularly the long bones when present, to establish the correct etiology.

Archaeological specimens of spongy hyperostosis attributed to thalassemia have come from various parts of the New World. Such cases have been found in prehistoric Florida Indians (Snow, 1962); prehistoric and early historic Pueblo Indians of the Southwest (Hooton, 1930; Zaino, 1967); the ancient Maya in Yucatan (Hooton, 1930); prehistoric Indians in Alabama (Wakefield et al., 1937); and ancient Peruvians (Williams, 1929). However, modern blood group studies of the modern Maya, South American Indians, and North American Indians have revealed very few cases of abnormal hemoglobins, particularly thalassemia. The few cases of abnormal hemoglobins found in Indian populations studied can reasonably be attributed to recent gene flow from immigrant populations, particularly Negro (Livingstone, 1967). Most workers today believe that malaria is a recent arrival to the New World brought along with the slaves from Africa (Bruce-Chwatt, 1965; Dunn, 1965). Thus, the existence of thalassemia or any other abnormal hemoglobins in these prehistoric and early historic American Indians is highly improbable and one must therefore consider iron deficiency anemia as the probable disorder present in these Indians. Recent evidence presented here further suggests that this is the case.

IRON DEFICIENCY ANEMIA AND AMERICAN INDIANS

Active lesions of spongy hyperostosis were noted by Hooton in the skulls of immature subjects found in the Sacred Cenote of Chichen Itza in the Yucatan. In fact, in twenty-one well-preserved skulls of these ancient Maya between the ages of about six and twelve years, marked spongy hyperostosis was observed in fourteen crania or 66 percent (Hooton, 1930). Using perhaps the largest and best preserved series of ancient Maya skeletal remains from Altar de Sacrificios in Guatemala, Saul found active and healed lesions of spongy hyperostosis present in 36 percent of the ninety individuals (Saul, 1972). This figure is much lower than Hooton's 66 percent because Saul includes adults in which the lesions may be completely healed. Incomplete skulls may also be a factor. Nevertheless, both 36 and 66 percent indicate a very high incidence of some form of chronic anemia in these ancient Maya. This incidence is much higher than the gene frequencies of thalassemia and abnormal hemoglobins in modern populations (Livingstone, 1967), particularly when one considers that many heterozygotes do not exhibit anemic bone changes (Moseley, 1963).

In Guatemala and Mexico today there are several populations of unmixed Maya Indians. Iron deficiency anemia, vitamin B_{12} and folic acid deficiency anemia, protein deficiency anemia, and to a lesser extent vitamin C deprivation (ascorbic acid deficiency or scurvy) are common in these Maya and other populations in this area at the present time (Shattuck, 1933; Scrimshaw and Tejada, 1970; *World Health Statistics Annual*, 1973). The underlying causes of these anemias are likely to have been present in ancient times (Saul, 1972, 1973). For example, infants and small children have increased iron requirements, particularly when the prenatal stores of iron are inadequate because of nutritional iron deficiency in the mother. This is further exacerbated by an extended milk-feeding period which is a poor source of iron and malabsorption due to chronic gastrointestinal disease, particularly parasitic infestation. In fact, gastrointestinal infection accounted for 39 percent of all deaths in Yucatan, particularly among children under five years of age (Shattuck, 1933). Even today bacterial, helminthic, and hookworm infection are common. (For paleopathological evidence of such organisms in America see Pizzi and Schenone, 1954; Samuels, 1965; Moore et al., 1969; and Allison et al., 1974). Not only is iron absorption inhibited by such disease, but chronic intestinal bleeding often occurs thus making the iron deficit even worse (Layrisse and Roche, 1964).

By far the major source of food for both ancient and modern Maya is corn prepared as tortillas or pozole. Beans and to a lesser extent squash are also staples. Other vegetables, fruit, and meat are rarely consumed (Gann, 1918; Shattuck, 1933). Corn and beans are high in carbohydrate content, low in protein, and completely lacking in ascorbic acid. Corn contains appreciable amounts of iron, but the food iron absorption is very low for both corn and beans (Layrisse et al., 1969). This is due to the high phytic acid content which inhibits the intestinal absorption of iron. Both corn and beans are usually cooked in water for long periods of time. This destroys as much as 90 percent of the folic acid and vitamin B_{12} which are necessary for normal development of the red blood cells (Davidson and Passmore, 1969). Ascorbic acid is totally lacking in corn and beans and ascorbic acid-rich fruit is rarely eaten. Ascorbic acid facilitates the reduction of ferric ion to the more easily absorbed ferrous form and therefore ascorbutism (scurvy) further decreases iron absorption. Iron loss in the tropics is much greater than in temperate areas due to a higher loss of iron-rich sweat (Lawson and Stewart, 1967). Protein malnutrition must be both chronic and severe to affect the amino acid requirements in the synthesis of hemoglobin. When combined with deficiencies of any one of the compounds already mentioned, a severe iron deficiency anemia will result (Aksoy et al., 1966; Lanzkowsy, 1968; Viteri et al., 1968).

Thus, many factors besides a simple dietary insufficiency of iron operate synergistically to produce iron deficiency anemia (Scrimshaw, 1968). A tropical climate increasing iron loss; protozoal, helminthic, and bacterial infections of the intestine preventing adequate iron absorption and indeed, causing chronic loss of blood; cooking methods which destroy folic acid and vitamin B_{12} in the food; and dietary insufficiencies of iron, ascorbic acid, and protein—all contribute to produce the high frequency of iron deficiency anemia in modern-day Maya and presumably in the ancient Maya as well. Further work by archaeologists, radiologists, medical anthropologists, and epidemiologists is needed in this area to confirm the thesis the author presented above. Only then will the widespread distribution and high frequency of spongy hyperostosis lesions in both space and time be correctly explained.

CRIBRA ORBITALIA (USURA ORBITAE, HYPEROSTOSIS SPONGIOSA ORBITAE)

Another problem still lacking a definite solution is the etiology of cribra orbitalia, also known as usura orbitae (Møller-Christensen, 1961) or hyperostosis spongiosa orbitae (Hengen, 1971). The lesion is almost always bilateral and presents as many small apertures in the anterior portion of the orbital roofs. The pathogenesis involves both bone lysis and new bone formation. The hypertrophy of the underlying diploic bone produces pressure atrophy of the thin cortical bone layer composing the orbital roof. This increase in spongy bone results in increased thickness of the orbital plate, often several times its normal thickness. The excess spongy or cribrous bone protrudes onto the orbital surface and does not invade the frontal sinus or endocranial cortex of the orbital roof. A vascular net of blood vessels appears to occupy the spaces in the spongy bone connecting the veins of the diploë with those of the orbit (see Hengen, 1971; Møller-Christensen and Sandison, 1963; Toldt, 1886; Wolff, 1954).

Classification

The orbital lesions may be classified according to their morphological appearance into three basic types representing different degrees of development (Nathan and Haas, 1966a). The *porotic type* is characterized by scattered fine openings affecting the roof of the orbit. In the *cribrotic type*, the openings are larger and more numerous, tending to coalesce into larger apertures. In the *trabecular type*, the small openings have lost all individuality by coalescing into large, irregular apertures often arranged in radiating patterns from one or more centers in the orbital roof. Resorption of the cortical bone is the predominant feature of the first two stages while the trabecular type shows marked hypertrophy of the underlying diploë.

The frequencies of the three degrees of cribra orbitalia are

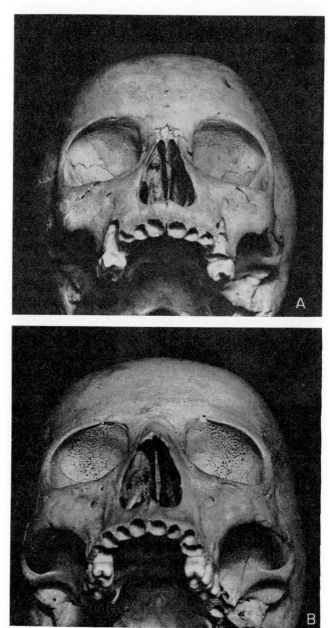

Figure 93. Cribra orbitalia in predynastic Egyptians from Siwa. (A.) A ten-year-old child with spongy and highly vascularized bone tissue protruding into the orbital cavity. This individual also has early signs of spongy hyperostosis affecting both parietals. (B.) A young adult showing large portions of the orbits affected by the cribrous perforations and vascular grooves.

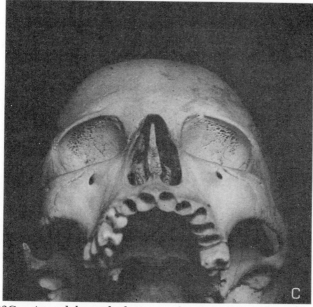

Figure 93C. An adult with large perforations located mainly in the anterior region of the orbital roofs.

similar in both children and adults. In sixty-seven affected subadult crania, 68.6 percent showed the porotic type of pitting, 17.9 percent the cribrotic type, and 13.4 percent the trabecular type. In the 115 affected adult skulls, 52.3 percent were of the porotic type, 28.6 percent of the cribrotic type, and 19.2 percent the trabecular type. The lesions occurred bilaterally in almost 90 percent of the affected skulls. In only a few cases were the lesions in the two orbits of different types (Nathan and Haas, 1966a).

Previous Studies

Cribra orbitalia was first described by Welcker who found it in 3.7 percent of German crania (Welcker, 1888). He considered it to be of racial significance, and this was accepted until the cribriform lesions were found in 68 percent of 100 leprotic crania, or 134 percent more often than in nonleprotic crania (Møller-Christensen, 1961, 1965). Most anthropologists regard cribra orbitalia as further evidence of spongy hyperostosis, because it is presumably the initial lesion found in many cases of spongy hyperostosis (Angel, 1966; Carlson and Van Gerven, 1974; Hengen, 1971; Hrdlička, 1914; Zaino, 1967). However, pathologists and radiologists make no reference to these lesions in their descriptions of hemolytic and iron deficiency anemias (Moseley, 1963). The exact etiology or etiologies of this condition are therefore uncertain.

In a study of 743 predominantly Scottish crania dating from the 18th century, orbital lesions were noted in 8.0 percent of the entire series. When broken down into two age groups, children and adults, cribra orbitalia was found to be much more common in children. Thirteen of twenty-five crania (52.0 percent) were affected in the young age group while only forty-seven of 718 crania (6.6 percent) exhibited such lesions in adults (Møller-Christensen and Sandison, 1963). A similar pattern was found in 182 affected crania from a collection of 718 crania representing various American Indian, European, and Asiatic groups. Of the subadult skulls 64.4 percent exhibited cribra orbitalia compared to 25.3 percent of the adult skulls (Nathan and Haas, 1966a). Many of these cases of childhood cribra orbitalia must heal with advancing age or else lead to an early death in order to explain the lower incidence among adults.

Of the sixty Scottish crania affected by cribra orbitalia, only one skull had associated lesions of cranial spongy hyperostosis. A similar lack of spongy hyperostotic lesions was noted in 285 prehistoric Nubian crania. Of the Nubian specimens 21.4 percent had lesions of cribra orbitalia while only one skull had spongy hyperostosis in addition to the orbital pitting (Carlson, et al., 1974). In fifty-three crania of Hawaiian aboriginal infants and children, 22.8 percent exhibited cribra orbitalia. Of these cases two crania had associated evidence of spongy hyperostosis (Johnson and Kerley, 1974; Zaino and Zaino, 1975).

If both cribra orbitalia and spongy hyperostosis are caused by an acute or chronic anemia, then the data above suggest that cribra orbitalia is the initial manifestation of the disorder. Very few cases progress to spongy hyperostosis of the cranial vault which presumably represents a more pronounced or chronic manifestation of the anemia. Some workers have questioned why the marrow hyperplasia begins initially in the orbital roof rather than in other more important sites of hematopoietic tissue (Blumberg and Kerley, 1966). Actually, the marrow hyperplasia probably occurs in various parts of the skeleton simultaneously, but only the orbital roof reveals the process because its thin external table is easily eroded by the underlying hypertrophy and hyperplasia of the diploë.

Possible Causes of Cribra Orbitalia

Several hypotheses have been put forward regarding the etiology of cribra orbitalia. It is important to note that just as several conditions may produce the lesions of spongy hyperostosis, cribra orbitalia may also have multiple etiologies.

Lacrimal Gland Irritation

Inflammation of the lacrimal gland has been blamed as the cause of cribra orbitalia by irritating the periosteum. The lacrimal gland may become inflamed by leprotic infection (Hogan and Zimmerman, 1962), and this may account for the high incidence of cribra orbitalia (69.7 percent) in ninety-nine leprous skeletons from a medieval leprosarium cemetery in Denmark (Møller-Christensen, 1961). However, the many cases of cribra orbitalia from all over the world are certainly not all caused by leprosy. It has even been suggested that mumps may cause cribra orbitalia (Møller-Christensen and Sandison, 1963). Mumps fulfills many of the criteria because it is a common disease with a wide distribution and affects children more often than adults. However, this is only speculative because little is known about lacrimal gland involvement in mumps. It is important to note that in many cases the orbital lesions are far removed from the fossa of the lacrimal gland and therefore could not be produced by inflammation of this gland.

Trachoma

A high incidence of trachoma (granular opthalmia and conjunctivitis) has been reported in leprosy patients (Chatterjee and Chaudhury, 1964), and others have suggested that nonspecific trachoma is the cause of the cribra orbitalia (Blumberg and Kerley, 1966). Local infections of the eye such as viral trachoma may affect large numbers of individuals living under primitive conditions. It is calculated that at least 50 percent of the world's population becomes infected, particularly in the poverty-stricken countries of Africa and Asia (Henschen, 1966). Such infections are usually bilateral, affect children as well as adults, and are more closely related anatomically to the anterior portion of the orbits (Nathan and Haas, 1966a). However, the conjunctival sac also adjoins the medial and lateral walls of the orbit, yet these regions are never involved with cribra orbitalia. Therefore, trachoma and other forms of conjunctivitis do not appear to qualify as causes of cribra orbitalia.

Comparative Primatology

Cribrous perforations of the orbit have been reported in skulls of other primates and are identical with the lesions in man. The condition was present in 14.1 percent of the total series which included gorillas, orangutans, chimpanzees, baboons, and macaques (Nathan and Haas, 1966b). Cribra orbitalia in other primates all but eliminates such infections as leprosy and mumps as major causes of the orbital lesions because these diseases do not even affect the species of primates examined. More importantly, the occurrence of cribra orbitalia in animals provides a means of attacking the problem by experimental research.

Nutritional Deficiency

All the information at the present time suggests that some type of nutritional deficiency is the most probable cause of cribra orbitalia. A nutritional disorder agrees well with the occurrence of cribra orbitalia in populations from all periods and places. It may also explain why children are affected more often than adults since the growing skeleton is more susceptible to nutritional deficiency. The high incidence of cribra orbitalia in the crania of lepers for example may be due to the combination of poor nutrition in the leprosarium and nutritional disturbances produced by the disease itself (Henschen, 1956, 1961). Similarly, early Greek skeletal remains from periods of poor living condi-

Figure 94. Cribra orbitalia in a child from Early Bronze Age Corinth in Greece (around 2500 B.C.).

tions have a greater frequency of cribra orbitalia. The lesion was attributed to thalassemia or response to a childhood vitamin deficiency or both (Angel, 1964). It is interesting to note that not a single case of cribra orbitalia was found in over 2000 recent autopsies of Swedes, possibly reflecting the better living conditions and improved nutrition (Henschen, 1961).

In a study of cribra orbitalia in different skeletal populations, higher incidences of the lesion were found in those groups which had lived under conditions of inadequate food supply (Nathan and Haas, 1966a). The prevalence of orbital pitting was particularly high in ancient Jewish skeletons from certain caves used for refuge during the War of Bar-Kockba against the Romans (132-135 A.D.). 95.2 percent of the children were affected by cribra orbitalia which may have been produced by starvation during the Roman seige.

Protein deficiency may be ruled out as the cause of cribra orbitalia as this deficiency affects the bones by retarding the bone remodeling process and certainly does not stimulate marrow hyperplasia.

Vitamin Deficiencies

Hypovitaminoses have been suggested as possible producers of cribra orbitalia. Vitamin C deficency or scurvy may produce orbital lesions due to orbital hemorrhage in about 10 percent of all cases of infantile scurvy (Still, 1915). The hemorrhages occur between the periosteum and the anterior bony roof of the orbit and are usually bilateral (Rodger and Sinclair, 1969). Although the bone changes produced by the orbital hemorrhages have not been described in humans, it has been briefly reported that monkeys kept on a vitamin C-deficient diet develop bone changes of the skull, especially pitting of the orbits (Howe cited by Hooton, 1930; see also Zilva and Still, 1920). Scurvy is certainly not the major cause of cribra orbitalia because the orbital lesions occur in only a small proportion of all cases of scurvy, and postcranial scorbutic lesions would be evident in such cases as well (see Chapter VIII). Lack of vitamin A and pantothenic acid also affect the skeleton, but neither produces bone changes in the orbital roof.

Iron Deficiency

As discussed earlier, iron deficiency anemia appears to be a major cause of spongy hyperostosis of the cranial vault and recent work suggests that this deficiency may also produce cribra orbitalia (Carlson et al., 1974; Hengen, 1971; and Moseley, 1965). Compiling the frequencies of cribra orbitalia for various peoples of the world, Hengen found that the frequencies increased to a maximum near the equator. This correlates well with the worldwide distribution of iron deficiency anemia (WHO report, 1968). The high incidence of iron deficiency anemia in the tropics is chiefly caused by parasitic infestations such as by *Entamoeba histolytica, Balantidium coli, Strongyloides, Ascaris lumbricoides, Giardia intestinalis, Trichuris trichuris,* and particularly hookworms (*Ancylostoma duodenale* and *Necator americanus*). In fact, an early pathological description of cribra orbitalia came from a dissected case of severe ancylostomiasis (Koganei, 1912). Other factors important in causing iron deficiency anemia include dietary insufficiencies of iron, presence of certain substances such as phytic acid which inhibits intestinal iron absorption, malabsorptive diseases such as sprue, prolonged suckling of infants on milk which has a low iron content, and iron loss through sweating (see the previous section of this chapter).

The greater incidence of cribra orbitalia in children may be easily explained by the nutritional trauma (known as weanling diarrhea) experienced by the infant when switching from milk to other foods, the greater need for iron during the growing years, and the greater frequency of parasitic infestation in children (May, 1958; Trowell and Jelliffe, 1958). Many of the primates are frequently infested by parasites, explaining the occurrence of cribra orbitalia in these animals. Improved nutrition and hygiene has resulted in the dramatic decrease of cribra orbitalia in Europe and other civilized areas. Furthermore, a deficiency of iron agrees well with the examples cited previously of people starving while under siege and patients suffering from a chronic, destructive disease in a wretched medieval leprosarium.

Additional evidence from studies of skeletal populations and their environments reinforces the belief that iron deficiency anemia causes cribra orbitalia. In ancient Nubia, the high frequency (21.4 percent) of cribra orbitalia in 285 individuals probably resulted from the diet of milled cereal grains which contain little iron and the prevalence of parasitic infection (Carlson et al., 1974). Iron deficiency anemia is still common in young children and multiparous females of this region. In 118 individuals from Nusplingen, Germany around 400-800 A.D., 60 percent exhibited cribra orbitalia. Analysis of the water and soil in this region revealed hardly a trace of iron which may partially explain the high incidence (Hengen, 1971).

As discussed earlier, iron deficiency anemia is probably the cause of cranial spongy hyperostosis in 36 percent of ninety ancient Maya individuals (Saul, 1972). Only one of these individuals had associated cribra orbitalia (Saul, personal communication, 1974), but the very fragmentary nature of the remains makes this low figure questionable. Upon reexamining thirty-six Maya crania from the Sacred Cenote in Yucatan (see Hooton, 1930; 1940), the author found that 25 percent of the crania had cribra orbitalia, which provides a more accurate indication of the prevalence of orbital lesions among the Maya.

More comparative studies of skeletal populations with differing nutritional backgrounds will hopefully provide further support to the argument of iron deficiency anemia as the cause of cribra orbitalia. More importantly, experimental studies using restricted diets in monkeys should provide the information necessary to solve the interesting and important riddle of cribra orbitalia.

ANEMIA BIBLIOGRAPHY

Aksoy, M., N. Camli, and S. Erdem: "Roentgenographic bone changes in chronic iron deficiency anemia," *Blood, 27*:677-685, 1966.

Allison, A. D.: "Protection afforded by the sickle-cell trait against subtertian malarial infection," *Brit. Med. J., 1*:1187-1190, 1954.

—————: Genetic factors in resistance to malaria," *Ann. N.Y. Acad. Sci., 91*:710-724, 1961.

Allison, M. J., A. Pezzia, I. Hasegawa, and E. Gerszten: "A case of hookworm infestation in a pre-Columbian American," *Amer. J. Phys. Anthrop., 41*:103-106, 1974.

Anderson, J. E.: "Late Paleolithic skeletal remains from Nubia," pp. 996-1040 in F. Wendorf (ed.): *The Prehistory of Nubia,* vol. 2, Southern Methodist Univ. Press, Dallas, 1968.

Angel, J. L.: "Osteoporosis: thalassemia?" *Amer. J. Phys. Anthrop., 22*:369-374, 1964.

—————: "Porotic hyperostosis, anemias, malarias, and marshes in the prehistoric eastern Mediterranean, *Science, 153*:760-763, 1966.

—————: "Porotic hyperostosis or osteoporosis symmetrica," in D. R. Brothwell and A. T. Sandison (ed.): *Diseases in Antiquity,* Thomas, Springfield, 1967.

—————: "Early Neolithic skeletons from Catal Huyuk: demography and pathology," *Anatolian Studies, 21*:77-98, 1971a.

—————: *Lerna, The People,* vol. 2, American School of Classical Studies at Athens and Smithsonian Institution Press, Washington, 1971b.

Armelagos, G. J.: "Future work in paleopathology," in W. D. Wade (ed.): *Miscellaneous Papers in Paleopathology, Museum of Northern Arizona Technical Series, 7*:1-8, 1967.

Bohrer, S. P. and G. E. Connah: "Pathology in 700 year old Nigerian bones," *Radiology, 98*:581-584, 1971.

Boyd, M. F. (ed.): *Malariology,* 2 vols., Saunders, Philadelphia, 1949.

Blumberg, J. M. and E. R. Kerley: "A critical consideration of roentgenology and microscopy in paleopathology," in S. Jarcho (ed.): *Human Palaeopathology,* Yale Univ. Press, New Haven, 1966.

Bruce-Chwatt, L. J.: "Paleogenesis and paleo-epidemiology of primate malaria," *Bull. World Health Org.*, *32*:363-387, 1965.
Caffey, J.: "The skeletal changes in the chronic hemolytic anemias," *Amer. J. Roentgenol.*, *37*:293-324, 1937.
—————: "Cooley's erythroblastic anemia: some skeletal findings in adolescents and young adults," *Amer. J. Roentgenol.*, *65*:547, 1951.
Carlson, D. S., D. P. Van Gerven, and G. J. Armelagos: "Factors influencing the etiology of cribra orbitalia in prehistoric Nubia," *J. Hum. Evol.*, *3*:405-410, 1974.
Carroll, D. S.: "Roentgen manifestations of sickle cell disease," *Southern Med. J.*, *50*:1486-1490, 1957.
Cavalli-Sforza, L. L. and W. F. Bodmer: *The Genetics of Human Populations*, W. E. Freeman, San Francisco, 1971.
Chatterjee, S. and D. S. Chaudhury: "Pattern of eye diseases in leprosy patients of North Ghana," *Internat. J. Leprosy*, *32*:53-63, 1964.
Cooley, T. B., E. R. Witwer and P. Lee: "Anemia in children with splenomegaly and peculiar changes in the bones," *Amer. J. Dis Child.*, *34*:347-363, 1927.
Cule, J. and I. L. Evans: "Porotic hyperostosis and the Gelligaer skull," *J. Clin. Path.*, *21*:753-758, 1968.
Currarino, G. and M. E. Erlandson: "Premature fusion of the epiphyses in Cooley's anemia," *Radiology*, *83*:656-664, 1964.
Davidson, S. and R. Passmore: *Human Nutrition and Dietetics*, Williams and Wilkins, Baltimore, 4th ed., 1969.
Dunn, F. L.: "On the antiquity of malaria in the Western Hemisphere," *Hum. Biol.*, *37*:385-393, 1965.
Eng, L. L.: "Chronic iron deficiency anemia with bone changes resembling Cooley's anemia," *Acta Haemat.*, *19*:263-268, 1958.
Gann, T. W. F.: "The Maya Indians of southern Yucatan and southern British Honduras," *Bur. Amer. Ethnol. Bull.*, 64, 1918.
Gejvall, N-G.: *Westerhus: Medieval Population and Church in the Light of Skeletal Remains*, Boktryckeri, Lund, 1960.
Gray, P. H.: "Bone infarction in antiquity," *Clin. Radiol.*, *19*:436-437, 1968.
Hamperl, H. and P. Weiss: "Über die spongiose Hyperostose an Schadeln aus Alt-Peru," *Arch. Path. Anat.*, *327*:629-642, 1955.
Hengen, O. P.: "Cribra orbitalia: pathogenesis and probable etiology," *Homo*, *22*:57-75, 1971.
Henschen, F.: "Zur Palaopathologie des Schadels—uber die sog. Cribra Cranii," *Verh. dtsch. Ges. Path.*, 39 Tag, Stuttgart, 1956.
—————: "Cribra cranii, a skull condition said to be of racial or geographical nature," *Path. Microbiol.*, *24*:724-729, 1961.
Hogan, M. J. and L. E. Zimmermann: *Opthalmic Pathology*, Saunders, Philadelphia, 1962.

Hooton, E. A.: *Indians of Pecos Pueblo,* Yale Univ. Press, New Haven, 1930.

———: "Skeletons from the Cenote of Sacrifice at Chichen Itza," in: *The Maya and Their Neighbors,* Appleton-Century, New York, 1940.

Hrdlička, A.: "Anthropological work in Peru in 1913, with notes on pathology of the ancient Peruvians," *Smithson. Misc. Coll., 61:*57-59, 1914.

Jarcho, S., N. Simon, and H. L. Jaffe: "Symmetrical osteoporosis in a prehistoric skull from New Mexico," *El Palacio, 72:*26-30, 1965.

Johnson, L. C. and E. R. Kerley: "Report on pathological specimens from Mokapu," Appendix B in C. E. Snow: *Early Hawaiians, an Initial Study of Skeletal Remains from Mokapu, Oahu,* Univ. Ky. Press, Lexington, 1974.

Koganei, Y.: "Cribra cranii und cribra orbitalia," *Mitt. Med. Fak. Tokyo, 10:*113-154, 1912.

Lanzkowsky, P.: "Radiological features of iron deficiency anemia," *Amer. J. Dis. Child, 116:*16-29, 1968.

Lawson, J. B. and D. B. Stewart: *Obstretics and Gynecology in the Tropics and Developing Countries,* Arnold, London, 1967.

Layrisse, M., J. D. Cook, C. Martinez, M. Roche, I. N. Kuhn, R. B. Walker, and C. A. Finch: "Food iron absorption: a comparison of vegetable and animal foods," *Blood, 33:*430-443, 1969.

Layrisse, M. and M. Roche: "The relationship between anemia and hookworm infection," *Amer. J. Hyg., 79:*279-301, 1964.

Livingstone, F.: *Abnormal hemoglobins in human populations,* Aldine, Chicago, 1967.

Loiacono, P. L. and M. M. Reeder: "An exercise in radiologic-pathologic correlation," *Radiology, 92:*385-394, 1969.

May, J. M.: *The Ecology of Human Disease,* MD Publ., New York, 1958.

McLaren, D. S.: *Malnutrition and the Eye,* Academic Press, New York, 1963.

Møller-Christensen, V.: *Bone Changes in Leprosy,* Munksgaard, Copenhagen, 1961.

Møller-Christensen, V. and A. T. Sandison: "Usura orbitae (cribra orbitalia) in the collection of the University of Glascow," *Path. Microbiol., 26:* 175-183, 1963.

Moore, J. G., G. F. Fry, and E. Englert: "Thorny-headed worm infection in North American prehistoric man, *Science, 163:*1324-1325, 1969.

Moore, S.: "Bone changes in sickle anemia with a note on similar changes in skulls of ancient Mayan Indians," *J. Missouri Med. Assn., 26:*561-564, 1929.

Moseley, J. E.: *Bone Changes in Hematologic Disorders,* Grune and Stratton, New York, 1963.

―――――: "The paleopathological riddle of 'symmetrical osteoporosis,'" *Amer. J. Roentgenol.*, 95:135-142, 1965.

―――――: "Radiographic studies in hematologic bone disease: implications for paleopathology," in S. Jarcho (ed.): *Human Palaeopathology*, Yale Univ. Press, New Haven, 1966.

Motulsky, A. G.: "Metabolic polymorphisms and the role of infectious disease in human evolution, *Hum. Biol.*, 32:28-61, 1960.

Nathan, H. and N. Haas: "Anthropological data on the Judean Desert skeletons," in E. Goldschmidt (ed.): *Genetics of Migrant and Isolate Populations*, Williams and Wilkins, Baltimore, 1963.

―――――: "Cribra orbitalia, a bone condition of the orbit of unknown nature," *Israel J. Med Sci.*, 2:171-191, 1966a.

―――――: "On the presence of cribra orbitalia in apes and monkeys," *Amer. J. Phys. Anthrop.*, 24:351-360, 1966b.

Pizzi, T. and H. Schenone: "Hallazzo de heuvos de *Trichuris trichiura* en contenido intestinal de un cuerpo arqueologico incaico," *Boletin Chileno de Parasitologia*, 9:73-75, 1954.

Reynolds, J.: "A re-evaluation of the 'fish vertebra' sign," *Amer. J. Roentgenol.*, 97:696-707, 1966.

Rodger, F. C. and H. M. Sinclair: *Metabolic and Nutritional Eye Diseases*, Thomas, Springfield, 1969.

Samuels, R.: "Parasitological study of long-dried fecal samples," *Mem. Soc. Amer. Arch.*, 19:175-179, 1965.

Saul, F.: "The human skeletal remains of Altar de Sacrificos," *Papers of the Peabody Mus. of Arch. and Ethnol.*, vol. 63. no. 2, Harvard University, 1972.

―――――: "Disease in the Maya area: the pre-Columbian evidence," in T. P. Culbert (ed.): *The Classic Maya Collapse*, Univ. of New Mexico Press, Albuquerque, 1973.

―――――: Personal communication, 1974.

Scrimshaw, N. S.: "An epidemiologic approach to the causes and control of the nutritional anemias," *Vitamins and Hormones: Advances in Research and Applications*, 26:705-716, Academic Press, New York, 1968.

Scrimshaw, N. S. and C. Tejada, C.: "Pathology of living Indians as seen in Guatemala," in T. D. Stewart (ed.): *Physical Anthropology*, vol. 9, of *Handbook of Middle American Indians*, Univ. of Texas Press, Austin, 1972.

Shattuck, G. C.: *The Peninsula of Yucatan; Medical, Biological, Meteorological, and Sociological Studies*, Carnegie Institution, Washington, D.C., 1933.

Smith, E. G. and W. Jones: "Report on the human remains," *Archaeological Survey of Nubia, Report of 1907-1908*, Cairo, Ministry of Finance, 1910.

Snow, C. E.: "Indian burials from St. Petersburg, Florida," *Contrib. of the Florida State Museum,* 8, 1962.

Still, G. F.: *Common Disorders and Diseases of Childhood,* Oxford Univ. Press, London, 1915.

Toldt, C.: "Über Welcker's cribra orbitalia," *Mitt. Anthrop. Ges. Wien.,* 16:20, 1886.

Trowell, H. C. and D. B. Jelliffe: *Diseases of Children in the Subtropics and Tropics,* E. Arnold, London, 1958.

Virchow, R.: "Über die puerperalen Krankheiten," *Verh. Ges. Geburtsh.,* 3:151, 1848; cited by Henschen, 1961.

Viteri, F. E., J. Alvarado, D. G. Luthringer, and R. P. Wood: "Hematological changes in protein calorie malnutrition," *Vitamins and Hormones: Advances in Research and Applications,* 26:573-615, Academic Press, New York, 1968.

Wakefield, E. G., S. C. Dellinger, and J. D. Camp: "Study of osseous remains of mound builders of eastern Arkansas," *Amer. J. Med. Sci.,* 193:488-495, 1937.

Welcker, H.: "Cribra orbitalia, ein ethnologisch-diagnostisches Merkmal am Schädel mehrerer Menschenrassen," *Arch. Anthrop.,* 17:1-18, 1888.

Williams, H. U.: "Human paleopathology with some original observations on symmetrical osteoporosis of the skull," *Arch. Path.,* 7:839-902, 1929.

Wolff, E.: *The Anatomy of the Eye and Orbit,* H. K. Lewis, London, 1954.

World Health Organization Technical Report Series, No. 405: "Nutritional anemias: report of a WHO scientific group," Geneva, 1968.

World Health Statistics Annual for the year 1970, World Health Organization, Geneva, 1973.

Zaino, D. E. and E. C. Zaino: "Cribra orbitalia in the aborigines of Hawaii and Australia," *Amer. J. Phys. Anthrop.,* 42:91-94, 1975.

Zaino, E. C.: "Paleontologic thalassemia," *Ann. N.Y. Acad. Sci.,* 119:402-412, 1964.

————: "Symmetrical osteoporosis, a sign of severe anemia in the prehistoric Indians of the Southwest," in W. D. Wade (ed.): *Miscellaneous Papers in Paleopathology; Museum of Northern Arizona Technical Series,* 7:40-47, 1967.

Zilva, S. S. and G. S. Still: "Orbital hemorrhage with proptosis," *Lancet,* 1:1008, 1920.

Chapter VIII

METABOLIC BONE DISEASE

METABOLIC DISEASES OF bone refer to disorders producing a reduction in bone mass due to inadequate osteoid production, inadequate osteoid mineralization, or excessive deossification of normal bone tissue. The factors that maintain the normal osteoid balance are complex and poorly understood (see Sissons, 1963). Particular emphasis will be placed on the nutritional deficiencies causing a disturbance in normal bone metabolism as such diseases have important implications in paleopathology.

DISEASES DUE TO INADEQUATE OSTEOID SYNTHESIS—OSTEOPENIA

Dietary Osteopenia-Scurvy

Osteoid is the organic matrix upon which the inorganic bone minerals are deposited. Collagen, a fibrous structural protein, makes up about 90 to 95 percent by weight of the dry, fat-free organic matrix (McLean and Urist, 1968). Both proline and hydroxyproline are important amino acids in the triple helical structure of collagen, although only proline is incorporated in the protocollagen molecule. The hydroxylation of proline to hydroxyproline requires vitamin C (ascorbic acid). People lacking vitamin C cannot synthesize adequate amounts of osteoid and the result is scurvy.

The richest natural sources of vitamin C are citrus fruits and uncooked green vegetables. Potatoes contain a smaller amount but are an important source when eaten in large quantities as in parts of Europe. Cereals totally lack vitamin C and meat contains very little. In Eskimos, whose diet is almost completely carnivorous, the liver and other glandular tissues of the animals they eat will provide adequate amounts of vitamin C (Davidson and Passmore, 1969).

Historically, scurvy has been the scourge of sailors on long voyages, soldiers in the field, and people under siege. It was not unusual for 50 to 80 percent of an entire crew to perish from scurvy during a single long voyage (Hess, 1920). During the famine years of 1846-47 in Ireland, many died from scurvy (Henschen, 1961). During periods of prolonged drought, epidemics of scurvy occurred among the Australian aborigines (Basedow, 1932). Ossified subperiosteal hemorrhages probably due to scurvy have been found in 27 percent of sixty-three adult ancient Maya skeletons and may indicate famine and disease as prominent factors in the collapse of the Mayan civilization (Saul, 1972, 1973).

Infantile Scurvy

Because of its similarity in both symptoms and age of onset, infantile scurvy was grouped with rickets until Barlow differentiated it pathologically from rickets and showed its close association with adult scurvy (Barlow, 1883). Most cases occur in infants between the ages of six months and one year. Scorbutic infants are highly susceptible to intercurrent infections which often prove fatal.

Subperiosteal hemorrhage is the most important single sign of scurvy (Kato, 1932). In the healing stage or in cases of moderate scurvy, ossification or calcification of the subperiosteal hematoma occurs providing valuable evidence of the disease in archaeological material. In the dried bone specimen they may be seen along the diaphyses of the long bones as irregular areas of subperiosteal bone apposition. In advanced cases, the entire shaft may show swelling as a result of this subperiosteal apposition. The long bones of the lower limb are most often involved (see Fig. 95).

If the subperiosteal hemorrhages are resorbed rather than calcified, the hyperemia and increased osteoclastic activity

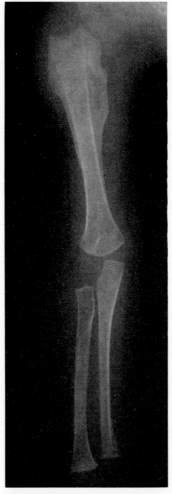

Figure 95. Clinical roentgenogram of an infant with severe scurvy. A large, partially calcified subperiosteal hemorrhage is evident in the proximal humerus. The cortices in the bones are unusually thin. (Courtesy of Dr. E. B. D. Neuhauser.)

causes weakening of the cancellous bone particularly in the metaphyses. Fractures through the metaphyses are common near the junction between the metaphysis and epiphysis. Small calcified spurs may protrude from the lateral border at these sites indicating callus formation. As a result of bone resorption and deficient periosteal new bone formation, the cortices become very thin and the sparse trabeculae may present a "ground glass" appearance in the roentgenogram. In severe cases of scurvy the entire skeleton undergoes deossification due to disuse atrophy of the immobilized infant.

Scorbutic beading of the ribs occurs at their costochondral junctions. The beading is similar to that characteristic of rickets but is more angular and less knobby. In addition, subperiosteal hemorrhages often occur along the rib shaft.

Very little has been written about cranial changes in scurvy. In some cases there is subperiosteal hemorrhage on the ophthalmic surface of the orbital portion of the frontal bone (Jaffe, 1972). Perhaps organization and resorption of the hematoma produces the lesions of cribra orbitalia, but this is not known at present. It is noteworthy that monkeys kept on a scorbutic diet exhibited marked cribriform changes in the orbital roofs (Howe cited by Hooton, 1930).

Adult Scurvy

Adult scurvy is similar in appearance to the infantile form except for the absence of involvement at the epiphyseal ossification centers which have closed. Pathological fractures are common in the diaphyses rather than the metaphyses and are caused by the generalized deossification without new bone formation (see Figs. 96 & 97).

Endocrine Osteopenia—Senile and Postmenopausal Osteoporosis

The sex hormones, estrogen and androgen, are necessary for normal osteoblastic activity. With the onset of old age (and menopause in females), a hormonal imbalance is created by an increase in the antianabolic adrenal glucocorticoid hormones relative to the anabolic sex hormones. The result of this imbalance is osteoporosis which is a reduction in the amount of osseous tissue per unit of bone volume. Other factors may be involved and there still remains a large group of cases of idiopathic osteoporosis (see McLean and Urist, 1968).

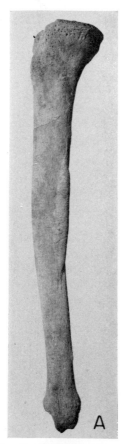 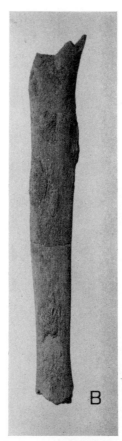

Figure 96. Ossified subperiosteal hemorrhages in ancient Mayans from Guatemala (500-1000 A.D.). (A.) Medial view of right tibia of an elderly male. An elongated, vermiform ossified subperiosteal hemorrhage is present along the shaft. Similar lesions are present along the linea aspera of both femora. (B.) Medial view of right tibia of an adult male. Several irregular areas of ossified subperiosteal hemorrhages are evident along the shaft.

Age Incidence

Senile osteoporosis does not affect all old people or women after menopause. However, the incidence of osteoporosis rises steadily after the age of forty years. A study of 218 ambulatory women aged forty-five to seventy-nine years with no clinical symptoms of osteoporosis gives some idea of the actual frequency

Figure 96C. Lateral view of left tibia of an elderly male showing the remains of a massive subperiosteal hemorrhage. This individual also has severe periodontal degeneration and resorption suggestive of scurvy. (Courtesy of Dr. Frank P. Saul, 1972.)

of osteoporosis in females. Between the ages of forty-five to forty-nine years, 7 percent had radiographic evidence of osteoporosis. This incidence jumped to 46 percent in those sixty to sixty-four years of age and 78 percent in females over seventy-five years old (Smith et al., 1960). In a study of the incidence of bone fractures due to senile osteoporosis, it was found that

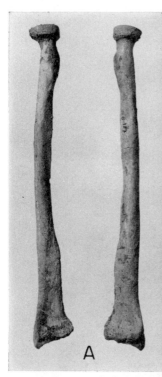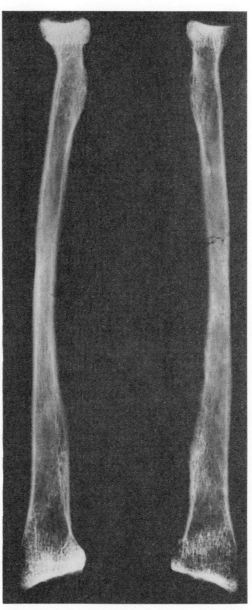

Figure 97. Ossified subperiosteal hemorrhages in an ancient Maya of the Plancha Phase (300 B.C.-200 A.D.). (A.) Radii of an adult with ossified subperiosteal hemorrhages in the distal portions of the radii producing a clubbed effect. (B.) Roentgenogram showing the new bone along the medial surfaces of the distal radii and merging with the cortex. (Courtesy of Dr. Frank P. Saul, 1972.)

populations which continue vigorous physical activity into old age have a lower incidence of osteoporosis than populations in which physical exericse is reduced (Chalmers and Ho, 1970).

Bone Changes

Senile osteoporosis first appears in the vertebral column, sternum, ribs, and pelvis. The changes are most apparent in the vertebral bodies with their abundance of spongy bone and thin cortical margins. The spongy trabeculae become thinned, particularly those which are transversely oriented. The weakened vertebrae often are compressed or wedged by weight-bearing pressure. Fractures may occur and callus formation should be evident in such vertebrae.

As osteoporosis progresses, the long bones may also become involved. The trabeculae become sparse initially and finally the cortex itself is thinned from the medullary surface outward (Jaffe, 1972). Fractures in the brittle bones are common, particularly in the neck of the femur (Chalmers and Ho, 1970).

Although the skull is rarely involved in senile osteoporosis, it is thought that the parietal thinning found in occasional excavated crania is evidence of this disorder (Epstein, 1953; Lodge, 1967). The thinning is usually bilateral in the posterior portions of the parietals. The outer table and the internal diploë may be completely resorbed leaving only the inner tables as shallow depressions on both sides of the vault. Most of the archaeological cases described have been in older individuals. The majority of such specimens come from ancient Egypt with fewer found in European material (Smith, 1906-07). Roentgenographic examination of the postcranial remains should be undertaken to determine if osteoporotic changes are present. Only then can one correctly assume an association between senile osteoporosis and bilateral parietal thinning. Moreover, one should be wary of soil conditions which leach the postcranial bones and make them very light and brittle as in osteoporosis.

Stress Deficiency Osteopenia—Atrophy

Infections, metabolic disturbances, growth, and senility all alter the physical properties and architecture of bone. However, one of the most important factors affecting the bone is mechanical stress. This has been formulated into the classic but still controversial Wolff's Law which states that every change in the function and form of a bone is followed by certain definite changes in the internal architecture and external form in accordance with mathematical laws (Wolff, 1892). Thus, in the normal long bone the maximal thickness of the cortex corresponds to the maximal bending stress in this area. The main distribution of cancellous bone is near the bone ends where axial pressure is greatest (Weinmann and Sicher, 1947; Frost, 1973). As pressure trajectories are altered old laminae are resorbed and replaced by new ones arranged in the new stress lines.

Disuse atrophy is interesting because osteoblasts do not lay down osteoid without the stress stimulus. After normal catabolic lysis the laminae are not reformed and osteopenia develops. Disuse atrophy can be apparent after only a few weeks of immobilization. In archaeological material bone atrophication would probably be found associated with severe injury, congenital deformity, or some form of paralysis. The diagnostic features are similar to those of senile osteoporosis except that only the specific bones involved in immobilization would be osteoporotic and atrophied.

Congenital Osteopenia—Osteogenesis Imperfecta

Osteogenesis imperfecta, also known as "brittle bones," is a relatively uncommon disease. It is transmitted as an autosomal dominant gene or may appear by spontaneous mutation. The abnormality apparently involves the mesenchyme because other connective tissues besides the osseous tissue are affected. A defect in the formation and chemical composition of collagen is common to all of these tissues. In bone, the result is a reduction in the quantity and quality of bone formation (Jaffe, 1972).

Osteogenesis imperfecta may occur *in utero* or may not be evident until years later. The prognosis is very poor for infants born with the disorder. The outlook is much better when an adolescent or adult is affected. Initially the limb bones appear long and gracile with thin cortices and sparse spongiosa. However, numerous fractures followed by malalignment and callus formation cause dwarfing and deformity in the delicate bones. The skull may appear normal except for numerous Wormian bones composing much of the vault.

The bones in osteogenesis imperfecta are so fragile that their chance of preservation is scant although two cases have been reported. An infant from the 21st Dynasty of Egypt (around 1000 B.C.) provides the classic findings of osteogenesis and dentinogenesis imperfecta (Gray, 1969). The skull vault is made up of a very large number of Wormian bones. The teeth have short roots and microscopic examination of the dentine reveals abnormal tubular structure. All of the postcranial bones present are deformed, particularly those of the lower extremities with marked anterolateral bowing of the femurs. Roentgenograms show cortical thinning and greatly decreased amounts of spongiosa in all of the bones.

An isolated left femur from a Saxon burial ground (650-850 A.D.) shows the characteristic features of osteogenesis imperfecta (Wells, 1965). The femur is from a sixteen- to seventeen-year-old individual and the distal two-thirds of the bone makes an angle of 75 degrees with the proximal one-third. There is evidence of several previous fractures in the region of greatest angulation; the shaft is very slender with expanded metaphyses and a roentgenogram shows the thinned cortex and diminished spongiosa. Rickets might also be considered a possible cause although rachitic changes are absent in hundreds of Saxon skeletons from the same period (see Fig. 98).

INADEQUATE OSTEOID MINERALIZATION

Rickets

Rickets is a disturbance in the formation of bone in the growing skeleton caused by a failure to deposit calcium and phosphorus within the organic matrix of cartilage and osteoid. Furthermore, in order to maintain normal plasma calcium levels, parathyroid hormone is secreted, leading to the mobilization of calcium and phosphorus from bone and thereby weakening the

Figure 98. Osteogenesis imperfecta in a left femur of an early Saxon. (Courtesy of Dr. Calvin Wells, 1965.)

previously normal bone tissue. The most common cause of rickets is a deficiency of vitamin D causing inadequate absorption of calcium and phosphorus from the intestine. Other causes of this disorder are chronic intestinal diseases, insufficient amounts of calcium and phosphorus in the diet, and chronic renal tubular malfunction (for an excellent review see Mankin, 1974).

History

Vitamin D occurs naturally only in the fat of certain food products of animal origin. It is also synthesized in the body when ultraviolet rays from sunlight convert ergosterol in the skin to vitamin D. Rickets is often referred to as a "disease of civilization." It became very common during the Industrial

Revolution of the 18th and 19th centuries, when large numbers of laborers and their families had poor diets and insufficient sunlight. This was particularly true in the more temperate regions such as Britain and Northern Europe (Hess, 1929). Social customs can also be a major factor in producing rickets. In India many Moslem girls keep their faces and bodies heavily covered and rarely venture outside. Among 1482 Moslem girls between the ages of five and seventeen, skeletal rachitic lesions were evident in 40 percent (Wilson, 1931).

Although the archaeological evidence of rickets is scarce, it should not be considered a disease unique to modern man. In the large amount of ancient Egyptian skeletal material, no definite cases of rickets were found although a few long bones exhibited bowing deformities very suggestive of rickets (Smith and Jones, 1910). A similar paucity of rachitic bones may be noted in other areas where sunlight is abundant and can rectify any deficiencies of vitamin D in the diet.

In more temperate regions possible cases of rickets are more numerous. Nine individuals from a series of 800 Medieval skeletons in Denmark exhibit rachitic lesions (Møller-Christensen, 1958). Two possible cases have been reported from a Medieval graveyard in Sweden (Gejvall, 1960). Earlier specimens of rickets come from the Neolithic and Iron Age of Denmark and Norway (Nielsen cited by Møller-Christensen, 1958; Sigerist, 1961). Several unequivocal literary references to rickets may be found in China dating back to 300 B.C. (Lee, 1940). Similar references come from Europe around the 2nd century A.D. (Foote, 1927b).

The lack of skeletal evidence of rickets in far northern areas such as Alaska is probably due in large part to a meat-eating diet high in vitamin D. Pre-Columbian rickets was practically nonexistent in North and South America (Hrdlička, 1907). Two possible cases have been reported in Archaic Indians from Kentucky around 3300 B.C. (Foote, 1927; Snow, 1948). One of these individuals is a twelve- to fifteen-month-old infant with beading of the ribs at the costochondral junctions. The distal and proximal portions of the tibiae and distal portions of the femora and ulnae appear swollen and spongy. This is particularly evident in the tibiae. The hypertrophy evident in these limb bones is characteristic of advanced rickets in infancy. Rickets has also been reported in more recent American Indian skeletal material from New Mexico and Utah (Hooton, 1930; Brues, 1946).

It should be noted that pseudopathological lesions simulating rickets may occur where acidic conditions in the soil lead to decalcification of the long bones. These long bones may then become warped by earth pressure and resemble rachitic bowing (Wells, 1967). However, other features of rickets will be absent such as cranial bossing, flaring of the metaphyses, and beading of the ribs at the costochondral junctions.

Age

Most cases of rickets occur between the ages of six months and two years—a period of rapid bone growth, often little exposure to sunlight, and poor diet. During the years preceding puberty there is also a high incidence of rickets. This vitamin D deficiency is known as late rickets when occurring between the ages of six and fifteen years. During adult life after epiphyseal ossification centers have closed, there is a different response of the mature bone to vitamin D deficiency. This disorder is called osteomalacia and will be described in a following section.

Bone Changes

The formation of organic matrix (osteoid) by the osteoblasts is undisturbed in rickets. However, the mineralization of this matrix does not occur because of the lack of vitamin D. The osteoid is resistant to resorption and therefore persists in excessive amounts. Depending upon the phase of rickets at death, the osteoid may or may not be evident in the skeletal remains. In florid or active rickets, the osteoid will not be mineralized and therefore not preserved although expanded and "cupped" metaphyses may be present. In partially healed rickets, the osteoid may be moderately well mineralized although the bone contours will still be abnormal. In both cases the bones may feel extraordinarily light and of brittle texture.

Skull

The skull is almost always affected in rickets and the changes may be thinning (craniotabes) or thickening or a combination of the two processes in different areas of the vault. Craniotabes occurs during the period of rapid brain enlargement and consequent growth of the cranial vault. This usually disappears by the ninth month when growth has slowed. The cranial thinning is rarely symmetrically distributed and usually found on the side of the head on which the infant habitually lies. Thinned areas are therefore most often found in the posterior parts of the parietal bones and the upper part of the occipital bone. These thinned areas may be 2 to 4 cm in diameter (Weinmann and Sicher, 1947; Jaffe, 1972).

Thickening of the vault also occurs in rickets, particularly at two to three years of age. The thickening is not evenly distributed and is more pronounced at the eminences of the frontal and parietal bones. The excessive deposition of new bone (and osteoid) is almost entirely on the outer surface of the vault. In advanced rickets the result is a "squared head" because of bone deposition on the four frontal and parietal bosses. During healing, part of the excessive osteoid is absorbed, but most of it ossifies and remains permanently as evidence of rickets (Luck, 1950).

Long Bones

Rachitic changes in the long bones will differ according to the amount of resorption of osteoid and demineralization of the bone taking place. Excessive demineralization of the bone in order to maintain the normal plasma calcium levels will cause thinning of the cortex and the spongy trabeculae become sparse and slender. In the majority of cases, such demineralization does not occur and the "hypertrophic" form of rickets is seen. The excessive amounts of osteoid and cartilage cause marked expansion of the metaphyses often described as "trumpeting." There is also osteoid apposition subperiosteally along the diaphysis. The cortical layer is thickened and the marrow cavity narrowed. Similar changes may occur in the metacarpals and metatarsals.

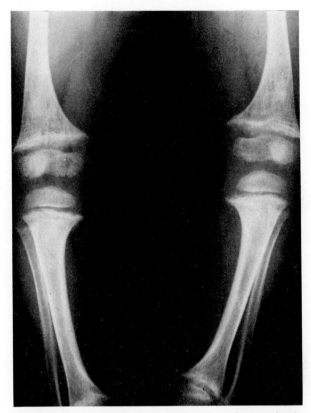

Figure 99. Clinical roentgenogram of infantile rickets. The bones of the lower extremities have flared metaphyses due to excessive osteoid synthesis. (Courtesy of Dr. E. B. D. Neuhauser.)

Fractures often occur in the shafts of the weakened long bones and in the ribs. There is rarely any displacement of the fractured ends. Callus formation will take place, but the deposited osteoid will not provide much support. Bowing of the long bones often occurs, mainly resulting from such fractures and weight placed on the weakened shafts. The long bones of the arms are usually less involved than those of the leg. The normal anterior bend of the femur becomes more marked and there may be lateral bowing of this bone as well. The cortices become wider on the concave side in response to the need for mechanical reinforcement. The neck of the femur may be bent downwards

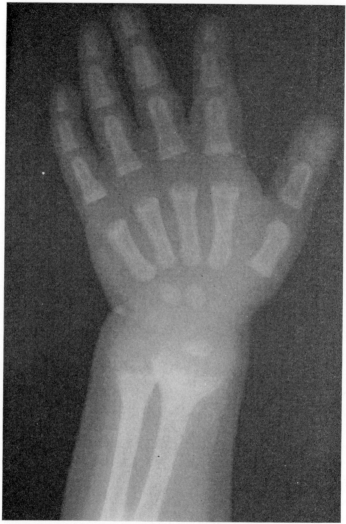

Figure 100. Clinical roentgenogram of infantile rickets. There is widening and cupping of the distal metaphyses of both ulna and radius. The bones of the left hand appear slightly swollen with coarsening of the trabeculae. (Courtesy of Dr. E. B. D. Neuhauser.)

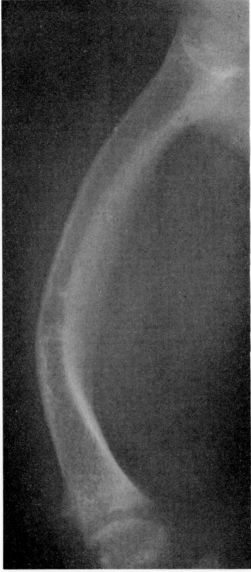

Figure 101. Clinical roentgenogram of rickets. There is lateral bowing of the right femur with thickening of the cortex along the concave side. (Courtesy of Dr. E. B. D. Neuhauser.)

until it forms a right angle with the shaft and produces coxa vara. The tibia and fibula may be curved anteriorly and are usually markedly curved laterally (producing bowlegs). Other curves are sometimes present and may be explained by postural habits (Knaggs, 1926). In severe rickets, deformities of the arms occur in children who use their arms as props to help the weakened spine support the head. The distal portions of the radius and ulna are curved posteriorly and the humerus is bent outward near the deltoid insertion.

Ribs, Spine, and Pelvis

The beading of the ribs at the costochondral junctions is due to the same process that causes the flaring of the metaphyses in the long bones. This beading (rachitic rosary) is often more pronounced on the inner surface at the ends of the ribs. The fifth and sixth ribs are most often affected. Lateral to these enlargements, the softened ribs sink into the chest forming a groove on both sides of the sternum. This is known as Harrison's groove or sulcus.

Lateral curvature or scoliosis frequently occurs in rickets. Demineralization of the vertebral bodies results in compression and kyphosis may be present as well as scoliosis. The pelvis may also be compressed by stress from the weight of the trunk above and pressure from the femora upon the acetabula. The resultant narrowing of the pelvic canal may seriously interfere with childbirth later on.

Dentition

Rickets retards the eruption of the deciduous teeth although the great variability in time of eruption for deciduous teeth should be noted. When the teeth do erupt they usually fail to appear in the proper sequence. Of more importance are the hypoplastic defects in the enamel, particularly of the upper incisors. The permanent teeth may also exhibit this enamel hypoplasia which occurs as horizontal grooves or depressions along the surface of the tooth.

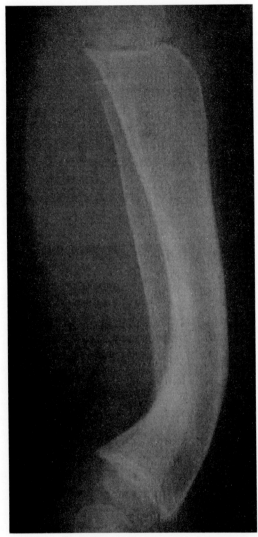

Figure 102. Clinical roentgenogram of severe rickets in an infant. The right leg shows extreme bowing deformity. (Courtesy of Dr. E. B. D. Neuhauser.)

Skeletal Dwarfism

Permanent skeletal dwarfism occurs in severe cases of rickets mainly due to the disturbance in enchondral bone growth of the long bones. Other factors producing the stunted growth or actual dwarfism are the bowing deformities of the long bones of the legs and pronounced scoliosis of the vertebral column.

Duration of Lesions

If rickets is treated before severe deformities have developed, the skeletal remodeling over a period of months or even years may obliterate any sign of the disease. However, when marked bowing of the long bones, cranial bossing, and dental hypoplasia occur, such lesions may be partially healed but never completely obliterated and remain as permanent evidence of this disorder.

Osteomalacia

Osteomalacia in adults is the sequel of rickets in infants and adolescents. As in rickets, the chief factors are deficient calcium, intestinal disease, and inadequate amount of sunlight. Of further importance to females are multiple pregnancies in rapid succession and prolonged lactation, both of which severely drain the skeletal stores of calcium and phosphorus. Indeed, in 131 cases collected from the literature, eighty-five occurred in pregnant or puerperal women, thirty-five in nonpregnant women, and eleven in men (Litzman, 1861). Very young women who become pregnant are especially prone to osteomalacia because of their own increased mineral requirements. In earlier cultures with marriage during adolescence and numerous pregnancies often the rule, one would expect to find a number of individuals exhibiting the skeletal deformities of osteomalacia.

The differences between rickets and osteomalacia are the result of differing responses of the growing and the mature skeleton to deficiencies of vitamin D or calcium and phosphorus. Osteoid formation in rickets is most active in the metaphyses and in the outer layers of diaphyseal cortex. In osteomalacia the process is diffused throughout the bones. The osteoid is formed to replace bone resorbed by normal catabolic lysis. The resultant deformities in the softened bones are most severe in

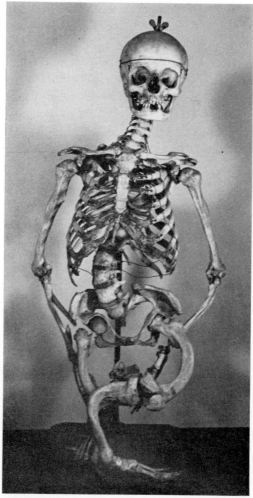

Figure 103. Osteomalacia in an adult female. There is extensive deformation of the weight-bearing bones such as the pelvis, femora, and tibiae. Scoliosis of the vertebral column is apparent. (WAM 1544)

the vertebral column, pelvis, ribs, sternum, and femora. The skull is rarely involved.

Vertebral involvement in osteomalacia is characterized by kyphosis, particularly in the lower thoracic region, and scoliosis. As in rickets, the forward protrusion of the sacral promontory and compression dorsally of the acetabula produce a severe deformation of the pelvis. Pelvic distortion is very common in puerperal osteomalacia and also very dangerous in childbearing (see Fig. 103).

The long bones are not markedly affected except in severe cases. Flaring of the metaphyses is absent because the epiphyseal ossification centers have already closed in adults, although such widening may be present in late rickets of adolescents. Bowing of the weakened long bones and collapse of the ribs may occur. Compacted trabeculae and thickened cortex will be present on the weight-bearing concave side while the convex side undergoes disuse atrophy. Fractures in these bones are common and the osteoid callus formed is excessive.

In severe osteomalacia with deformation of the vertebral column, pelvis, and long bones, the articular cartilage surfaces are placed under severe stress. The degenerative changes in these surfaces produces a traumatic osteoarthritis (Aegerter and Kirkpatrick, 1968). Complete inactivation of the person may occur due to the pain of arthritis and fracture and immobility from skeletal deformity. Such immobility leads to disuse osteopenia, a disorder in the production of osteoid, and this osteoporosis will be evident in the final stages of severe osteomalacia.

METABOLIC BONE DISEASE BIBLIOGRAPHY

Aegerter, E. and J. Kirkpatrick: *Orthopedic Diseases,* W. B. Saunders, Philadelphia, 2nd ed., 1968.

Barlow, T.: "On cases described as 'acute rickets' which are probably a combination of scurvy and rickets, the scurvy being an essential, and the rickets a variable, element," *Med.-Chir. Trans., 66:*159-184, 1883.

Basedow, H.: "Diseases of the Australian aborigines," *J. Trop. Med. Hyg., 35:*247-250, 1932.

Brues, A.: "Alkali Ridge skeletons, pathology and anomaly," Appendix B in J. O. Brew: *Archaeology of Alkali Ridge, Southeastern Utah, Papers of the Peabody Museum, 21:*327-329, 1946.

Chalmers, J. and K. C. Ho.: "Geographical variations in senile osteoporosis," *J. Bone Joint Surg., 52B:*667-675, 1970.

Davidson, S. and R. Passmore: *Human Nutrition and Dietetics,* Williams and Wilkins, Baltimore, 4th ed., 1969.

Epstein, B. S.: "The concurrence of parietal thinness with postmenopausal, senile, or idiopathic osteoporosis," *Radiology, 60:*29-35, 1953.

Foote, J. A.: *Disease of the Bones, Joints, Muscles, and Tendons,* Appleton, New York, 1927a.

————: "Evidence of rickets prior to 1650," *Amer. J. Dis. Child.,* 34:443-452, 1927b.

Frost, H. M.: *Orthopaedic Biomechanics,* Thomas, Springfield, 1973.

Gejvall, N-G.: *Westerhus: Medieval Population and Church in the Light of Skeletal Remains,* Boktryckeri, Lund, 1960.

Gray, P. H.: "A case of osteogenesis imperfecta, associated with dentinogenesis imperfecta, dating from antiquity," *Clin. Radiol.,* 21:106-108, 1969.

Henschen, F.: *The History and Geography of Diseases,* Delacorte Press, New York, 1966.

Hess, A. F.: *Scurvy Past and Present,* Lippincott, Philadelphia, 1920.

————: *Rickets Including Osteomalacia and Tetany,* Lea and Febiger, Philadelphia, 1929.

Hooton, E. A.: *Indians of Pecos Pueblo,* Yale Univ. Press, New Haven, 1930.

Hrdlička, A.: *Handbook of American Indians,* Bureau of American Ethnology Bulletin 30, pt. 1, 1907.

Jaffe, H. L.: *Metabolic, Degenerative, and Inflammatory Diseases of Bones and Joints,* Lea and Febiger, Philadelphia, 1972.

Kato, K.: "A critique of the roentgen signs of infantile scurvy," *Radiology,* 18:1096-1110, 1932.

Knaggs, R. L.: *Diseases of Bone,* William Wood and Co., New York, 1926.

Lee, T.: "Historical notes on some vitamin deficiency diseases in China," *Chinese Med. J.,* 58:314-323, 1940. (Also in D. R. Brothwell and A. T. Sandison (ed.): *Diseases in Antiquity,* Thomas, Springfield, 1967.)

Litzmann, C.: *Die formen des Beckens nebst einem Anhange über Osteomalacie,* Berlin, 1861.

Lodge, T.: "Thinning of the parietal bones in early Egyptian populations and its etiology in the light of modern observations," in D. R. Brothwell and A. T. Sandison (ed.): *Diseases in Antiquity,* Thomas, Springfield, 1967.

Luck, J. V.: *Bone and Joint Disease,* Thomas, Springfield, 1950.

Mankin, H. J.: "Rickets, osteomalacia, and renal osteodystrophy, part I," *J. Bone Joint Surg.,* 56A:101-128, 1974.

McLean, F. C. and M. R. Urist: *Bone: Fundamentals of the Physiology of Skeletal Tissue,* Univ. Chicago Press, Chicago, 1968.

Møller-Christensen, V.: *Bogen om Aebelholt Kloster,* Dansk Videnskabs Forlag, Copenhaegn, 1958.

Ruffer, M. A.: "On the diseases of the Sudan and Nubia in ancient times," *Mit. zur Geschichte der Medizin u. der Naturwissenschaften,* 13:453, 1914.

Sandison, A. T. and C. Wells: "Endocrine diseases," in D. R. Brothwell and A. T. Sandison (ed.): *Diseases in Antiquity*, Thomas, Springfield, 1967.

Saul, F.: "The human skeletal remains from Altar de Sacrificios," *Papers of the Peabody Mus. of Arch. and Ethnol.*, vol. 63, no. 2, Harvard University, 1972.

------: "Disease in the Maya area: the pre-Columbian evidence," in T. P. Culbert (ed.): *The Classic Maya Collapse*, Univ. of New Mexico Press, Albuquerque, 1973.

Sigerist, H. E.: *A History of Medicine, I. Primitive and Archaic Medicine*, Oxford Univ. Press, London, 2nd ed., 1961.

Sissons, H. A. (ed.): *Bone Metabolism*, Pitman Med. Publ., London, 1963.

Smith, E. G.: "The causation of the symmetrical thinning of the parietal bones in ancient Egyptians," *J. Anat. Physiol*. London, *41*:232-233, 1906-07.

Smith, E .G. and W. Jones: "Report on the human remains," *Archaeological Survey of Nubia*, Report of 1907-1908, Cairo, Ministry of Finance, 1910.

Smith, R. W., W. R. Eyler, and R. C. Mellinger: "On the incidence of senile osteoporosis," *Ann. Int. Med.*, *52*:773-781, 1960.

Snow, C. E.: "Indian Knoll skeletons of site Oh2, Ohio County, Kentucky," *Univ. Kentucky Reports in Anthropology*, *4*:371-554, 1948.

Weinmann, J. P. and H. Sicher: *Bone and Bones, Fundamentals of Bone Biology*, Mosby, St. Louis, 1947.

Wells, C.: "Osteogenesis imperfecta from an Anglo-Saxon burial ground at Burgh Castle, Suffolk," *Med. Hist.*, *9*:88-89, 1965.

------: "Pseudopathology," in D .R. Brothwell and A. T. Sandison (ed.): *Diseases in Antiquity*, Thomas, Springfield, 1967.

Wilson, D. C.: "Osteomalacia (late rickets) studies," *Indian J. Med. Res.*, *18*:951-978, 1931.

Wolff, J.: *Das Gesetz der Transformation der Knocken*, Berlin, 1892.

Chapter IX

ARTHRITIS

ARTHRITIS IS A general term used when the joints themselves are the major seat of rheumatic disease. There are many forms of arthritis, and clinical classifications are exceedingly lengthy and complex. However, limitations imposed by examining only the dried bone specimen greatly simplify the classification of arthritic disease in archaeological material (Morse, 1969). In a paleopathological study, a case of arthritis may be diagnosed as one of the following disorders:

1. Osteoarthritis (Degenerative Joint Disease)
2. Vertebral Osteophytosis
3. Traumatic Arthritis
4. Rheumatoid Arthritis
5. Ankylosing Spondylitis
6. Infectious Arthritis
7. Gout

Joints are divided into two fundamental types. A *synarthrosis* lacks a joint cavity and the two bones are connected by various types of connective tissue. There are several different types of synarthroses labelled according to the type of connective tissue uniting the two bones. This may be fibrous connective tissue as in syndesmoses and sutures. Hyaline cartilage temporarily unites two ossification centers in the embryonic skeleton forming a synchondrosis. This cartilage is gradually replaced by bone and growth ceases. When fibrocartilage separates two bones, this is known as an amphiarthrosis or symphysis. Examples of such joints are sutures of the skull and the pubic symphysis.

There is little if any movement between the two bones of an amphiarthrosis.

Joints in which a fibrous capsule connects two bones and defines a space between them that is lined with a synovial membrane are known as *synovial joints* or *diarthroses*. The articular surfaces of the bones are covered with cartilage and the bones are united by the fibrous capsule and ligaments. The synovial membrane produces hyaluronic acid which combines with the plasma transudate to lubricate the joint. A diarthrosis permits more or less free movement between the two bones and this type of joint is involved in the arthritic disorders. Examples of diarthroses are the hips, knees, shoulders, and elbows (see Fig. 104).

OSTEOARTHRITIS

Degenerative Joint Disease, Arthrosis, Arthritis Deformans, Hypertrophic Arthritis

Osteoarthritis is by far the most common form of arthritis. Visible bone changes are usually present after the age of fifty years in modern populations. Thus, the disease is one of adults past middle age. In most series, the incidence of osteoarthritis is slightly higher in males.

The term *osteoarthritis* implies an inflammatory process which is inaccurate in the early stages. A more accurate term for this condition is degenerative joint disease. Osteoarthritis or degenerative joint disease is a disorder of diarthrodial joints characterized by deterioration and abrasion of articular cartilage and formation of new bone at the joint surfaces. Several factors contribute to the degenerative changes in the articular cartilage. The major factor is simply biological aging with its concurrent decrease in bone vascularization and ability of the bone to repair itself (Sokoloff, 1963; 1969). Repeated minor trauma or "wear and tear" on the joints gradually produces the degenerative changes in the articular cartilage and subsequent bone reaction. Genetic factors may be important in the causation of osteoarthritis (see Stecher, 1955; Lawrence et al., 1962). Anthropometric studies indicate a correlation between physical type and degenerative

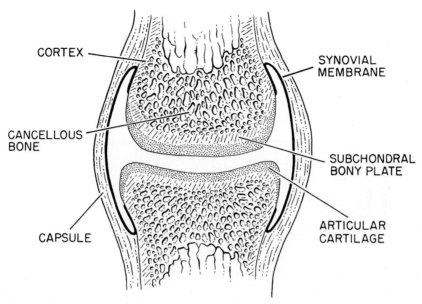

Figure 104. Diagram of a normal diarthrosis. (Modified from E. F. Traut: *Rheumatic Diseases*, St. Louis, 1952, The C. V. Mosby Co.)

joint disease (Seltzer, 1943). Victims of osteoarthritis tend to be well-muscled with a large trunk and tapering lower extremities.

Degenerative joint disease usually affects the weight-bearing joints such as the knees and hips. Other joints exposed to chronic trauma may also be involved, particularly the first metatarsophalangeal, temporomandibular, shoulder, and proximal interphalangeal joints of the fingers (Heine, 1926; Keefer and Myers, 1934). Although not a weight-bearing surface, the patella is subjected to enormous loads due to leverage when the knee is flexed in the squatting position. Thus, eroded areas are often seen on the central portions of the facets. Osteoarthritis is rarely confined to a single joint. On the other hand, generalized involvement is not as frequent as in rheumatoid arthritis.

Pathogenesis and Gross Morphology

Articular cartilage degeneration may occur by changes in the rate of production and chemical composition of the cartilage matrix (Mankin, 1968) or by changes in the structure of the subchondral bone altering the articular contours (Johnson, 1959). In either case, many tiny, irregular pits may be seen on the surface of the articular cartilage which gradually develop into irregular crevices and fissures. These zones of injury occupy principally the central portions of the joint surface where the

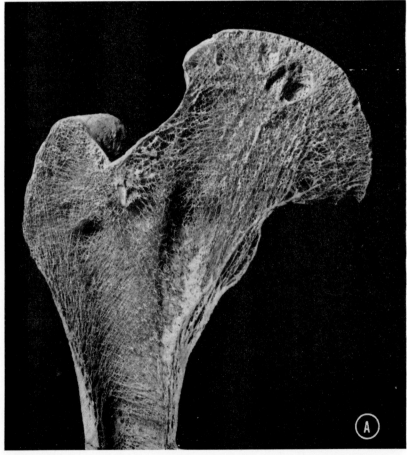

Figure 105. Osteoarthritis affecting the femur. (A.) Sectioned left femur showing several important features of osteoarthritis. In the femoral head are several cystic defects surrounded by sclerosis. The femoral neck is widened by subperiosteal new bone. The original cortical outline is still visible beneath the new bone.

stress of weight-bearing is greatest. The surface gradually becomes more irregular and the cartilage thickness decreases. Simultaneous changes occur in the subchondral bone with "cystic" areas of rarefaction surrounded by rims of sclerotic new bone. The sclerosis is most marked in the weight-bearing areas. In some cases, particularly the femoral head, the cystic areas may

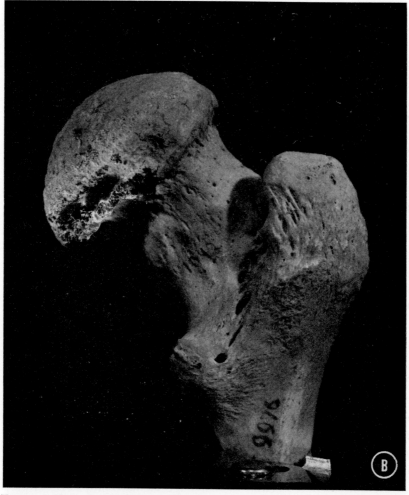

Figure 105B. Externally, the femoral head appears flattened or compressed. The normally smooth surface is roughened by irregular growth of bone. A large fissure is evident in the head anteriorly. (WAM 9976)

coalesce to form large bone defects greater than 2 cm in diameter.

In moderately advanced cases new bone formation takes place at the margins of the articular cartilage as well as in the subchondral spongiosa (Harrison et al., 1953). The marginal lipping or osteophytes are produced by normal enchondral bone replacement of the cartilage spurs at the junction of the perichondrium and periosteum (Fig. 106). The direction of the osteophyte is

Figure 106. Sectioned femur with a large cystic defect immediately below the articular surface. The subchondral sclerosis and thickening of the trabeculae are quite apparent. Note the marginal lipping produced by new bone formed at the edges of the articular cartilage. The articular surface is pitted, particularly in the central area of greatest weight-bearing. (WAM 7641)

governed by the lines of mechanical force exerted on the area of growth which generally corresponds to the contour of the joint surface. For example, in the distal interphalangeal joints of the fingers the marginal projections at the base of the phalanges extend along the dorsal surface of the joint capsule and are known as Heberden's nodes. In the hip and knee the osteophytic growths may interfere with motion and cause further reactive bone changes.

In more advanced cases the cartilage surface completely erodes and the underlying bone is exposed. With continued friction, the articular surface of the sclerotic subchondral bone becomes

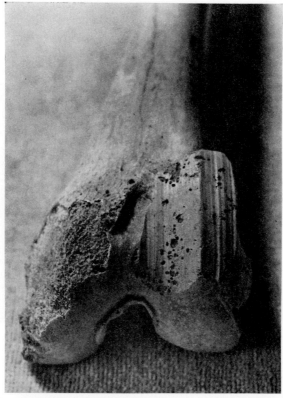

Figure 107. Distal end of left femur from an ancient Peruvian. The lateral condyle is eburnated and has an ivory-like appearance. The shallow grooves produced by bone-on-bone friction are particularly evident. (USNM 265,331)

eburnated and polished. Shallow grooves may appear on the eburnated surface and irregular bony spurs arise from the periphery of the eburnated bone (see Fig. 107 & 108). With progressive marginal hyperplasia and enlargement, the osteophytes may become true exostoses having cancellous bone continuous with the marrow cavity.

Although the eburnated bone is dense and hard, it is a very thin layer and the bone beneath is weakened by the presence of many cystic defects. Such cysts may be formed by the hydraulic action of synovial fluid forced by pressure

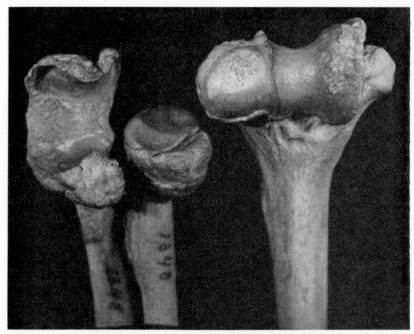

Figure 108. Left humerus, radius, and ulna with osteoarthritis. Marginal hyperplasia is particularly evident in the ulna and radius. The capitulum of the humerus has a well-defined area of eburnated bone where it articulates with the radius. (WAM 1348)

through the degenerated articular cartilage into the cancellous tissue (Landells, 1953). Others believe that the cysts result from localized hemorrhages which become encapsulated (Pommer, 1913). In any case, the weakened bone structure cannot withstand the weight-bearing pressure and gross deformity of the bone contour occurs. Further alteration in the stress trajectories is produced by the deformation creating a vicious cycle of degenerative changes. This is well illustrated by the increased lipping of the acetabulum and mushrooming of the femoral head (Fig. 109). Despite the hypertrophic bone proliferation, ankylosis is very rare.

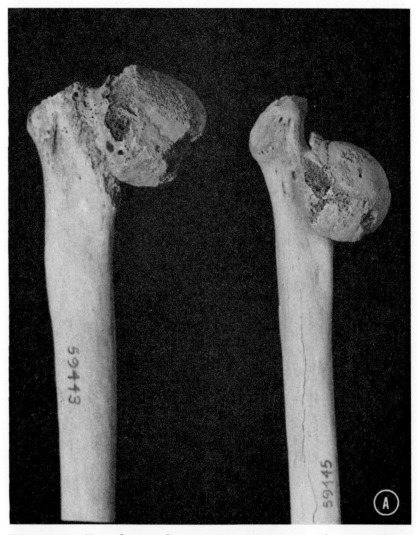

Figure 109. Two femora from ancient Peruvians with osteoarthritis. (A.) Externally, the superior surfaces of the femoral heads are severely eburnated. Both heads have been compressed with subsequent loss of the necks producing a coxa vara.

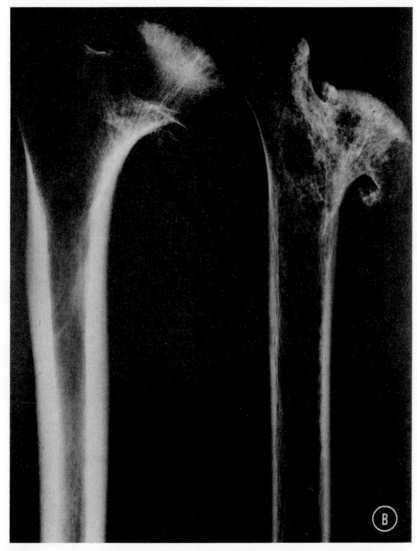

Figure 109B. Roentgenogram showing areas of sclerosis. The right femur has a mushroom head from compression and marginal hyperplasia. (PM 59443; PM 59445)

VERTEBRAL OSTEOPHYTOSIS

Spondylosis Deformans, Hypertrophic Spondylitis

Degenerative joint disease of the spine involves two separate intervertebral articular systems. The apophyseal joints or facets of the vertebrae are true synovial joints and therefore the degenerative changes in these joints are included under osteoarthritis. The intervertebral joints are secondary cartilaginous joints without a synovial membrane. For this reason degenerative changes of the intervertebral joints are defined as a separate entity—vertebral osteophytosis (Collins, 1949). It should be noted that other workers do not regard vertebral osteophytosis as distinct from osteoarthritis on anatomic and pathologic grounds (Jaffe, 1972).

Vertebral osteophytosis is a very common condition being present in almost all persons over sixty years old. Degenerative changes may be present as early as thirty years of age. Modern series show a higher incidence among males, but this is possibly due to differences in amount of physical work. Vertebral osteophytosis may affect any portion of the spine. Marked changes occur with greatest frequency in the lower thoracic and lumbar regions reflecting the weight-bearing function of this area.

Pathogenesis and Gross Morphology

The initial lesion of vertebral osteophytosis occurs in the intervertebral disk so we shall briefly review its anatomy. The interior of the disk is known as the *nucleus pulposus*. In young persons this portion is semigelatinous tissue with considerable amounts of amorphous mucoprotein. In the adult the nucleus pulposus is very much more collagenous and reduced in fluid. The peripheral portion of the intervertebral disk is called the *annulus fibrosus* and consists of collagenous fibrous tissue. The vertebral bodies are firmly united along their marginal ridge by fibers from the intervertebral disk and also by the anterior and posterior longitudinal ligaments. The fibers of the annulus fibrosus intermesh with the fibers of the anterior and posterior ligaments where these ligaments attach to the margins of the vertebral disks.

Following degenerative tearing of some of the fibers of the annulus from the marginal ridge, the nucleus pulposus compresses and protrudes against the anterior longitudinal ligament. The resultant pressure provokes subperiosteal new bone formation where the ligament is firmly attached at the anterior margin of the vertebrae (see Fig. 110).

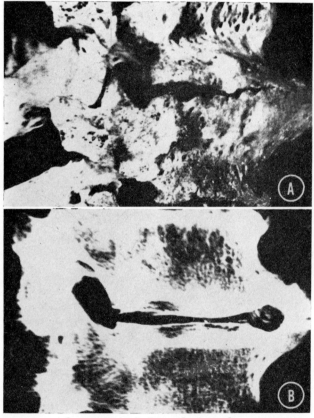

Figure 110. Osteophytosis. (A.) Lateral view showing origin of osteophytes from midvertebral level. The new bone formation is produced by the flowing anteriorly of disk material raising the periosteum. The osteophytes extend horizontally forming a "parrot beak" projection. (B.) Lateral roentgenogram of the vertebrae pictured above. The osteophytes appear anterior in this view, but actually lie anterolateral. A thick, inelastic anterior ligament prevents a true anterior formation of osteophytes. The cancellous bone of the osteophytes is continuous with that of the adjacent vertebral bodies. (Courtesy of Dr. T. Dale Stewart from Tobin and Stewart, 1953.)

The marginal osteophytes are thus produced by the periosteum. They are usually located anterolaterally because the thick, inelastic anterior ligament prevents true anterior formation. Osteophytes rarely form on the posterior surface because the intervertebral disk material is compressed anteriorly. Marginal osteophytes of adjacent vertebral bodies may become so large that they impinge on each other. The space between the two "kissing" osteophytes usually contains disk material. Following complete destruction of this material, the two osteophytes often fuse producing bony anklyosis. Marginal osteophytes are usually multiple and predominantly unilateral. Although ankylosis may often occur, the "flowing" ossification of the anterior and posterior ligaments as seen in the "bamboo" spine of ankylosing spondylitis is not found in vertebral osteophytosis (see Fig. 111).

TRAUMATIC ARTHRITIS

Traumatic arthritis is a secondary form of osteoarthritis initiated by trauma. The degenerative changes in the articular tissues are secondary to alterations in the joint anatomy produced by injury. The injury causing such changes is usually a sprain or fracture into a joint.

The joints most often affected are the joints of the lower extremity: the hip, knee, and ankle. Osteoarthritis affects several joints, often symmetrically, whereas traumatic arthritis usually involves only one joint. While osteoarthritis is a disease of the elderly, traumatic arthritis may affect persons of any age. Furthermore, severe joint injury may produce bony ankylosis which is very rare in osteoarthritis (see Fig. 112).

RHEUMATOID ARTHRITIS

Etiology

Rheumatoid arthritis is a systemic disorder which may involve all of the connective tissues, particularly the joints. The etiology of this widespread condition is not known. Almost all patients with rheumatoid arthritis are found to have a specific "rheumatoid factor" antibody in their serum. The antibody is

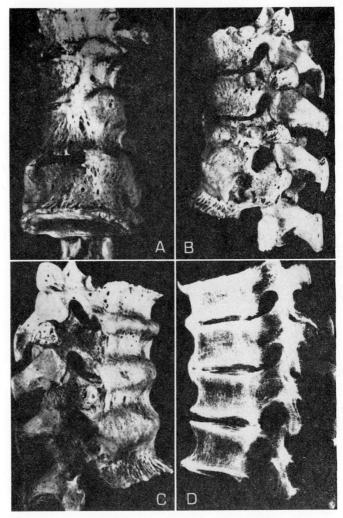

Figure 111. Osteophytosis of the lower thoracic vertebrae. (A.) Anterior portion of the spine showing predominantly unilateral distribution and anterolateral position of the osteophytes. Fusion of adjacent osteophytes has occurred. (B.) Lateral view of another specimen showing singly fused osteophytic outgrowths. The apophyseal joints are not involved. (C.) Opposite view of same specimen showing fusion of all the osteophytes. The horizontal bulging is the result of the degeneration and flowing anteriorly of disk material. (D.) Lateral roentgenogram showing some narrowing of the intervertebral spaces. (Courtesy of Dr. T. Dale Stewart from Tobin and Stewart, 1953.)

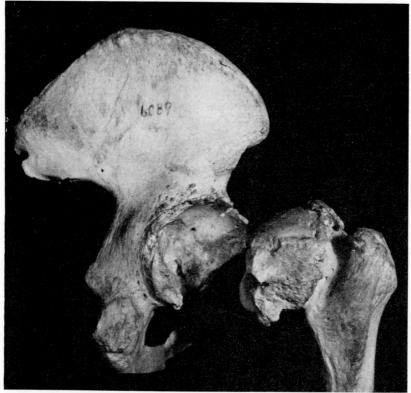

Figure 112. Traumatic arthritis of the right hip following severe injury fourteen years previously. (WAM 6089)

produced by the sensitized plasma cells in the inflamed synovial membrane and in lymph nodes near the affected joints. The antigen responsibile for the antibody formation has not been identified. It is speculated that rheumatoid arthritis develops by an abnormal response of the host's connective tissue cells, particularly the lining cells of the synovial membrane, to a component of bacterial or viral cell walls. This same bacterial or viral component alters the host's own gamma globulin resulting in an autoimmune reaction by production of the rheumatoid factor (Hamerman, 1966; 1968). Other factors involved in producing rheumatoid arthritis such as genetic predisposition, endocrine disorder, personality traits, and metabolic errors have been excellently reviewed elsewhere (Robinson, 1966).

Incidence

In most series the incidence of established rheumatoid arthritis is about three times greater in females, perhaps due to a genetic or endocrine factor. Although the disease may occur during adolescence, it mainly affects adults between the ages of thirty and forty-five years. Since rheumatoid arthritis is chronic, its incidence in a given population rises with increasing age of the subjects. Rheumatoid arthritis is more common in temperate regions than in warmer climates. The occurrence of the disease within a given population is not correlated with any particular occupation or physical activity (Kellgren et al., 1953).

Distribution

Rheumatoid arthritis is a systemic disorder and therefore affects many joints in a symmetrical manner. In the hand, the arthritic changes are usually located in the proximal interphalangeal joints, metacarpophalangeal joints, and less frequently the intercarpal joints. Other common sites of rheumatoid arthritis are the wrist, elbow, shoulder, temporomandibular joint, tarsal joints, and knees. The apophyseal joints of the vertebral column, particularly the cervical region, are often affected. The hip is usually unaffected except in advanced cases (Garrod, 1890).

Pathogenesis

The initial lesion of rheumatoid arthritis is inflammation of the synovial membrane with synovial hypertrophy, edema, hyperemia, and large numbers of plasma cells, lymphocytes, and macrophages. Fibrin deposition and hyperplastic overgrowth of the synovia forms a layer of granulation tissue covering the articular surfaces. This layer of granulation is known as a *pannus* and causes destruction of the underlying articular cartilage. This same sequence of events takes place beneath the articular plate by the inflammatory reaction in the subchondral bone. Eventually the joint space becomes filled with the thickened synovial tissue and granulation tissue. The joint surfaces are completely destroyed and distorted.

When the inflammatory process subsides, organization of the granulation tissue occurs with a proliferation of fibrous tissue. As part of this reparative process, fibrous adhesions join both epiphyses producing fibrous ankylosis. This often proceeds to bony ankylosis by calcification and then ossification of these adhesions. The distorted joint is now immobile and functionless.

Gross Morphology and Roentgenographic Appearance

In the early stages of rheumatoid arthritis, the major roentgenographic change is osteoporosis in the bones of the affected joints. Both the factors of disuse atrophy and inflammatory hyperemia are responsible for this osteoporosis. The marked trabecular thinning produces a "ground glass" appearance in the roentgenogram. The articular cortex is also thinned with irregular contours.

As the disease progresses, inflammation of the subchondral bone produces further destruction of the cortex and irregular, cyst-like areas of destruction in the cancellous tissue. Subluxation and distortion of the joint often occurs due to muscle contractures. In the final stages the fibrous adhesions are ossified producing bony ankylosis. The bones are usually ankylosed in a flexed position.

Differential Diagnosis

The differential diagnosis of rheumatoid arthritis and osteoarthritis in archaeological material is generally difficult because many cases of rheumatoid arthritis may be complicated by degenerative joint changes. Rheumatoid arthritis usually affects a younger age group, although many specimens will undoubtedly be over forty-five years of age due to the chronicity of the disease. Rheumatoid arthritis commonly involves many joints in a symmetrical manner. Osteoarthritis affects only a few joints and with less symmetry. The distal interphalangeal joints of the hands are less commonly involved by rheumatoid arthritis while this is the location of the Heberden's nodes frequently found in

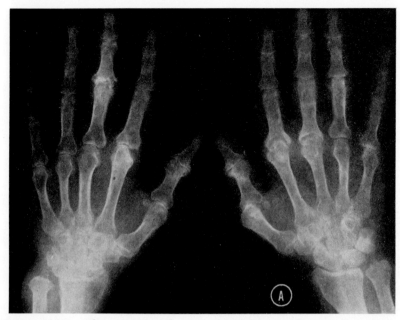

Figure 113A. Rheumatoid arthritis. The metacarpophalangeal joints are particularly involved and several are subluxed. Several carpal joints and the left wrist are also affected. (Courtesy of Dr. Charles Short.)

osteoarthritis. Rheumatoid arthritis is often described as atrophic arthritis because of the marked osteoporosis in the epiphyses. In contrast, osteoarthritis is referred to as hypertrophic arthritis because of the subchondral sclerosis and marginal spurring. Subchondral cysts are present in both diseases. Bony ankylosis is very common in rheumatoid arthritis but distinctly rare in osteoarthritis. Eburnation rarely occurs in the immobilized rheumatoid joint but is much more common in osteoarthritis.

ANKYLOSING SPONDYLITIS

Rheumatoid Spondylitis, Bamboo Spine, Poker Spine, Marie-Strümpell Disease

Ankylosing spondylitis is a chronic and usually progressive disease principally affecting the vertebral column. It frequently begins in the sacro-iliac joints and spreads upward to involve

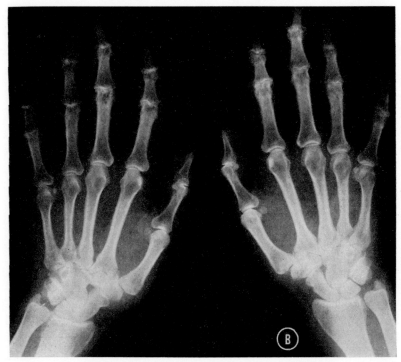

Figure 113B. Osteoarthritis. The interphalangeal joints are prominently affected while the metacarpophalangeal joints appear normal. (Courtesy of Dr. Charles Short.)

the synovial joints of the vertebrae (apophyseal and costovertebral joints) and ossification beneath and within the spinal ligaments. The pubic symphysis is also a frequent site of ligamentous ossification. Peripheral joints such as the hips, knees, and shoulders may become affected in about 20 percent of all cases. The microscopic pathologic changes in the vertebral and peripheral joints are identical to those seen in rheumatoid arthritis.

Ankylosing spondylitis has been classified as a distinct disease entity separate from rheumatoid arthritis for various reasons. The relative sex incidence of ankylosing spondylitis greatly favors males by a ratio of about 7 or 9 to 1. Rheumatoid arthritis in contrast is more common among females. Ankylosing spondylitis has an inheritance pattern distinctly different from rheumatoid arthritis (DeBlecourt et al., 1961). Various differences in clinical characteristics and soft tissue pathology also separate the two diseases.

Incidence

Ankylosing spondylitis chiefly occurs between the ages of sixteen and forty years (Boland and Present, 1945). As mentioned above, males comprise almost 90 percent of most series. In the general population today, about one person in 2000 is afflicted with the disease (West, 1949)

Pathogenesis and Gross Morphology

The initial lesion of ankylosing spondylitis is often a rheumatoid inflammation of the sacro-iliac joints followed by bony ankylosis, usually bilateral. A similar process causes erosion and irregularities in the apophyseal and costovertebral joints. Ankylosis of these joints develops, often involving the entire vertebral column (see Fig. 114).

Paravertebral ossification starts in the annulus fibrosus and neighboring connective tissue. In advanced cases ossification of the longitudinal ligaments and ligamenta flava may occur. The ossification is probably caused by an inflammation of the fibrous tissue similar to that which occurs intraarticularly. If viewed laterally, the vertebrae often appear squared due to loss of normal of anterior concavity. In advanced cases, the ossified ligaments appear as a smooth, undulating surface of bone connecting the vertebrae and creating the appearance of a "bamboo" spine. The ossification occurring in the outer fibers of the annulus fibrosus may spread to the nucleus pulposus so that the entire intervertebral disk is composed of spongy bone. When the lumbar region of the spine is affected, there is loss of normal lordosis and the individual develops a "poker back."

Differential Diagnosis

A differential diagnosis of ankylosing spondylitis and vertebral osteophytosis in archeological material is often possible. Advanced cases of the latter condition may appear very similar to ankylosing spondylitis because of advanced osteophytic formation causing

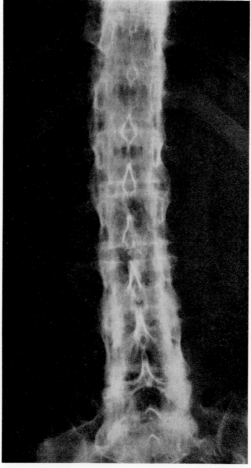

Figure 114. Clinical roentgenogram of ankylosing spondylitis showing extensive ossification and fusion of the vertebrae. (Courtesy of Dr. Charles Short.)

ankylosis of several vertebrae. In general, the osteophytes develop at the edges of the vertebral bodies and tend to extend horizontally to form the characteristic "parrot beak" bony projections. When fused, these osteophytes are thick and situated several millimeters out from the vertebral bodies. In ankylosing spondylitis the bony bridging is situated very close to the

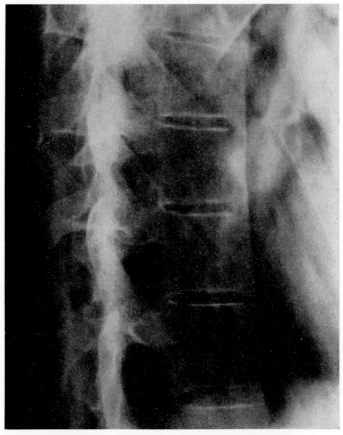

Figure 115. Ankylosing spondylitis causing squaring of the vertebrae. (Courtesy of Dr. Charles Short.)

vertebral bodies. The syndesmophytes between the vertebrae often form a solid, smooth sheet of bone in contrast to the roughened bony bridges of vertebral osteophytosis. The vertebral osteophytes are also fewer in number and often unilateral. Furthermore, involvement of the sacro-iliac joint is rare in vertebral osteophytosis while practically pathognomonic of ankylosing spondylitis. Rheumatoid changes may be present in the latter condition.

INFECTIOUS ARTHRITIS

Inflammation of the joints occurs in several infectious diseases, notably pyogenic osteomyelitis, tuberculosis, and syphilis. The joint changes produced by these diseases have been fully described in previous chapters but will be briefly reviewed here.

Almost 15 percent of all cases of pyogenic osteomyelitis are complicated by pyogenic arthritis. They comprise the majority of all cases of infectious arthritis. The disease mainly involves the knee or hip in children below the age of sixteen years. Multiple joint involvement occurs very rarely. The bone changes reflect the chronicity of the infection and may resemble osteoarthritis. However, bony ankylosis often occurs in pyogenic osteomyelitis while this is rare in osteoarthritis. Furthermore, sinus tracts in the epiphyses and diaphyses reveal the pyogenic nature of the arthritis.

Tuberculous arthritis resembles pyogenic arthritis in several respects. The hip and knee are the most frequent sites of involvement and children below the age of sixteen years are predilected. Only one joint is affected in most cases. At times the tubercle bacillus stimulates a strong leukocyte response with production of an exudate similar in all respects to the pus produced by pyogenic infections (Collins, 1949). The sinus tracts created in such cases will be indistinguishable from the bone changes of pyogenic osteomyelitis. Tuberculous arthritis is characterized by identical lesions in the two opposing bones and "kissing sequestra," particularly in the knee. Furthermore, many cases of tuberculous arthritis have concomitant involvement of the vertebral column.

Syphilitic arthritis is uncommon and occurs in three forms which in order of frequency are: Charcot's joint in tabes dorsalis, Clutton's joint in late congenital syphilis, and gummatous osteitis eroding the articular surfaces. Together they comprise about 12 percent of all cases of osseous syphilis. Refer to Chapter IV for a complete discussion of syphilitic joint disease.

GOUT

Primary gout is an inborn disorder in the metabolism of purines resulting in an abnormally high production of uric acid. The genetic determinants of the hyperuricemia in gout are multifactorial. Environmental factors such as diet are also involved in producing the high levels of uric acid. Hyperuricemia produced in association with other conditions such as chronic renal insufficiency is referred to as secondary gout.

Primary gout is a disease of the adult male. Females are rarely afflicted. The first joint affected is usually the metatarsophalangeal joint of the big toe. Other common sites are the joints of the lower extremity and hand.

The essential pathology in gout is the deposition of uric acid crystals in the cartilages and epiphyseal bone. These crystals provoke an inflammatory response, and the released leukocytic enzymes destroy the cartilage and bone. Large deposits of the crystals together with the foreign body reaction of fibrosis form nodules called *tophi* in the soft tissues and external surface of the bone.

The degenerative changes produced by gout are very similar to osteoarthritis. A differential diagnosis of the two disorders in archaeological material would be presumptuous. When mummified soft tissues are also present, a possible diagnosis may be made as discussed in the next section.

PALEOPATHOLOGY OF ARTHRITIC CONDITIONS

Osteoarthritis

Osteoarthritis or degenerative joint disease is one of the oldest and most common pathological lesions found in animal and human remains. Such lesions have been reported in dinosaurs of the Cretaceous and cave bears of the Pleistocene (Moodie, 1923). Osteoarthritis may be found today in many mammalian and reptilian species regardless of their phylogenetic position (Fox, 1923; 1939). The condition is present in human skeletal remains from every region and period. Neanderthal individuals,

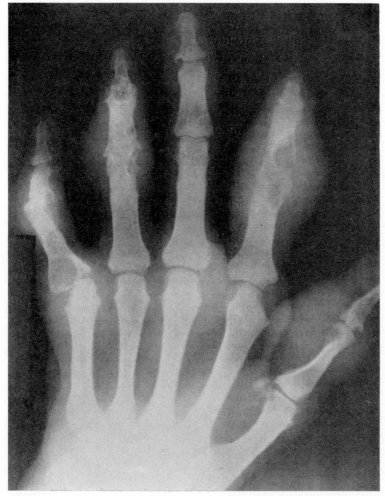

Figure 116. Gout affecting the hand. Many of the joints are affected by the lytic lesions produced by uric acid crystals. (Courtesy of Dr. Charles Short.)

particularly from La Chapell-aux-Saints, exhibit osteoarthritic lesions in the hip joint, temporomandibular joint and phalanges (Vallois, 1949; Straus and Cave, 1957). Gross osteoarthritic lesions have also been reported in upper Paleolithic and Neolithic remains (Vallois, 1949; Ackerknecht, 1953). Ancient Egyptians,

Peruvians, pre-Columbian American Indians and Anglo-Saxons, to name but a few, were afflicted with degenerative joint disease (Hooton, 1930; Hrdlička, 1914; Ruffer, 1918; Ruffer and Rietti, 1912; Smith and Dawson, 1924; Wells, 1962).

Osteophytosis

Osteophytosis or degenerative arthritis of the vertebral column is also a widespread pathological condition, often being included with osteoarthritis in descriptions of skeletal material. Osteophytic growths and complete fusion of two vertebrae have been reported in caudal vertebrae from six different specimens of *Diplodocus longus*, a large Jurassic dinosaur. Many of these cases were probably the result of a single severe injury to the particular region affected (Blumberg and Sokoloff, 1961). Vertebral osteophytosis has also been reported in a Miocene crocodile and Eocene and Pleistocene mammals (Moodie, 1923). The condition almost always accompanies the lesions of osteoarthritis found in human skeletal remains as shown by the references previously cited. Indeed, when cases of vertebral osteophytosis are included under osteoarthritis, they make up the great majority of lesions (see also Roche, 1957).

Racial variations in the region of the vertebral column most affected by osteophytosis have been reported. Early Eskimo, various American Indian, ancient Egyptian, and Iron Age British skeletal series have the vertebral lipping confined mainly to the lumbar region (Stewart, 1947; 1966; Chapman, 1964; Bourke, 1971). The findings in these early populations differ markedly from the high incidence of cervical osteophytosis found in a modern autopsy room series of skeletons (Stewart, 1947). A higher mean age in this latter series may account for some of the difference because significant lumbar and thoracic osteophytosis was also present. However, cultural and environmental factors should also be considered (see Wells, 1964).

When such variation in osteophytosis is recognized and examined within the specific population, it is possible to obtain a rough age estimate from the degree of osteophytosis in older age groups where other criteria for age estimation are not available (Stewart, 1957). Vertebral osteophytosis may be graded according to an arbitrary scale from 0 to 4 as described below:

Degree 0   No lipping present.
Degree 1   Slight lipping at the inferior and superior margins of the centra.
Degree 2   More pronounced lipping at the margins.
Degree 3   Extensive lipping often resembling a mushroom-like eversion with bony spurs.
Degree 4   Actual ankylosis or bony union between two or more vertebrae.

Rheumatoid Arthritis

Rheumatoid arthritis has received very little attention in paleopathology, perhaps due to its infrequency in archaeological material. Indeed, some workers find no convincing evidence of the disease prior to Sydenham's clinical description in 1676 (Short, 1974). The paucity of reported cases of rheumatoid arthritis is perhaps due to its inclusion under cases of osteoarthritis. This is understandable because severe cases of rheumatoid arthritis have concomitant bone changes of osteoarthritis resulting from the articular destruction and mechanical changes in the joint articulation. Many of the less severe cases involve only the soft tissues of the joint leaving minimal alterations in the bones. Many cases of rheumatoid arthritis involve only the phalanges and metacarpals which are frequently not preserved in archaeological material.

The earliest reported case of possible rheumatoid arthritis comes from the 3rd Dynasty of Egypt around 2700 B.C. (Ruffer, 1918). The distal and middle phalanges of one finger are fused in a flexed position. The absence of other remains makes a definite diagnosis impossible although definite involvement of the spine makes this a possible case of ankylosing spondylitis with peripheral joint disease. Another possible case from the same period is an ankylosed knee with patella fused to the femur.

A frequently cited case of possible rheumatoid arthritis comes from the 5th Dynasty of Egypt around 2400 B.C. (May, 1897; Karsh and McCarthy, 1960). The mummified hands of a fifty to sixty-year-old male are deformed by joint swelling, ulnar deviation of the fingers, hyperextension at the metacarpophalangeal joints, and flexion at the interphalangeal joints. The "clawing" presented in this specimen is not entirely typical of rheumatoid arthritis. A more recent Egyptian specimen around 330-30 B.C. consists of ankylosis and deformity of the wrist and intercarpal joints.

Only one case of rheumatoid arthritis has been described in American Indian skeletal material (Morse, 1969). The individual is an elderly male probably about seventy-five years of age from the Mississippian Culture of Illinois. During excavation it was noted that the knees were in a flexed position, differing from the extended position of the other 250 burials from the same mound. Almost all of the joints are affected by slight lipping and the flexure of the legs may have been the result of rheumatoid arthritis affecting the soft tissues of the knee joints.

Rheumatoid arthritis may be more prevalent in archaeological material than these few cases would indicate. Avoiding the confusing terminology of arthritic conditions found in the earlier literature and careful examination of the smaller joints will probably result in many new cases of rheumatoid arthritis coming to light. Another possibility is that rheumatoid arthritis is a recent variant of ankylosing spondylitis (Short, 1974).

Ankylosing Spondylitis

In contrast to rheumatoid arthritis, many cases of the related rheumatic disorder, ankylosing spondylitis, have been reported. This is probably due to the easily recognizable and often striking appearance of the disease rather than reflecting a higher prevalence in ancient times. However, it is worth mentioning that population studies of modern American Indian populations have revealed a high prevalence of ankylosing spondylitis in certain groups such as the Haida Indians. About 10 percent of all male Haida Indians over the age of twenty-five years have radiological evidence of ankylosing spondylitis (Gofton et al., 1968).

Several specimens characteristic of ankylosing spondylitis have been found in Egyptian skeletal material (Bourke, 1967).

One such spine is completely fused from the second cervical vertebra to the fifth lumbar (the atlas and sacrum are missing). Other diagnostic signs include squaring of the vertebral bodies and ossification of the spinal ligaments and apophyseal joints. Earlier reported cases of ankylosing spondylitis from the same area were reexamined and found to be simply severe cases of osteophytosis (Zorab, 1961). Possible specimens have also been noted in Neolithic France and Denmark and Bronze Age and Saxon Britain (Snorrason, 1942; Zorab, 1961; Wells, 1962).

Several cases of ankylosing spondylitis come from the New World. The earliest case comes from Mayan skeletal remains in Guatemala around 500 A.D. (Saul, 1972). The fragmentary skeleton of an elderly male shows ankylosis of the right sacro-iliac joint and varying degrees of degeneration and new bone formation in various other joint surfaces. Three individuals from Hopewell and Mississippian sites in Illinois have characteristic lesions of the disease (Morse, 1969). At the Harvard Peabody Museum are the fragmentary remains of an adult male from a Mississippian site in Ohio. Besides bilateral sacro-iliac fusion, vertebrae from the three different regions are solidly fused at the apophyseal joints without any sign of osteophytic lipping or ligamentous ossification.

Another previously unreported case is at the Smithsonian Institution. The remains are those of an adult male Pueblo Indian collected by the Hemenway Expedition of 1887-88 near Zuni, New Mexico. Almost the entire vertebral column is fused although postmortem breakage makes the exact extent uncertain. The first three cervical vertebrae are ankylosed to the skull. There are four sections of vertebrae which are still fused involving twenty-two out of a possible twenty-four vertebrae. The areas of fusion are the apophyseal joints and the vertebral bodies (see Fig. 118).

A recently excavated Pueblo Indian skeleton from northern New Mexico has evidence of severe ankylosing spondylitis involving many peripheral joints as well as the spine. The individual is a thirty-six- to fifty-five-year-old male from the Historic Period around 1550-1672 A.D. (Turner, personal communication,

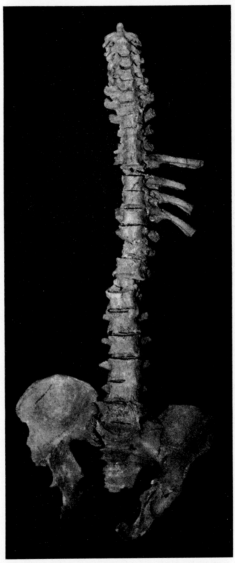

Figure 117. Ankylosing spondylitis in an adult male Indian from the Hopewell Culture of Illinois. There is fusion of both sacro-iliac joints, several costovertebral joints, and many apophyseal joints. (Courtesy of Dr. Dan Morse, 1969.)

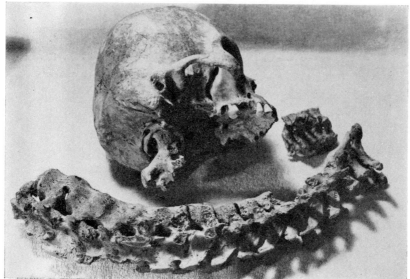

Figure 118. Anklyosing spondylitis in an adult male Pueblo Indian. (USNM 239,208)

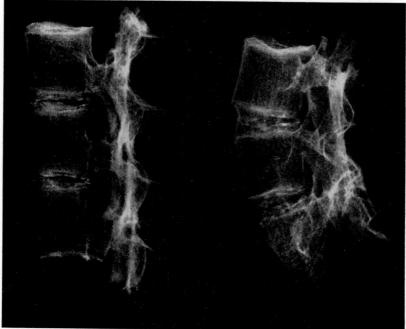

Figure 119. Fused sections of thoracic and lumbar vertebrae from an adult male Pueblo Indian with severe ankylosing spondylitis. (Courtesy of Dr. Christy Turner.)

308 Paleopathological Diagnosis and Interpretation

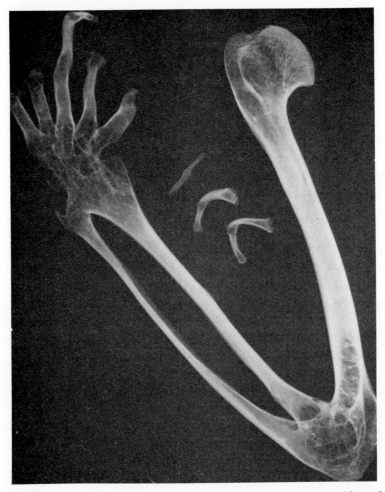

Figure 120. Severe ankylosing spondylitis in an adult male Pueblo Indian affecting the peripheral joints in a manner similar to rheumatoid arthritis. The elbow is ankylosed in acute flexion and all joints of the hand are fused. (Courtesy of Dr. Christy Turner.)

1975). The lumbar and thoracic regions of the spine are completely ankylosed at the apophyseal joints and spongy bone is evident in the intervertebral spaces (see Fig. 119). Ankylosis has occurred at the hips, knees, elbows, wrists, intercarpal joints, and fingers. The presence of severe bone atrophy and extensive bone changes following ankylosis indicate a childhood onset of the disease in this adult male Indian. It should be noted that very similar bone changes may occur in juvenile rheumatoid arthritis.

Two possible cases of ankylosing spondylitis have been reported from an early historic Huron ossuary in Ontario (Kidd, 1954). In summary, at least thirty possible cases of ankylosing spondylitis have been reported in the Old World, particularly in Egyptian skeletal material. In addition to three cases described here for the first time, about six cases of ankylosing spondylitis have been reported in skeletal material from the New World.

Gout

The highly characteristic clinical features of gout have been recognized since earliest times. As early as the 5th century B.C., gout was described as podagra, cheiagra, or gonagra by Hippocrates depending on whether the big toe, wrist, or knee was involved. Several interesting articles discuss the history of gout, particularly as it affected famous individuals of recent centuries (Bywaters, 1962; Gaidner, 1849; Rodnan, 1961; Rodnan and Benedek, 1963). Benjamin Franklin, King George IV and many others of the English nobility suffered from gout.

The unusual frequency of gout among the Georgian English (18th and 19th centuries) was blamed on their diet of alcohol and rich foods (see Garrod, 1859). Recent evidence suggests that a high level of lead in the liquor, particularly port wine, created the "epidemic" of gout. The distillation of the alcohol in leaden vessels and addition of lead intentionally to soften the taste and retard fermentation resulted in unusally high concentrations of lead. Chemical analysis of four old wines (1770-1805) revealed levels of lead comparable to that found in known cases of gout following "moonshine" lead poisoning (Ball, 1971). The chronic exposure to small amounts of lead produces renal nephritis interfering with the excretion of uric acid and secondary gout develops.

As mentioned earlier, the soft tissues must be present for a definite diagnosis of gout in archaeological material. An Egyptian mummy from the Christian period exhibits lesions characteristic of gout. Large tophi were found on the metatarsals of the big toes. The metatarsophalangeal joints were also affected. Similar tophi were present on the tarsal bones, in the knee joints and distal ends of the tibiae and fibulae. Chemical analysis of the chalk-like tophi revealed a high uric acid content (Smith and Dawson, 1924).

A diagnosis of gout is sometimes possible in the dried bone specimen if certain criteria are met. When the urates of the gouty tophi are deposited on the external surface of the bone within the joint space or extrarticularly, they produce sharply circumscribed defects in the affected regions of the bone. As discussed earlier the major joints affected are the metatarsophalangeal joint of the big toe, other joints of the lower extremities, and the hands. One very interesting case which meets these criteria comes from Gloucestershire, England during the Romano-British period or about 150 A.D. (Wells, 1973). The well-preserved remains are of a male aged forty-five to sixty-five years and exhibit small, sharply defined lesions in the bones mentioned above as well as both elbow joints. Of the twenty-two surviving bones of the hand, at least fifteen are affected, particularly at their articular surfaces. Although the urate tophi were not recovered at excavation, the specimen most likely represents a case of severe gouty arthritis (see Fig. 121).

It is interesting to note that the affected individual was one of two individuals buried in handsome stone sarcophagi while the other 266 burials had only wooden coffins or shrouds (Wells, 1973). Persistent overeating of foods such as meat which are rich in nucleoproteins and therefore produce large amounts of purines, contributes to the onset of gout. Furthermore, it has been determined that very high levels of lead are present in wine manufactured according to the directions of the 1st century A.D. Roman, Pliny (Kobert, 1909). While only speculative, overindulgence in rich foods and leaded wines could have been important factors producing such a severe case of gout.

It might prove interesting to closely examine ancient Roman and Greek skeletal remains for evidence of gout, because both

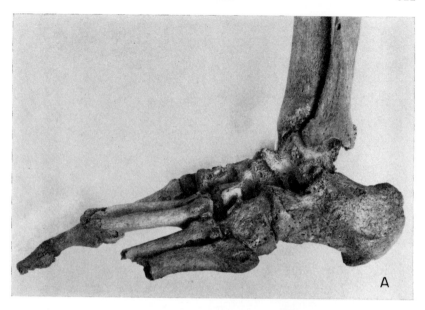

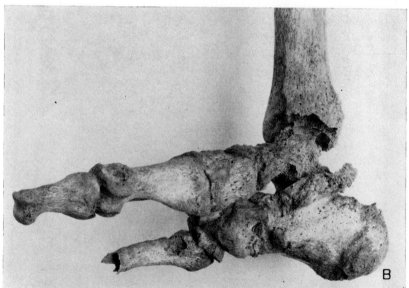

Figure 121. Gout in an adult male from the Romano-British period. (A.) Left foot with several well-defined lytic lesions in metatarsals, tarsals, and fibula. (B.) Right foot with similar lesions, particularly in the joints. (Courtesy of Dr. Calvin Wells, 1973.)

used lead extensively in water piping, cooking vessels, and in making wine. Indeed, Paul of Aegina described an epidemic of lead poisoning throughout the Roman world in the 7th century A.D. due to ingestion of lead from leaden water pipes and leaded wines (Waldron, 1973).

ARTHRITIS BIBLIOGRAPHY

Ackerknecht, E. H.: "Paleopathology," in A. L. Kroeber (ed.): *Anthropology Today*, Univ. of Chicago Press, Chicago, 1953.

Ball, G. V.: "Two epidemics of gout," *Bull. Hist. Med.*, 45:401-408, 1971.

Bennett, G. A., H. Waine, and W. Bauer: *Changes in the Knee Joint at Various Ages*, Commonwealth Fund, New York, 1942.

Blumberg, B. S. and L. Sokoloff: "Coalescence of caudal vertebrae in the giant dinosaur *Diplodocus*," *Arthritis Rheum.*, 4:592-601, 1961.

Boland, E. W. and A. J. Present: "Rheumatoid spondylitis: a study of 100 cases with special reference to diagnostic criteria," *J. Amer. Med. Assn.*, 129:843-849, 1945.

Bourke, J. B.: "A review of the palaeopathology of the arthritis diseases," in D. R. Brothwell and A. T. Sandison (ed.), *Diseases in Antiquity*, Thomas, Springfield, 1967.

————: "The palaeopathology of the vertebral column in ancient Egypt and Nubia," *Med. Hist.*, 15:363-375, 1971.

Bywaters, E. G. L.: "Gout in the time and person of George IV: A case history," *Ann. Rheum. Dis.*, 21:325-338, 1962.

Chapman, F. H.: "The incidence and age distribution of osteoarthritis in an archaic American Indian population," *Proc. Indiana Acad. Sci.*, 72:64-66, 1964.

Collins, D. H.: *The Pathology of Articular and Spinal Diseases*, E. Arnold, London, 1949.

Copeman, W. S. C. (ed.): *Textbook of Rheumatic Diseases*, E. and S. Livingstone, London, 1964.

DeBlecourt, J. J., A. Polman, and T. DeBlecourt-Meindersman: "Hereditary factors in rheumatoid arthritis and ankylosing spondylitis," *Ann. Rheumat. Dis.*, 20:215-223, 1961.

Fisher, A. G. T.: *Chronic (Non-tuberculous) Arthritis*, Macmillan, New York, 1929.

Fox, H.: *Disease in Captive Wild Mammals and Birds*, F. A. Davis Co., Philadelphia, 1923.

————: "Chronic arthritis in wild animals," *Trans. Amer. Philosoph. Soc.*, 31:73, 1939.

Gaidner, W.: *On Gout, Its History, Its Causes, and Its Cure*, London, 1849.

Garrod. A. E.: *The Nature and Treatment of Gout and Rheumatism,* Walton and Maberly, London, 1859.
Gilfillan, S. C.: "Lead poisoning and the fall of Rome," *J. Occup. Med.,* 7:53-60, 1965.
Gofton, J. P., J. S. Lawrence, P. H. Bennett, and T. A. Burch: "Sacroiliitis in eight populations," *Excerpta Medica,* #148:293-298, 1968.
Hamerman, D.: "New thoughts on the pathogenesis of rheumatoid arthritis," *Amer. J. Med.,* 40:1-9, 1966.
―――: "Views on the pathogenesis of rheumatoid arthritis," *Med. Clin. N. Amer.,* 52:593-605, 1968.
Harrison, M. H. M., F. Schajowicz, and J. Trueta: "Osteoarthritis of the hip: a study of the nature and evolution of the disease," *J. Bone Joint Surg.,* 35B:598-626, 1953.
Heine, J.: "Über die arthritis deformans," *Virchow's Arch. Path. Anat.,* 260:521-663, 1926.
Hollander, J. L.: "Environment and musculoskeletal diseases," *Arch. Environ. Health,* 6:527-536, 1963.
Hooton, E. A.: *The Indians of Pecos Pueblo,* Yale Univ. Press, New Haven, 1930.
Hrdlička, A.: "Anthropological work in Peru in 1913, with notes on pathology of the ancient Peruvians," *Smithson. Misc. Coll.,* 61:57-59, 1914.
Jaffe, H. L.: *Metabolic, Degenerative and Inflammatory Diseases of Bone and Joints,* Lea and Febiger, Philadelphia 1972.
Johnson, L. C.: "Kinetics of osteoarthritis," *Lab. Invest.,* 8:1223-1238, 1959.
Karsh, R. S. and J. D. McCarthy: "Archaeology and arthritis," *Arch. Intern. Med.,* 105:640-644, 1960.
Keefer, C. S. and W. K. Myers: "The incidence and pathogenesis of degenerative arthritis," *J. Amer. Med. Assoc.,* 102:811-813, 1934.
Kellgren, J. H., J. S. Lawrence, and J. Aitken-Swan: "Rheumatic complaints in an urban population," *Ann. Rheum. Dis.,* 12:5-15, 1953.
Kidd, K. E.: "A note on the palaeopathology of Ontario," *Amer. J. Phys. Anthrop.,* 12:610-615, 1954.
Knaggs, R. L.: "Report on the Strangeways Collection of rheumatoid joints in the Museum of the Royal College of Surgeons," *Brit. J. Surg.,* 20:113-129; 309-330; 425-443, 1932-1933.
Kobert, R.: "Chronische bleivergiftung im klassischen altertume," in P. Diergart (ed.): *Beitrage aus der Geschichte der Chemie dem Gedachtnis nov Georg Kahlbaum,* Leipzig u. Wien, 1909.
Landells, J. W.: "The bone cysts of osteoarthritis," *J. Bone Joint Surg.,* 35B:643-649, 1953.

Lawrence, J. S., R. De Graaff, and V. A. Laine: *The Epidemiology of Chronic Rheumatism*, Blackwell, Oxford, 1962.

Mankin, H. J.: "The effect of aging on articular cartilage," *Bull. N.Y. Acad. Med., 44*:545-552, 1968.

May, W. P.: "Rheumatoid arthritis (osteitis deformans) affecting bones 5500 years old," *Brit. Med. J., 2*:1631-1632, 1897.

Moodie, R. L.: *Paleopathology: An Introduction to the Study of Ancient Evidence of Disease*, Univ. of Illinois Press, Urbana, 1923.

Morse, D.: *Ancient Disease in the Midwest*, Illinois State Mus. Rep. of Invest., #15, 1969.

Pommer, G.: *Mikroskopische Befunde bei Arthritis Deformans*, Wien, 1913.

Robinson, W. D.: "The etiology of rheumatoid arthritis," in J. L. Hollander (ed.): *Arthritis and Allied Conditions*, Lea and Febiger, Philadelphia, 1966.

Roche, M. B.: "Incidence of osteophytosis and osteoarthritis in 419 skeletonized vertebral columns," *Amer. J. Phys. Anthrop., 15*:433-434, 1957.

Rodnan, G. P.: "A gallery of gout being a miscellany of prints and caricatures from the 16th century to the present day." *Arthritis Rheum., 4*:27-46, 1961.

Rodnan, G. P. and T. G. Benedek: "Ancient therapeutic arts in the gout," *Arthritis Rheum., 6*:317-340, 1963.

Ruffer, M. A.: "Arthritis deformans and spondylitis in ancient Egypt," *J. Path. Bact., 22*:152-196, 1918.

Ruffer, M. A. and A. Rietti: "On osseous lesions in ancient Egyptians," *J. Path. Bact., 16*:439-465, 1912.

Saul. F. P.: "The human skeletal remains of Altar de Sacrificios," *Papers of the Peabody Museum*, vol. 63, #2, 1972.

Seltzer, C. C.: "Anthropometry and arthritis," *Medicine, 22*:163-203, 1943.

Short, C. L.: "The antiquity of rheumatoid arthritis," *Arthritis Rheum., 17*:193-205, 1974.

Smith, E. G. and W. R. Dawson: *Egyptian Mummies*, Allen and Unwin, London, 1924.

Snorrason, E. S.: "Rheumatism past and present," *Canad. Med. Assoc. J., 46*:589-594, 1942.

Snow, C. E.: *Early Hawaiians, an Initial Study of Skeletal Remains from Mokapu, Oahu*, Univ. Ky. Press, Lexington, 1974.

Sokoloff, L.: "The biology of degenerative joint disease," *Perspect. Biol. Med., 7*:94-106, 1963.

―――――: *The Biology of Degenerative Joint Disease*, Univ. of Chicago Press, Chicago, 1969.

Stecher, R. M.: "Heberden's nodes. A clinical description of osteoarthritis of the finger joints," *Ann. Rheum. Dis., 14*:1-10, 1955.

Stewart, T. D.: "Racial patterns of vertebral osteoarthritis," *Amer. J. Phys. Anthrop.*, 5:230-231, 1947.
――――: "The rate of development of vertebral hypertrophic arthritis and its utility in age estimation," *Amer. J. Phys. Anthrop.*, 15:433, 1957.
――――: "Problems in human palaeopathology," in S. Jarcho (ed.): *Human Palaeopathology*, Yale Univ. Press, New Haven, 1966.
Straus, W. L. and A. J. Cave: "Pathology and the posture of Neanderthal man," *Quart. Rev. Biol.*, 32:348-363, 1957.
Tobin, W. J. and T. D. Stewart: "Gross osteopathology of arthritis," *Clin. Orthop.*, 2:167-183, 1953.
Traut, E. F.: *Rheumatic Diseases*, C. V. Mosby, St. Louis, 1952.
Vallois, H. V.: "Paléopathologie et paléontologie humaine," *Homenaje a Don Luis de Hoyos Sainz*, 1:333-341, 1949.
Waldron, H. A.: "Lead poisoning in the ancient world," *Med. Hist.*, 17:391-399, 1973.
Wells, C.: "Joint pathology in ancient Anglo-Saxons," *J. Bone Joint Surg.*, 44B:948-949, 1962.
――――: "Hip disease in ancient man. Report of three cases," *J. Bone Joint Surg.*, 45B:790-791, 1963.
――――: *Bones, Bodies and Disease*, Thames and Hudson, London, 1964.
――――: "Diseases of the knee in Anglo-Saxons," *Med. Biol. Illus.*, 15:100-107, 1965.
――――: "Abnormality in mediaeval femur," *Brit. Med. J.*, 1:504-505, 1971.
――――: "A palaeopathological rarity in a skeleton of Roman date," *Med. Hist.*, 17:399-400, 1973.
West, H. F.: "The aetiology of ankylosing spondylitis," *Ann. Rheum. Dis.*, 8:143-148, 1949.
Zorab, P. A.: "The historical and prehistorical background to ankylosing spondylitis," *Proc. Roy. Soc. Med.*, 54:415-420, 1961.

Chapter X

TUMORS AND TUMOR-LIKE PROCESSES OF BONE

MODERN CLASSIFICATIONS of bone tumors and tumor-like processes list approximately forty different tumors of bone and its associated cartilage and fibrous connective tissue. Paleopathology is severely limited in diagnosing each distinct neoplastic process, because histological sections and biochemical data are not available for the dried bone specimen. Thus, the following discussion will be limited to fifteen different skeletal tumors and bone disorders which closely resemble neoplastic processes. Even with this simplification, a differential diagnosis of archaeological specimens exhibiting neoplastic lesions may be quite difficult, particularly if there is poor preservation or an incomplete skeleton.

A tumor may be classified as either *benign* or *malignant*. Such an identification is rather arbitrary with certain benign tumors such as giant-cell tumor which may become malignant. A benign tumor grows slowly and is usually localized in one area. A malignant tumor grows rapidly and may invade the adjacent soft tissue. Certain malignant tumors may cause widespread destruction of both bone and soft tissue through metastasis. In this process, malignant cells are liberated from the primary tumor and travel to different parts of the body through the lymphatics and bloodstream. After being deposited in another location, these malignant cells may then set up a secondary tumorous growth.

Tumors are classified according to the tissue in which they

originate. Tumors originating in the bone (osteogenic series) include osteomas, osteosarcoma, and osteochondroma. Chondrogenic tumors such as enchondroma begin in the cartilage. Fibrous tissue or collagen may be the site of tumors such as angioma and nonosteogenic fibroma. Ewing's sarcoma and multiple myeloma originate in certain cells in the bone marrow which are in turn derived eventually from the marrow reticulum. Malignant tumors of soft tissues are called carcinomas, and these neoplasms may invade the bone secondarily through metastasis. Metastatic invasion of contiguous bone tissue occurs in meningioma while widespread metastatic dissemination through the bloodstream occurs in carcinoma of the breast, prostate, and lung.

Table XI gives a relatively complete classification of primary benign and malignant tumors of bone and associated tissues

TABLE XI

CLASSIFICATION OF PRIMARY BENIGN AND MALIGNANT TUMORS OF BONE*

Tissue Origin	*Benign Tumor*	*Malignant Tumor*
Cartilage	Chondroblastoma (rare) Chondromyxoid fibroma Osteochondroma Chondroma	Chondrosarcoma
Bone	Osteoma Osteoid-osteoma Osteoblastoma	Osteosarcoma
Marrow elements		
Hematopoietic cells		Multiple myeloma Ewing's sarcoma Reticulum cell sarcoma
Fat cells	Lipoma	Liposarcoma
Fibrous connective tissue	Desmoplastic fibroma Periosteal desmoid fibromyxoma	Fibrosarcoma
Blood vessels	Hemangioma Hemangiopericytoma	Angiosarcoma (rare)
Neurogenous tissue	Neurilemoma Neurofibromatosis	
Notochord		Chordoma
Uncertain origin	Giant cell tumor	Giant cell tumor Adamantinoma

* Modified from AFIP classification: Spjut, et al., 1971.

(Spjut et al., 1971). This classification is based on predominant matrix formation and cell of origin. Neoplasms of bone are modified not only by the cell of origin but by the metabolic field in which they arise: the location in the bone, which bone is affected, and the age of the host at the time of growth. Thus, the neoplastic entities listed are not always well-defined and may merge in many cases when modified by the above conditions. This very interesting concept may be used to dynamically examine bone tumors rather than regard them as static clinicopathologic entities (Johnson, 1953).

TABLE XII

MAJOR TUMORS AND TUMOR-LIKE PROCESSES
SIMULATING PRIMARY BONE TUMORS

Fibrous cortical defect
Solitary bone cyst
Aneurysmal bone cyst
Fibrous dysplasia
Localized myositis ossificans
Histiocytosis X
Meningioma
Metastatic carcinoma

BONE TUMOR PREVALENCE AND INCIDENCE

Incidence figures are available for the primary malignant bone tumors, although such figures may be greatly modified by race, environment, age, and sex. Valid information on the asymptomatic benign tumors is unobtainable, although these tumors evidently outnumber the malignant ones. It should be noted that primary malignant tumors of bone account for only 0.5 percent of all malignant neoplasms (*End Results in Cancer* #4, 1972). This figure does not include multiple myeloma, a primary tumor of hematopoietic tissue, which accounts for about 0.7 percent of all malignant neoplasms. Looking at it another way, 300,000 people die in the United States each year from malignancies. Out of this large number only about 3000 deaths are due to primary malignant neoplasms of bone and bone marrow.

The age-adjusted incidence for primary bone tumors is approximately 1 per 100,000 people. The peak incidence occurs during the second decade of life and is around 3 per 100,000 during adolescence (Coley, 1960). Even during this peak, bone tumors comprise only 3.2 percent of all childhood malignancies below the age of fifteen years (Stradford, 1971). The incidence rate decreases to a low of 0.2 per 100,000 between the ages of thirty and thirty-five. Thereafter it slowly rises until at sixty years of age the incidence equals that of the adolescent (Coley, 1960).

Carcinomas often produce metastatic lesions in bone. These secondary malignancies of bone are much more common than primary malignant neoplasms, and this point should be kept in mind by the paleopathologist. Approximately two thirds of all malignant neoplasms metastasize to the skeleton, predominantly to the spine and pelvis. This figure may reach as high as 85 percent for the most common carcinomas of the breast, lung, prostate, kidney, and thyroid (Jaffe, 1958). Many of these lesions may not be evident on gross examination of the bones unless they are sectioned and radiographed.

BENIGN TUMORS OF BONE

Osteochondroma (Solitary Osteocartilaginous Exostosis)

Incidence

Osteochondroma is the most common of the benign bone tumors. In the large Mayo Clinic series, osteochondroma comprised 45 percent of the benign bone tumors and 12 percent of the total series (Dahlin, 1967). Because many of these tumors are asymptomatic, their actual incidence is much greater than the surgical case figures indicate. Osteochondromas occur primarily in persons between the ages of ten and twenty-five years. Many of the lesions are apparently present at birth but do not become clinically evident until years later.

Location

The major sites of skeletal involvement in osteochondroma are shown in Figure 122. Osteochondromas may occur on any bone that develops by enchondral ossification. The metaphyses

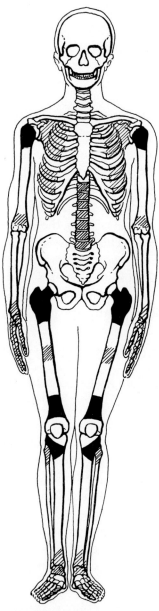

Figure 122. Skeletal distribution of osteochondroma. The solid black areas indicate the most frequent sites and the diagonal lines mark occasional sites.

of the long bones are most commonly involved. Over 50 percent involve either the femur or tibia. Other sites of importance are the humerus, ilium, and small bones of the hand and foot (Geschickter and Copeland, 1949; Dahlin, 1967).

Appearance

An osteochondroma always arises at or near an epiphyseal line and protrudes at a right angle to the long axis of the bone. The base or pedicle protrudes directly from the cortex of the underlying bone without an intervening area of abnormal osseous tissue. The "metaphysis" of the exostosis is usually fin-shaped leaning towards the diaphysis of the host long bone and away from the epiphysis. The ossified exostosis is capped by a large layer of cartilage undergoing calcification which is not evident in the archaeological specimen. The age at which an osteochondroma ceases to grow varies. However, it coincides roughly with the end of the individual's growth period. Thus, further growth stops when ossification ceases at the nearest epiphyseal line (Aegerter and Kirkpatrick, 1968).

The bone substance formed in osteochondroma is normal in every respect. It is produced by growth of aberrant foci of cartilage on the surface of bone. Thus, one could consider this tumor as a congenital anomaly or an enchondral hyperplasia rather than a true neoplasm.

Osteochondroma is almost always a solitary lesion. If multiple osteochondromas are present in a skeleton, this is due to hereditary multiple exostosis (also known as diaphyseal aclasis). The gross appearances of the two different tumors are almost identical except that the multiple exostoses, when viewed individually, show much greater bone involvement. Bilateral symmetry is also evident in the numerous exostoses of affected individuals. In advanced cases of hereditary multiple exostosis, there may be shortening and bowing of the long bones, particularly of the lower extremities.

Osteochondromas may closely resemble the ossification of muscle and tendons at their site of insertion which occurs in myositis ossificans circumscripta. This condition almost always

Figure 123. Small osteochondroma projecting from the medial surface of the right tibial metaphysis. Its growth ceased with the fusion of the epiphysis to metaphysis. (WAM 5287)

has an etiology of trauma, and the exostosis produced is usually more pointed than that of osteochondroma. The remarkable bony growth on the first discovered *Homo erectus* femur (formerly *Pithecanthropus*) is very likely an example of myositis ossificans. The proximal half of the femoral shaft has a prominent medial exostosis which is probably the result of trauma and subsequent ossification of the vastus medialis muscle tendons or other muscle tendons inserting in this region. Further examples of myositis ossificans and ossifying hematoma are reviewed by Brothwell, 1967.

Paleopathology

A possible osteochondroma of the distal femur comes from the 5th Dynasty of Egypt and was originally described as an

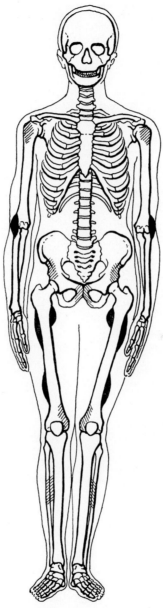

Figure 124. Skeletal distribution of traumatic myositis ossificans. Solid black areas indicate the most frequent sites and diagonal lines mark occasional sites.

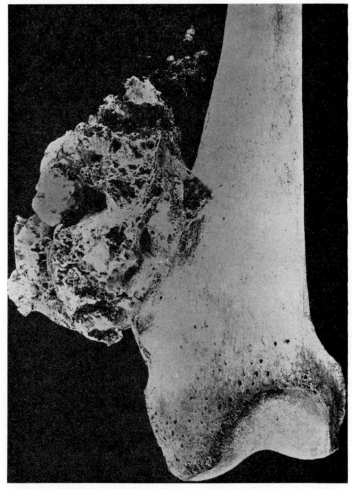

Figure 125. Anterior view of a left femur showing a large osteochondroma projecting from the medial surface. This specimen is from the 5th Dynasty of Egypt and originally diagnosed as an osteosarcoma. (From Smith and Dawson, 1924.)

osteosarcoma (Smith and Dawson, 1924). The large, irregular bone mass is situated very near the fused distal epiphysis on the medial side (see Fig. 125). Its location and general appearance are more suggestive of an osteochondroma (Rowling, 1961).

Chondroma (Particularly Enchondroma)

Incidence and Age

Chondromas make up over 10 percent of the benign tumor series at Mayo Clinic and 3 percent of the entire series. Persons of any age may be affected by the tumor which is almost always solitary. As Figure 126 illustrates, the major sites of involvement are the bones of the hands and feet, particularly the phalanges. Over 50 percent of the tumors are found in these bones. Other major sites in descending order are the humerus, femur, tibia, and radius (Jaffe, 1958; Dahlin, 1967).

Appearance

A chondroma is composed of mature hyaline cartilage which most commonly is centrally located in the bone. This subtype is known as enchondroma. Other chondromas are eccentric and bulge the overlying periosteum (periosteal chondroma). Chondromas begin in the metaphyseal region of the bone due to cartilage cells which have become isolated from the normal epiphyseal plate. Chondromas of the phalanges usually cause destruction of the cancellous tissue and bulging out and thinning of the cortex. This occurs to a much lesser extent in the thicker long bones. No cartilage will be apparent in the archaeological specimen, but a radiological examination of the tumor may give a lobulated appearance if ossification of the cartilage has occurred.

Osteoma (Ivory or Spongy Exostosis)

Classification

The term *osteoma* has been used to describe any protuberance of fibrous or cartilaginous tissue that becomes ossified. Thus, many types of tumors and tumor-like processes have been referred to as osteomas. The current literature on this entity is confusing and such all-inclusive classification should be discouraged in both clinical and archaeological material. For example, osteophytes are inflammatory cortical or supracortical deposits of bone from the periosteum. Exostoses of bone may be due to trauma and

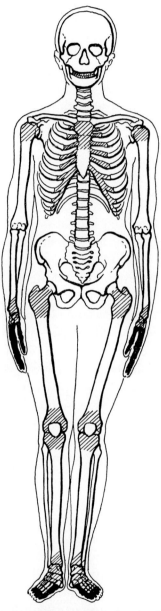

Figure 126. Skeletal distribution of chondroma. Black areas indicate the most frequent sites and diagonal lines mark occasional sites.

excess callus formation or may be osteochondromas and should not be included under osteoma.

Pathologists are notorious for developing their own classification of neoplasms based upon slightly different interpretation of the clinical material. In his study of the comparative pathology of animal and human bone tumors, Jacobson (1971) groups osteoma with ossifying fibroma (also called osteofibroma or fibrous osteoma). He describes benign osteoblastoma as a separate entity. Aegerter and Kirkpatrick (1968) on the other hand, list osteoma separately and group ossifying fibromas under benign osteoblastoma. Jaffe (1958) describes benign osteoblastoma as a distinct entity and groups both ossifying fibroma and osteoma under the term fibrous dysplasia, which in turn is given separate status by the previous authors. Amidst this confusion it has been decided here to keep the osteomas as a separate entity rather than place them under one of the other two groups.

Location

A true osteoma is usually classified as a bone-forming tumor. Actually it is a type of hamartoma in which cells of a circumscribed area of intramembranous bone grow more than the surrounding tissue. The cells eventually mature and therefore the tumor is not a progressive one. Thus, osteomas are benign and occur only in areas in which bone is normally formed by periosteum (intramembranous). Most osteomas are found on the inner and outer surfaces of the cranium and mandible, though a few may be located in bones of the postcranial skeleton.

Incidence

Accurate figures on the incidence and frequency of osteomas are not available due to confusion in classification of such tumors. Osteomas are usually symptomless and therefore not clinically apparent. Moreover, small osteomas of the cranium are often missed in routine autopsies and therefore not recorded. In general however, osteoma is one of the most common bone tumors or tumor-like processes in man. In 581 early Pueblo Indians (800-1800 A.D.), thirteen individuals had small "button" osteomas on their crania or a frequency of 2.2 percent (Hooton, 1930; see also Abbott and Courville, 1945).

Appearance

The most common type of osteoma is the button osteoma found as solitary or multiple projections on the bones of the cranium in the majority of cases (Bullough, 1965). Button osteomas are small, circular bone projections which may be dome-shaped or flattened. They are usually the size of a pea, and the bone texture may appear different from the underlying bone (see Figs. 127 and 128).

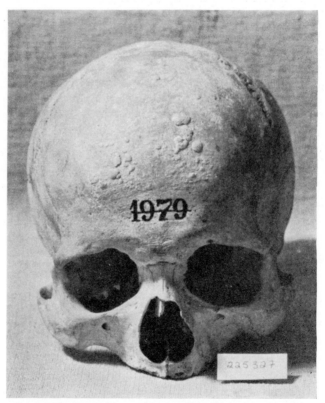

Figure 127. Button osteomas of the cranial vault. The small bony growths have an ivory-like appearance differing from the underlying bone. (USNM 225,327)

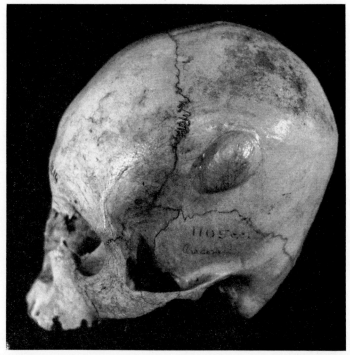

Figure 128. Large osteoma of the left parietal in an adult Indian from Ancon, Peru. (PM 7214)

Osteomas may be quite large and composed of cancellous as well as hard, dense bone. They are often located in the large sinuses of the skull. The skull of a twenty-five-year-old man from Medieval times in France has a very large frontoethmoidal osteoma as revealed by the roentgenogram (Béraud et al., 1961). An ancient Peruvian skull exhibits expansion of the medial portion of the left orbit and deflection of the median nasal septum to the right by a large cancellous osteoma (Moodie, 1926).

Ear Exostoses

Ear exostoses, also known as auditory tori or external auditory canal exostoses, are sometimes referred to as osteomas of the tympanic ring (Putschar, 1966). They are distinct bony excrescences within the external auditory canal, and may appear as

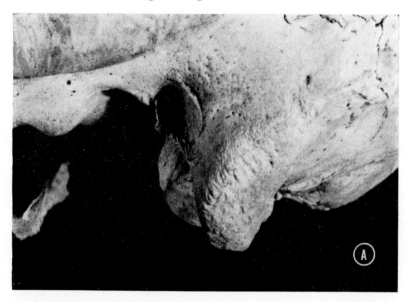

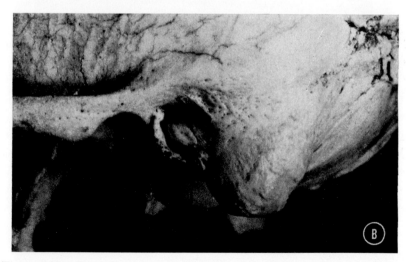

Figure 129. Ear exostoses in ancient Peruvians from the Chancay Valley of coastal Peru. Lateral views of the left side of the skull. (A.) Normal external auditory meatus. (B.) A ridge of eburnated sessile bone protruding into the external canal from the posterior canal wall. This is the most common form of ear exostosis.

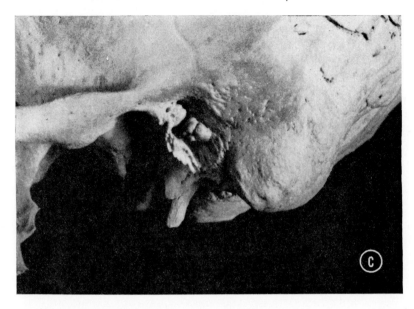

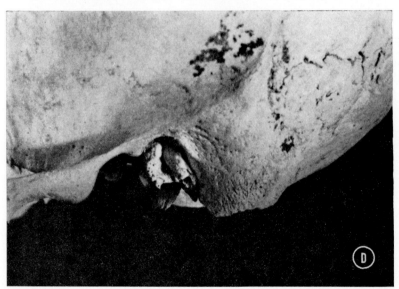

Figure 129. (C.) Rounded and pedunculated masses of bone protruding into the external canal from both the anterior and posterior walls. (D.) Large exostoses of bone almost completely occluding the external canal.

symmetrical "pearls" or as irregular bony masses. In some cases they may completely occlude the auditory canal (see Fig. 129).

Ear exostoses are probably not true osteomas. Definite racial differences in frequency have been noted and genetic factors are probably involved (Hrdlička, 1935; Roche, 1964; Berry and Berry, 1967). Table XIII lists the frequency of ear exostoses in various ancient American Indian populations in order of increasing frequency. There is a distinct sex difference with males having a greater frequency of ear exostoses than females. The exostoses rarely affect individuals below twenty-five years of age (Gregg and Bass, 1970). With advancing age, the exostoses become more common and are larger (Roche, 1964). They are usually bilateral.

TABLE XIII

INCIDENCE OF EAR EXOSTOSES IN VARIOUS AMERICAN INDIAN POPULATIONS LISTED IN ORDER OF INCREASING FREQUENCY[1]

Indian Population	No. Skulls Examined	No. Skull Exostoses	Percent Exostoses	Percent with Exos.: Male	Female
Pueblos*	500	12	2.4	2.8	2.0
Florida*	395	35	8.9	13.7	3.3
Texas†	348	36	10.3	17.9	1.9
California*	435	46	10.6	15.2	5.7
New England*	112	13	11.6	16.0	10.2
Peruvian*	3651	522	14.3	22.2	6.3
Virginia*	65	14	21.5	28.6	13.3
Louisiana*	61	15	24.6	46.4	6.1
Arkansas*	173	47	27.2	38.6	16.7
S. Dakota*	109	30	27.5	38.2	16.7
Indian Knoll‡	69	22	31.8	56.8	3.1
Indian Knoll§	405	141	34.8	56.7	8.1

[1] Incidence is computed for the separate sexes as well as as for the total population.
* Hrdlička, 1935; † Goldstein, 1957; ‡ this study; § Snow, 1948.

Almost all ear exostoses are located near the lateral end of the bony meatus and growing from the edge of the tympanic plate on the posterior wall of the canal. Less often the exostoses may occur in other locations and have different appearances. On this basis, Gregg and Bass have defined four different types of ear exostoses.

TABLE XIV

FREQUENCY OF EAR EXOSTOSES IN THE INDIAN KNOLL POPULATION ACCORDING TO THE SIDE INVOLVED, SEX AND TOTAL POPULATION

Ear Exostoses	Male	Percent	Female	Percent	Total	Percent
Left & Right	16/37	43.2	0/32	0.0	16/69	23.2
Left Only	2/37	5.7	1/32	3.1	3/69	4.3
Right Only	3/37	8.1	0/32	0.0	3/69	4.3
Total Present	21/37	56.8	1/32	3.1	22/69	31.8

Osteoid-Osteoma

Osteoid-osteoma is a relatively rare type of solitary, benign bone tumor first recognized forty years ago. It may be defined as an oval nidus composed of osteoid and trabeculae of newly formed bone deposited within a layer of highly vascularized osteogenic connective tissue (Lichtenstein, 1970). Even when fully developed, the nidus rarely exceeds a centimeter in its greatest dimension. The femur and tibia are most commonly involved. Less commonly the lesion may occur in the fibula, humerus, vertebrae, or bones of the feet. It may be located within the cortex or less commonly in the spongiosa. Almost all cases occur between the ages of ten and twenty-five years (see Fig. 130).

Appearance

Osteoid-osteoma has a very distinctive radiographic appearance. The small nidus of osteoid and bone trabeculae stands out clearly as a well-defined core from the surrounding tissue. The osteoid portion of the nidus may be poorly preserved in the archaeological specimen. If the bony portion of the nidus is also destroyed, one is left with a radiolucent defect within the bone. The bone tissue surrounding the nidus is usually modified by reactive thickening for a variable distance. The bone thickening is also accompanied by periosteal new bone formation if the nidus is located in the cortex. The shaft may be thickened for several inches in the area of the osteoid-osteoma, closely resembling a cortical bone abscess or a sclerosing nonsuppurative osteomyelitis. Indeed, many of the latter cases have been rediagnosed as osteoid-osteomas.

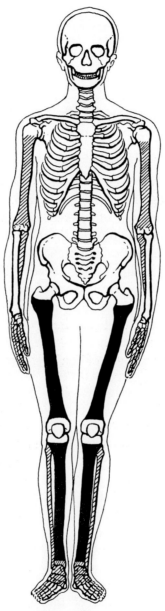

Figure 130. Skeletal distribution of osteoid-osteoma. Black areas indicate the most frequent sites and diagonal lines mark occasional sites.

Paleopathology

Three cases of possible osteoid-osteoma have been reported in the literature. The first comes from a 7th century Anglo-Saxon individual from Britain (Wells, 1965). The left femur has a localized region of cortical thickening along its shaft and

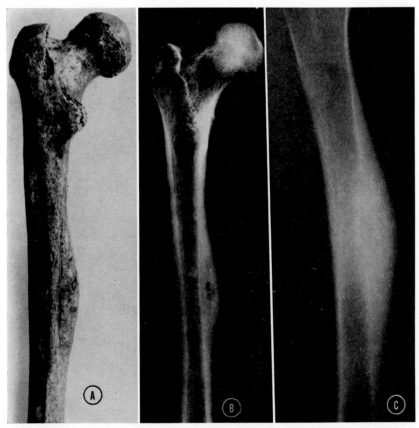

Figure 131. Osteoid-osteoma. (A.) Left femur of an adult male from Early Saxon times with a localized and elliptical growth of bone along the medial surface. (B.) Roentgenogram of the specimen reveals a well-defined radiolucent defect beneath the bone growth corresponding to the site of the nidus. (Courtesy of Dr. Calvin Wells, 1965.) (C.) Clinical roentgenogram of a left femur with an osteoid-osteoma for comparison. (Courtesy of Dr. E. B. D. Neuhauser.)

the roentgenogram reveals a small, well-defined radiolucent area corresponding to the site of the nidus (see Fig. 131). This cortical lesion was initially regarded as a subperiosteal ossified hematoma resulting from scurvy. A tibia of Saxon date from Britain has an elliptical swelling along the anterior border about 10 cm in length (Brothwell, 1961). Evidently the bone was not radiographed, but could well be an osteoid-osteoma or more likely a focal osteomyelitis.

The third reported case of osteoid-osteoma comes from Czechoslovakia during the 11th century A.D. (Vyhnánek, 1971). The right tibia of a forty- to fifty-year-old male has an area of localized cortical swelling along the medial aspect of the midshaft. Beneath this 8 cm area of swelling is an oval radiolucent defect presumably the site of an osteoid-osteoma nidus. It is worth noting that a proven case of focal osteomyelitis closely resembles this reported osteoid-osteoma (see Spjut et al., 1971, Figure 117).

Fibrous Dysplasia

Fibrous dysplasia is a nonneoplastic condition characterized by the presence of fibrous or fibro-osseous tissue within the affected part or parts of the bone. The lesions apparently result from a developmental defect in the bone-forming mesenchyme causing an arrest of bone maturation at the woven bone stage. Various extraskeletal aberrations may also occur such as blotchy pigmentation, premature sexual development, premature skeletal growth, and hyperthyroidism. These changes are referred to as Albright's syndrome.

Incidence

Fibrous dysplasia is not a rare disorder although relatively few cases progress to the syndrome described by Albright. It is about three times more common in females than in males. Most cases of fibrous dysplasia first occur in childhood or adolescence. The process usually abates by adulthood.

Location

Fibrous dysplasia may affect only one bone or several bones. Solitary lesions most often affect the femur, tibia, rib, or a

Tumors and Tumor-Like Processes of Bone

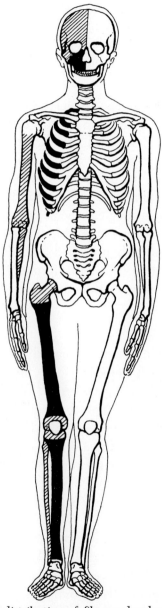

Figure 132. Skeletal distribution of fibrous dysplasia. The diagram above applies to both monostotic and polyostotic forms of the disease. Only half of the diagram is used to emphasize the predominantly unilateral distribution of lesions in polyostotic cases. Black areas indicate the most frequent sites and diagonal lines mark occasional sites.

facial bone. Polyostotic cases almost always affect the skeleton unilaterally, particularly the bones of the lower extremities. If the bones of the upper extremities are affected, the skull is also likely to be involved by the dysplasia (see Fig. 132).

Appearance

The portion of the bone involved by fibrous dysplasia is usually expanded due to erosion of the original cortex by the growing fibro-osseous tissue and the formation of subperiosteal new bone around the expanding tissue. The outer bone surface is usually smooth and the inner surface is often ridged. The ossified portion of the fibro-osseous tissue in the medullary cavity will persist in the archaeological specimen, usually as a mass of coarse trabeculae of woven bone. Many cyst-like areas will be present which were formerly occupied by fibrous tissue or islands of cartilage. In addition, true cysts formerly filled by blood or serous fluid will be present within the affected region or regions of the bone.

Paleopathology

Two very typical cases of fibrous dysplasia have been reported in archaeological material. The first comes from Early Saxon times in Britain around 650 A.D. (Wells, 1963). The left humerus of an adult male is bowed laterally and shows irregular expansion throughout much of the shaft. Longitudinal section and roentgenograms reveal the thinned cortex of the expanded area and the presence of spongy bone tissue composed of woven bone. Within the spongy tissue are several small, cystic areas (see Fig. 133). No other preserved bones are affected making this a probable monostotic case.

The second case of fibrous dysplasia comes from the Mississippian Culture of Illinois around 1000 A.D. (Denninger, 1931). The skeleton of this thirty-five-year-old male Indian has many bones involved, predominantly on the left side. These include the bones of the lower extremities, pelvic bones, several vertebrae, several ribs, and the left side of the face. The skeletal distribution of the lesions, bone expansion, and presence of abnormal spongy

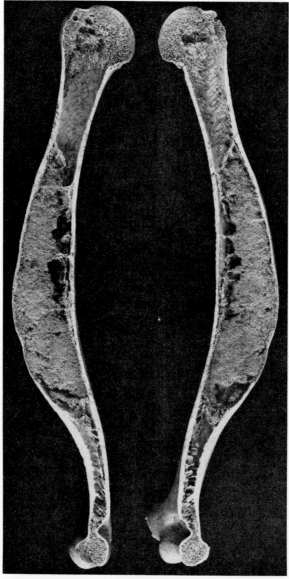

Figure 133. Fibrous dysplasia affecting the left humerus of an adult male from Early Saxon times. The sectioned bone reveals the medullary cavity filled with abnormal spongy tissue composed of woven bone. Several cystic defects are present within the abnormal tissue. The cortex in the expanded region is very thin. (Courtesy of Dr. Calvin Wells, 1963b.)

bone interspersed with cystic areas within the affected regions are characteristic of fibrous dysplasia (see Fig. 134).

Craniometaphyseal Dysplasia

It is worth mentioning here a rare familial dysplasia of bone totally unrelated to fibrous dysplasia, but resembling it in several respects. Craniometaphyseal dysplasia is a failure to absorb the secondary spongiosa in the growing metaphysis resulting in markedly expanded metaphyses and thinned cortices (Rubin, 1964). Unlike fibrous dysplasia, the bones are involved in a bilaterally symmetrical manner with expansion limited to the metaphyseal areas, and there are no cystic defects within the spongy tissue.

A case of this rare dysplasia has been reported in a Peruvian skeleton of the Mochica Culture around 200-800 A.D. (Urteaga and Moseley, 1967). The remains consist of a humerus, pelvic bones, femora, and tibiae, and all reveal metaphyseal expansion due to persistence of the secondary spongiosa through absence of funnelization. The long bones are normal in length although the tibiae have S-shaped bowing. The trabeculae of the excessive spongy bone are delicate and are not arranged in a normal stress pattern.

A femur and tibia from an ancient Nubian cemetery have metaphyseal expansion suggestive of craniometaphyseal dysplasia (Smith and Jones, 1910). However, the process was apparently unilateral and did not involve both metaphyses of each bone, perhaps indicating fibrous dysplasia as a more likely cause. A reexamination of these bones with longitudinal sectioning and radiography would be most helpful.

Nonossifying Fibroma and Fibrous Cortical Defect
(Synonyms and Related Terms: Nonosteogenic Fibroma, Metaphyseal Fibrous Defect, Fibroma, Xanthoma)

Classification

Nonossifying fibroma and fibrous cortical defect are almost certainly nonneoplastic lesions. Focal ischemia due to vascular occlusion by trauma or embolism has been suggested as the

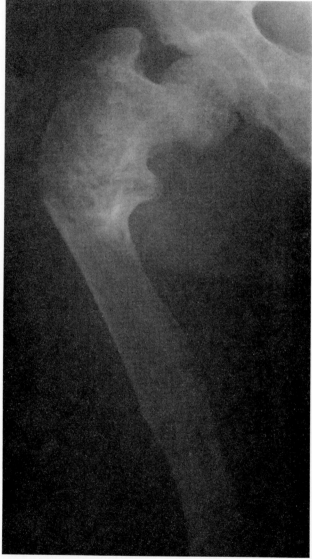

Figure 134. Polyostotic fibrous dysplasia in a nine-year-old girl. This individual is one of the five original cases described by Albright in 1937. The bone changes in the femur and other bones are identical to those present in the ancient Indian described by Denninger in 1931. (Courtesy of Dr. E. B. D. Neuhauser.)

initiating agent although the lesions differ from bone infarcts. The fibrous cortical defect represents the small, usually asymptomatic, manifestation of this disturbance of ossification. Nonossifying fibroma refers to the less common tumor-like form which involves the medullary cavity and may produce pain.

Incidence and Age

Surgical series do not provide accurate estimates of the prevalence of these lesions because the majority are asymptomatic. Radiological surveys of normal children provide much more accurate data. The occurrence of fibrous cortical defect in 151 apparently normal children was 27 percent (Selby, 1961). One or more defects were found in 35 percent of 154 children (Caffey, 1955). A definite familial tendency was also noted. Thus, it is clear that a large percentage of children (30-40 percent) develop one or more fibrous cortical defects. The incidence of nonossifying fibromas which only occasionally evolve from the cortical defects is much less.

Both lesions are almost exclusively found in children and adolescents. The fibrous cortical defect is most commonly found between the ages of four and eight years. The lesion is found in twice as many males as females. Nonossifying fibroma is usually found in persons over eight and under twenty years of age.

Location

Both lesions are found mainly in the metaphyseal portions of the femur, tibia, and fibula. The long bones of the arm are only occasionally involved and other areas not at all. The lesion first appears near the end of the shaft very close to the unclosed epiphyseal line. Multiple and bilateral cortical defects are nearly as frequent as the solitary lesion. One may encounter a nonossifying fibroma in one long bone and one or more fibrous cortical defects in other bones of the same individual. In the femur the lesions are most common posteriorly and medially and in the distal portion. In the tibia they are most common in the proximal end and may occur on any aspect. Fibular lesions may appear at either end of the bone (Caffey, 1955).

Gross Pathology

When the roentgenogram is examined showing the specimen in profile view, the cortical defect is found to be superficial and located in the cortical wall rather than in the interior of the bone. There is a border of bony sclerosis around the margins of the defect. In the majority of cases, the fibrous cortical defect heals after several years with normal architecture completely reconstituted. Occasional defects do not heal but progress to the state of nonossifying fibroma. The lesion usually remains eccentric, but may extend all the way across the shaft in thin

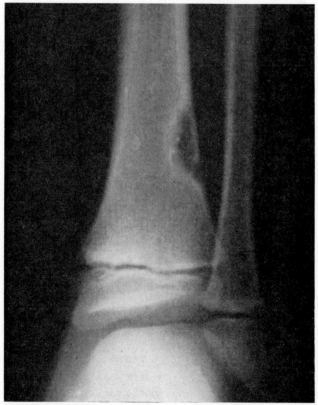

Figure 135. Clinical roentgenogram of nonossifying fibroma in the distal end of the left tibia. The lesion originated in the cortex and has partially invaded the medullary cavity producing a lobulated appearance. A thin rim of sclerosis surrounds the lesion. (Courtesy of Dr. E. B. D. Neuhauser.)

long bones such as the fibula. Some expansion of the cortex may occur as well as marked thinning of the cortex. There is often a multilocular appearance in the roentgenogram (see Fig. 135).

Solitary Bone Cyst (Unicameral Bone Cyst, Benign Bone Cyst, Juvenile Bone Cyst, Simple Bone Cyst)

Classification and Incidence

A solitary bone cyst is a benign, unicameral (single-chambered) cystic lesion of unknown etiology. (For a discussion of the possible etiologies see Coley, 1960). These cysts are not often clinically apparent except when a pathologic fracture is produced and their true anatomic incidence is therefore not known. The Netherlands Committee lists only sixty-six cases among their 1569 bone tumor cases or a frequency of 4.2 percent.

Solitary bone cysts are most commonly found in children between the ages of three and fourteen years. Because the cysts persist throughout life, archaeological specimens of solitary bone cyst may be found in individuals of any age. In most clinical series there is a male preponderance of 3 to 1.

Location

Solitary bone cysts are found almost exclusively in the long tubular bones of the skeleton. Two-thirds to three-fourths of the lesions occur in the proximal portion of the humerus or femur (Spjut et al., 1971). Indeed, the proximal portion of the humeral shaft may be involved in 50 percent of all cases (Jaffe, 1958). As Figure 136 illustrates, other sites of involvement are the tibia, fibula, radius, ulna and metatarsals.

Appearance

Located in the shaft of a long tubular bone, the cyst often lies near the epiphyseal plate in the young individual. In older individuals, the solitary bone cyst is found much farther away from the epiphysis due to growth of the epiphyseal plate away from the cyst. This difference in location with age (until maturity of the long bone involved), is of some importance in

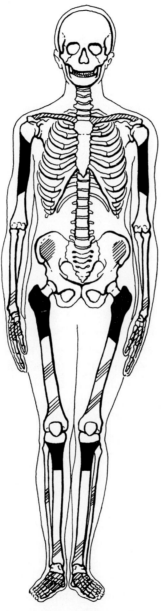

Figure 136. Skeletal distribution of solitary bone cyst. Black areas indicate the most frequent sites and diagonal lines mark occasional sites.

diagnosis of archaeological material. Very rarely will the cyst actually penetrate into the plate and this helps in differentiating the lesion from giant-cell tumor. The humeral lesion often extends down the shaft for several inches. There is usually slight expansion of the shaft as well as cortical thinning from the medullary surface. The periosteal surface of the cortex appears normal except in the region of a pathological fracture when present (see Fig. 137).

A roentgenogram of a typical solitary bone cyst reveals a circular or oval area in the medullary space. The wall is sharply defined and normal trabeculation is absent. The appearance may be very similar to Brodie's abscess except that the latter lesion is smaller and involves cortical thickening rather than cortical thinning and expansion.

Differential Diagnosis

A differential diagnosis between solitary bone cyst, giant-cell tumor, and Brodie's abscess has been briefly discussed above. With only the dry bone specimen available, other conditions such as a solitary focus of eosinophilic granuloma, enchondroma of a long bone, or an aneurysmal bone cyst may closely resemble the cystic lesion. An eosinophilic granuloma lesion is usually found in the middle of a shaft rather than towards the ends. There is also subperiosteal new bone formation extending beyond the defect. Most enchondromas are found in the bones of the hands and feet which are rarely the site of a solitary bone cyst. If ossification of the cartilage has occurred in the enchondroma, the lobulated appearance may help differentiate the lesion from solitary bone cyst. An aneurysmal bone cyst is often eccentrically located in the long bone rather than centrally placed. The characteristic extreme expansion of part of the contour of the affected area in an aneurysmal bone cyst usually makes a differential diagnosis possible.

Giant-Cell Tumor (Osteoclastoma)

Classification

A giant-cell tumor is a usually benign though often malignant neoplasm which develops within the bone and apparently arises

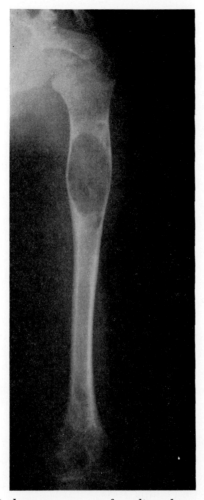

Figure 137. Clinical roentgenogram of a solitary bone cyst in the proximal shaft of a right humerus. The lesion appears as a sharply defined oval defect within the medullary cavity. The adjacent cortex is thinned and expanded. (Courtesy of Dr. E. B. D. Neuhauser.)

from the mesenchymal cells of the connective tissue framework. The multinucleated giant-cell (not present in the skeletal material) are a constant and prominent part of these tumors. Unfortunately, giant-cells occur in many other pathologic conditions of bone, thus accounting for the confusion in the early literature on these tumors. Such entities as nonossifying fibroma, aneurysmal bone cyst, and benign chondroblastoma have been erroneously labeled giant-cell tumors for this reason. Many neoplastic lesions of the maxilla and mandible have been classified as giant-cell tumors. Many if not all of these lesions are not true neoplasms and have been reclassified as giant-cell reparative granulomas (Jaffe, 1953).

Incidence and Age

When the other giant-cell lesions are correctly diagnosed, the true giant-cell tumor is not a common one. The Mayo series of 155 cases accounted for 3.9 percent of all bone tumors and 15.1 percent of the benign tumors (Dahlin, 1967). Giant-cell tumor occurs most frequently in young adults within ten to twenty years after fusion of the epiphysis of the site involved. Occurrence of giant-cell tumor is rare before the age of twenty or after the age of fifty-five years. The highest incidence is in the third decade of life.

Location

Giant-cell tumors are almost always found in the epiphyses of the long bones with a later secondary involvement of the metaphyses. Lesions located in the metaphyseal area of a bone without involving the epiphyses are almost certainly not true giant-cell tumors (Spjut et al., 1971). More than half of the giant-cell tumors in the Mayo series occurred about the knee. Other bones involved in descending order are the distal epiphyses of radius and ulna, sacrum, ilium, and ischium. Giant-cell tumors rarely involve the small bones of the hand or foot (Geschickter and Copeland, 1949; Dahlin, 1967).

Appearance

Roentgenograms and gross sections of giant-cell tumors reveal the characteristic rapidity of destructive activity. The lytic region is eccentric initially, but grows to involve the width of the bone. There is a lack of periosteal new bone formation and absence of reactive bony sclerosis in the advancing margins. Commonly, the bone is expanded and the cortex is thinned. While the tumor extends to the articular plate, it rarely destroys it. The aggressive nature of the giant-cell tumor accounts for their immense

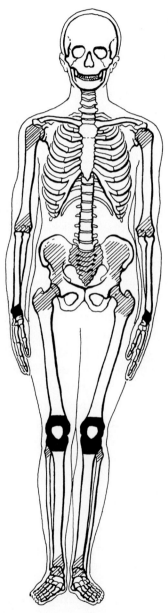

Figure 138. Skeletal distribution of giant-cell tumor. Black areas indicate the most frequent sites and diagonal lines mark occasional sites.

size when left untreated, as is presumably the case in archaeological specimens. In advanced cases, the thin shell of bone may be perforated, permitting the tumor to extend itself into the soft tissues and perhaps to another bone. Incomplete septa are often evident within the substance of the tumor. These are remnants of trabeculae and are not newly formed bone.

Differential Diagnosis

A differential diagnosis of giant-cell tumor may be difficult when confronted with the dry bone specimen. As mentioned earlier, most lesions of the jaws are due to giant-cell reparative granuloma rather than giant-cell tumor. A "brown tumor" focus of hyperparathyroidism may also simulate the appearance of giant-cell tumor. The entire skeleton should be radiographed for porosity of the bone and perhaps other circumscribed lytic lesions of hyperparathyroidism. Aneurysmal bone cyst resembles giant-cell tumor and may even be assoicated with it. Some workers regard giant-cell tumor as a common precursor of the nonneoplastic aneurysmal bone cyst (Aegerter and Kirkpatrick, 1968). In many cases, aneurysmal bone cysts do not involve the epiphyses of the bone and may be differentiated in this respect. Solitary bone cyst is sometimes confused with giant-cell tumor although it too does not often involve the epiphysis. Furthermore, the age of those with solitary bone cyst is substantially younger (3-15 years of age) than for those with giant-cell tumor (Jaffe, 1958). Nonossifying fibroma may be differentiated from giant-cell tumor for the same reason as solitary bone cyst. A fibrosarcoma of the epiphyseal region may closely resemble giant-cell tumor radiologically. Its incidence as to skeletal location and age of the affected person are closely comparable to that of giant-cell tumor. In most cases however, fibrosarcomas exhibit periosteal bone reaction and slight sclerosis around the margins of the tumor.

Hemangioma (Angioma)

Classification and Incidence

Hemangiomas are simple overgrowths of the blood vascular channels which cease growth at maturity of the surrounding tissue. These are true hamartomas and not neoplasms. The great majority of hemangiomas are asymptomatic and therefore make up only a small portion of surgical tumor series. The Mayo Clinic series included only forty-seven solitary hemangiomas in a total of 3907 bone tumors for a prevalence of 1.2 percent (Dahlin, 1967). On the other hand, the anatomic incidence of this lesion is quite common, being found in approximately 10 percent of all autopsies in which the vertebrae have been sectioned (Jaffe, 1958). According to most reports, hemangiomas are usually found in adults, particularly in the fifth decade. The most serious lesions, however, are often in younger adults.

Location

Hemangioma is found most often in the vertebrae, particularly the lower thoracic and upper lumbar segments. The cranium is also a common site of hemangioma, particularly in the parietal and frontal region (Sherman and Wilner, 1961; Jacobson, 1971). Hemangioma in other locations of the skeleton is much less common with the ribs, proximal femur, mandible, clavicle, and sternum involved in order of decreasing frequency. Almost all hemangiomas are solitary lesions, although several small lesions may be present in several vertebrae.

Gross Pathology

Hemangioma in a vertebra causes lysis of the trabeculae and if the lesion is adjacent to the cortex there may be cortical thinning or perforation. Sclerotic trabeculae may traverse the area of the lesion causing the "soap bubble" appearance seen in the roentgenogram. These trabeculae may also appear as parallel striations arranged vertically. The hemangioma is almost always restricted to the body of the vertebra although there is no predilected location within the vertebral body. The majority of the asymptomatic hemangiomas are less than 1.5 cm in their greatest dimension (Jaffe, 1958). A very large hemangioma causing symptoms in the affected person may fill out the entire vertebral body and cause partial expansion of the cortex. Although the lesion may be multiple and occur in several vertebrae, it will not cross the intervertebral disk and invade an adjoining vertebra.

A roentgenogram of a cranial hemangioma or the archaeological specimen itself can reveal a characteristic "sunray" appearance with the longitudinal spicules radiating from the center to the outer edges. This appearance may be confused with a sclerosing osteosarcoma and with the hyperostosis and spicule formation in meningioma. The lesion originates in the diploic spaces and can erode through either table. Expansion of the compact tables may occur, particularly toward the outside surface. The diameter of the lesion along the surface of the

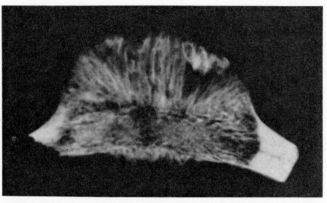

Figure 139. Roentgenogram of a macerated calvarial hemangioma measuring about 5 cm in diameter. (Courtesy of *Medical Radiography and Photography*.)

skull may vary between 2.5 cm and over 7 cm. The thickness of the lesion is usually greater than 1.5 cm and may exceed 7 cm (Jaffe, 1958; Dahlin, 1967).

Paleopathology

Very few possible cases of hemangioma have been reported in archaeological material. However, several of the specimens described as meningioma may have been the result of hemangiomas. A Late Dynasty Egyptian skull has a large crater some 8.0 cm in diameter and situated near the vertex. The external margins of the crater are raised above the level of the outer table and this bone has a highly vascular appearance. This lesion is possibly caused by a hemangioma or epidermoid cyst (Brothwell, 1967).

An upper lumbar vertebra from a Middle Kingdom Nubian site has a well-defined cavity within the spongy tissue of the vertebral body. It perforates the cortex anteriorly and several coarse trabeculae can be seen traversing the cavity (Smith and Dawson, 1924). There is no evidence of bony reaction or vertebral collapse, and the appearance suggests a hemangioma.

An adult skull from the Late Woodland Culture of Illinois around 900 A.D. has expansion of inner and outer tables in the midparietal region (Morse, 1969). There is an area of erosion affecting the inner table. The roentgenogram reveals a honeycomb pattern in the expanded region possibly due to hemangioma.

Meningioma

A meningioma is not a primary bone tumor but is briefly described here because of its close resemblance in many instances to primary bone tumors such as osteosarcoma. The tumor originates in the meninges, the outer membranous covering of the brain and spinal cord. The tumor may erode into the cranial bone and cause production of osteophytes in 10 to 25 percent of all cases of meningioma (Cushing, 1922; Geschickter and Copeland, 1949). The new bone formed is not tumorous but is merely a periosteal reaction to increased vascularization or to irritation by the invading tumor. Both tables of the skull may be affected and the slow-growing tumor may cause the bone forming process to persist for months or even years.

Location and Appearance

Most meningiomatous hyperostoses occur in the vault of the skull, although the orbital region and parts of the base may be affected. The new bone is laid down both parallel and perpendicular to the tables, perhaps depending on the extent of periosteal separation (see Abbott and Courville, 1939). The outer table is usually involved more extensively than the inner table. Destruction of bone also occurs either by direct action of the tumor or by pressure atrophy.

Differential Diagnosis

Osteosarcomas and meningiomas have often been confused in archaeological material because the pronounced and often radial bone formation combined with bone destruction of a meningioma resembles a malignant tumor arising in the bone tissue. However, the bone formed by a malignantly progressive osteosarcoma is rarely as heavily ossified and regularly ordered as that produced by a slow-growing meningioma. Furthermore, osteosarcoma rarely affects the cranium. In a series of 650 osteosarcomas, only thirteen (2.0 percent) were located in the vault area (Dahlin, 1967). A series of all types of primary sarcoma included only twelve calvarial lesions in 700 cases, or an incidence of 1.7 percent (Geschickter and Copeland, 1949). Finally, osteosarcoma usually occurs before the age of twenty-five and is rapidly fatal while meningioma predominantly affects the older age group.

Hemangioma of the calvarium may closely resemble a mengiomatous hyperostosis. The calvarial hemangiomas have longitudinal spicules of bone radiating from the center of the lesion to the outer edges, and some meningiomas produce an identical picture. Both lesions affect the same age group. Several of the ancient cases reported below could well be calvarial hemangiomas.

Paleopathology

At least twelve possible cases of meningiomatous hyperostosis have been recorded in the literature. Over half of these cases were described in a paper published only four years after Harvey Cushing's initial description of the meningioma (Moodie, 1926).

The earliest likely cases of meningioma come from Egypt (Rogers, 1949). A calvarium from the 1st Dynasty (around 3400 B.C.) shows a circumscribed hyperostosis affecting both tables of the right parietal. Large vascular channels are also evident near the hyperostosis. A skull from the 20th Dynasty (around 1200 B.C.) has bony spicules covering a large portion of both parietals. The spicules produce a honeycomb appearance and appear to radiate from a single site in the right parietal.

A Romano-British skull of an adult female has a cranial

hyperostosis of the right parietal originally diagnosed as an osteosarcoma. The diagnosis has been changed to meningioma because the heavy ossification in the specimen is unlike that occurring in osteosarcoma (Brothwell, 1967).

Another meningioma initially diagnosed as an osteosarcoma comes from Paucarcancha, Peru (MacCurdy, 1923; Moodie, 1926). The skull of an elderly male is of unknown date but is probably pre-Hispanic. The hyperostosis involves the left parietal and frontal covering an area 14 by 11 cm. The height of the growth of radiating spicules is 4.5 cm (see Fig. 140). The cranial wall is completely destroyed below the hyperostosis and roentgenograms support the diagnosis of a meningioma (Zariquiey, 1958).

The skull of an adult female from Chavina, Peru has been described in two separate reports (Moodie, 1926; Abbott and Courville, 1939). Bony hyperostosis covers the frontal and a large part of both parietals. It consists of many fused bony spicules radiating at right angles to the vault surface. According to the authors, the outer table remains intact beneath the growth, ruling out the possibility that this represents a severe case of spongy hyperostosis. Indeed, the authors had examined several cases of spongy hyperostosis and realized their nonneoplastic nature. A slight hyperostosis is also present on the inner table of this skull, which probably represents a meningioma.

The skull of an adult male Indian from an island off the coast of southern California may show a meningioma (Abbott and Courville, 1939). Hyperostosis mainly involves the right parietal and there is erosion of the inner table. The skull probably dates prior to Spanish contact.

Histiocytosis X (Reticuloendotheliosis)

Classification

Eosinophilic granuloma, Letterer-Siwe disease, and Schüller-Christian syndrome are considered by most authorities to be different clinical manifestations of the same condition known as histiocytosis X (Moseley, 1963; Lichtenstein, 1970). The basic pathology is a granulomatous proliferation of reticulum

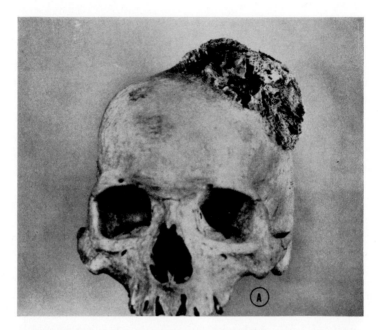
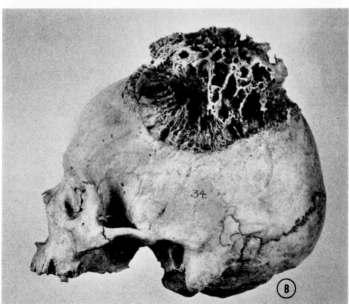

Figure 140. Meningioma in an elderly male from Paucarcancha, Peru. (A.) Frontal and (B.) lateral views of the large hyperostosis consisting of heavily ossified and radiating spicules of bone. (Courtesy of Dr. Manuel Zariquiey, 1958.)

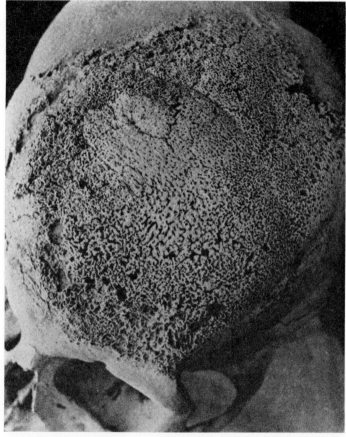

Figure 141. Meningiomatous hyperostosis in an adult female from Chavina, Peru. The close-meshed spicules of bone are perpendicular to the vault surface creating a coral-like appearance. (From Moodie, 1926.)

cells. According to Lichtenstein's classification, eosinophilic granuloma represents histiocytosis X localized in bone, Letterer-Siwe disease represents the acute disseminated stage of this disease, and Schüller-Christian syndrome represents a chronic disseminated form of histiocytosis X (Lichtenstein, 1953). It should be noted that some workers are still skeptical of this grouping (Lieberman et al., 1969).

Age and Location

Eosinophilic granuloma is a benign, nonneoplastic condition in which one or more focal areas of bone are destroyed by a granulomatous process and proliferation of the intraosseous reticulum. Most cases occur in children and adolescents with considerable predominance among males. Multiple lesions occur in almost half of the cases. The bones of the skull vault are most often involved, particularly the frontal bone. Other sites are the mandible, humerus, rib, and proximal metaphysis of the femur. When the lesion involves a long bone, it is usually located in the metaphysis or diaphysis with no apparent predilection for either area. When vertebrae are involved there is usually collapse of the vertebal body at its ossification center in the young child.

Eosinophilic granuloma is a benign osteolytic process with little cortical expansion and usually no subperiosteal bone apposition. The cortex may be completely destroyed with extension of the granuloma into the soft tissue. Letterer-Siwe disease, on the other hand, is a very serious disorder in which granulomatous lesions are widely disseminated in both hard and soft tissues. Infants and those under the age of two years are most often affected. Schüller-Christian disease lies between the previous two forms in its severity and is essentially a miscellaneous grouping of cases between the two extremes. Its age incidence and sites of involvement are identical to eosinophilic granuloma.

Appearance

The roentgenogram and dry bone specimen of all three forms of histiocytosis X show the characteristic round or oval lesion with no surrounding sclerosis. This "punched-out" defect represents a granuloma which has destroyed the bone. Flat bones are more frequently involved than long bones. Metastatic skull defects, multiple myeloma, and severe osteomalacia may present a somewhat similar appearance to these punched-out lesions. However, the older age groups favored by these conditions do not coincide with the much younger age group of histiocytosis X.

When the cortex of a long bone is destroyed by the lesion

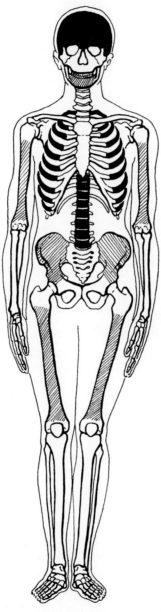

Figure 142. Skeletal distribution of histiocytosis X. Black areas indicate the most frequent sites and diagonal lines mark occasional sites.

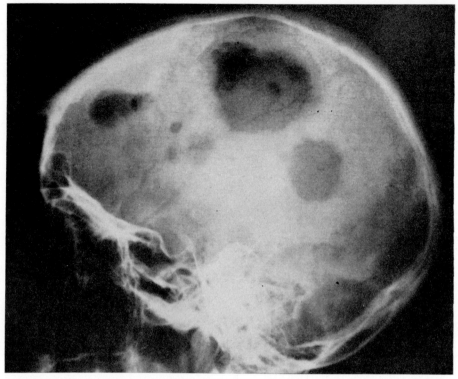

Figure 143. Clinical roentgenogram of histiocytosis X in an adult female who has had the condition for several years. (Courtesy of Dr. E. B. D. Neuhauser.)

originating in the medullary cavity, periosteal new bone formation may occur. This periosteal reaction may lead to confusion with Ewing's sarcoma and osteomyelitis.

Paleopathology

A possible case of eosinophilic granuloma comes from the Mississippian Culture of Illinois (1000-1600 A.D.) The skeletal remains of a 2½-year-old child exhibit lytic defects in two vertebrae and left ilium. The bodies of the 12th thoracic and 1st lumbar vertebrae are partially destroyed by a single granuloma which must have crossed the intervertebral disk. There is no evidence of bone regeneration. The left ilium has a large multilocular cavity involving the posterior surface and auricular facet with evidence of slight bone reaction (Morse, 1969).

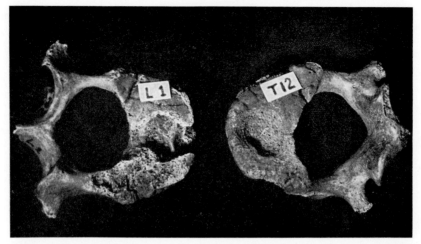

Figure 144. Histiocytosis X in a 2½-year-old child from the Mississippian Culture of Illinois. The lytic defect affecting both vertebrae is typical of an eosinophilic granuloma. (Courtesy of Dr. Dan Morse, 1969.)

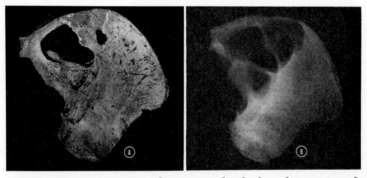

Figure 145. Histiocytosis X in the same individual as the previous figure. (A.) External view of the left ilium showing the large multilocular cavity. (B.) Roentgenogram of the specimen. (Courtesy of Dr. Dan Morse, 1969.)

Another likely case of histiocytosis X comes from New York State around 1200 A.D. and was originally described as a case of multiple myeloma (Williams et al., 1941). The individual is only ten years old, which coincides with the age group of histiocytosis X patients and is much younger than the vast majority of multiple myeloma patients. The skull, mandible, ribs, vertebrae, pelvis, and scapulae have numerous punched-out lesions from 3 to 10 mm in diameter. Other bones less severely affected are the femora, tibiae, fibulae, humeri, radii, ulnae, and clavicles. Roentgenograms reveal the presence of lesions within the marrow cavity in addition to those that have perforated the cortex. This closely resembles the Letterer-Siwe syndrome of histiocytosis X.

MALIGNANT TUMORS OF BONE

The subdivision of tumors into benign and malignant is not clear cut. Several benign bone tumors such as giant-cell tumor may evolve into a malignant, invasive neoplasm while still other benign tumors such as enchondroma may predispose to a malignant tumor at the same site. Malignant neoplasms grow rapidly, invade contiguous tissues, and inevitably proved fatal to the afflicted person of earlier times.

Osteosarcoma (Osteogenic Sarcoma)

Classification

Osteosarcoma is a malignant primary tumor of bone with evidence of malignant osteoid, bone, and frequently cartilage production. It is more generally called osteogenic sarcoma indicating that the tumor is of osseous origin and does not mean that the sarcoma necessarily produces bone. Osteosarcoma has been subdivided by earlier workers into several categories such as sclerosing, osteolytic, subperiosteal, medullary, and telangiectatic. In fact, there is a continuous graded series of osteosarcomas with wide variation in both the amount of osteoid tissue produced and the degree of calcification in it. Thus, most authorities currently avoid placing an osteosarcoma into one of the above categories (Aegerter and Kirkpatrick, 1968; Jacobson, 1971). These categories should not be confused with fibrosarcoma, chondrosarcoma, and parosteal osteosarcoma which have been classified as distinct tumors of bone.

Incidence

Osteosarcoma is the most common malignant bone tumor among younger people. Its frequency is exceeded only by multiple myeloma which occurs almost exclusively in later life. Osteosarcoma comprised 16 percent of the entire Mayo Clinic series and 21.9 percent of the malignant bone tumors (Dahlin, 1967). The Netherlands series has 246 osteosarcomas among 1402 primary tumors or a frequency of 17.5 percent. The actual prevalence of this tumor in the general population lies somewhere between 0.5 and 1.0 per 100,000 people (Coley, 1960; Spjut et al., 1971).

Age

A bimodal curve for peak incidence of osteosarcoma according to age has been reported by several workers (Price, 1962; Aegerter and Kirkpatrick, 1968). The first peak is in the second decade of life. In fact, three-fourths of the cases in a representative series are between ten and twenty-five years of age (Jaffe, 1958). The second and much smaller peak occurs after the age of fifty and is related to the increasing frequency of Paget's disease in these older individuals which may give rise to an osteosarcoma.

The combined age incidence and location of osteosarcoma is interesting and important to the paleopathologist. Osteosarcoma of the long bones occurs in patients under thirty in 80 percent of cases. Similar tumors of the flat bones occurred almost exclusively in individuals over forty years of age (Weinfield and Dudley, 1962; Aegerter and Kirkpatrick, 1968).

Location

The most common bone involved is the femur, particularly the distal portion. Nearly 50 percent of all osteosarcomas affect this bone. The upper end of the humerus and upper end of the tibia are next most commonly involved in the younger age group. It should be noted that these are the sites of principal growth in these bones. In the older age group, the flat bones such as the ilium, sacrum, sternum, ribs, and skull are most often involved. Osteosarcoma rarely affects the small bones of the hand and foot (Aegerter and Kirkpatrick, 1968). Osteosarcoma almost always begins in the metaphysis with the epiphyseal region often becoming involved secondarily. The tumor in both long and flat bones is almost always a single lesion and rarely

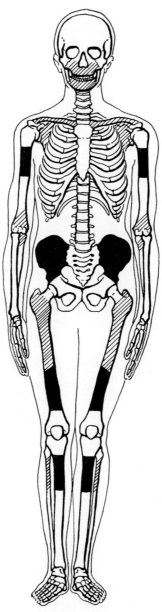

Figure 146. Skeletal distribution of osteosarcoma. Black areas indicate the most frequent sites and diagonal lines mark occasional sites.

metastasizes to other bones although metastasis to the lungs is common. If metastatic foci do appear, they usually involve the vertebrae, skull, or pelvic bones.

Size

Osteosarcomas exhibit rapid growth and may become quite large before causing death. This is particularly true in archaeological material in which the tumor has been allowed to run its full course uninhibited by radiation therapy or amputation. Radiographic measurement of 246 cases of osteosarcoma gave an average measurement of 10 by 3 cm (Spjut et al., 1971). In more advanced cases, the tumor is likely to measure between 10 and 20 cm in longitudinal extent.

Appearance

Osteosarcomas may vary widely in the amount of bone lysis, ossification, and calcification. The tumors may be completely lytic or predominantly sclerotic, but there is usually a combination of both features present. Beginning in the metaphysis, considerable destruction of the trabeculae occurs and then destruction of the cortex. There is a gradual transition from areas of marked lysis to areas of uninvolved bone making the borders of the lesion indistinct in the roentgenogram. The neoplastic cells usually produce osteoid or bone in appreciable amounts. In addition there is a nonneoplastic formation of periosteal new bone. As the periosteum is elevated by the tumor perforating through the cortex, several layers of this may be laid down creating an "onion peel" appearance very similar to Ewing's sarcoma. More commonly, a "sun burst" appearance may be produced by apposition of new bone along the vascular channels which have remained attached to both the cortex and elevated periosteum. At the boundaries of the tumor where the periosteum rejoins the cortex, there is localized cortical thickening from the subperiosteal apposition of bone.

Differential Diagnosis

A diagnosis of osteosarcoma is often not difficult except in separating this tumor from other primary malignant tumors such

Figure 147. Osteosarcoma affecting the distal metaphysis of a right femur. (A.) Externally, a profusion of bony spicules radiate from the metaphysis. The growth measures 12 cm in greatest dimension. (B.) The sectioned specimen reveals a large area of bone destruction and a pathologic fracture. Primitive bone fills part of the metaphysis. (WAM 8228)

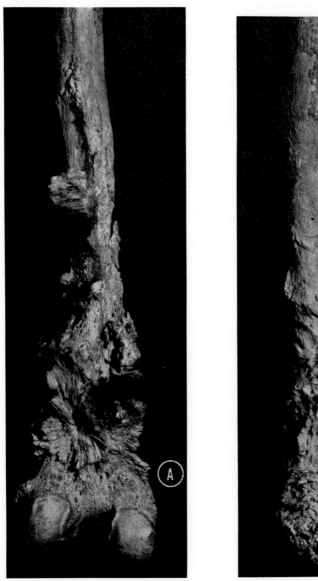

Figure 148. Osteosarcoma affecting the distal metaphysis of a left femur. (A.) Posterior view showing radiating spicules of bone in the metaphysis and along the shaft. (B.) Lateral view.

 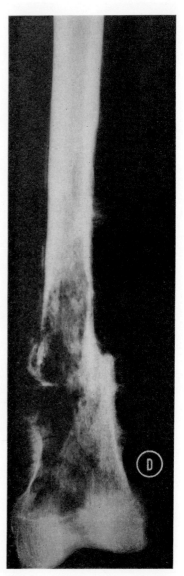

Figure 148 (C.) Anterior view showing large cortical perforations by bone lysis originating in the medullary cavity. (D.) Roentgenogram revealing the extent of bone destruction. (WAM 8273)

as chondrosarcoma and the less common fibrosarcoma. Chondrosarcomas are nearly as common as osteosarcomas, comprising 17 percent of the malignant tumors in the Mayo Clinic series. The majority of chondrosarcomas involve the pelvic bones and other common sites include the ribs, proximal femur, humerus, vertebrae, scapula, tibia, and fibula in that order (Spjut et al., 1971). This tumor is uncommon in children and adolescents which are by far the major victims of osteosarcomas. In the older age group with osteosarcoma, there is evidence of Paget's disease of bone in nearly 50 percent of the cases, and this may provide an important means of differentiating the tumors, particularly in the pelvis and femur.

Fibrosarcoma is much less common than either osteosarcoma or chondrosarcoma, comprising only 3.4 percent of the Mayo Clinic malignant tumor series (Dahlin, 1967). The major sites of skeletal involvement are closely similar to osteosarcoma. The age distribution of fibrosarcoma is fairly evenly spread between the ages of ten and sixty years which differs from the predominantly juvenile age group of osteosarcoma. The macerated specimens of fibrosarcoma may closely resemble a predominantly osteolytic osteosarcoma.

Parosteal osteosarcoma (juxtacortical osteosarcoma) is even less common than fibrosarcoma. It affects the same long bones as osteosarcoma, particularly the distal portion of the femur posteriorly. The average age of patients with the less malignant parosteal osteosarcoma is greater than for ordinary osteosarcoma. The lesion most often affects the metaphyseal region beneath the periosteum rather than within the affected bone as the conventional osteosarcoma does.

A solitary lesion of fibrous dysplasia may often be confused with osteosarcoma. Both lesions have similar sites of involvement and affect the same young age group. The lesion of fibrous dysplasia will usually be much smaller than the faster growing osteosarcoma. Furthermore, the area of bone rarefaction is much better defined than for osteosarcoma. Expansion with thinning of the cortex is not prominent except in smaller bones such as ribs.

Ewing's sarcoma may originate in the metaphyseal area of a long bone and create an appearance very similar to osteosarcoma. The age group is closely comparable to osteosarcoma. Ewing's sarcoma may produce the "sun burst" spiculation commonly seen in osteosarcoma just as osteosarcoma may produce the "onion peel" appearance associated with Ewing's sarcoma.

A solitary focus of metastatic carcinoma which is predominantly osteoblastic may simulate osteosarcoma. Metastatic lesions from carcinoma are by far the most common malignant tumors affecting the skeleton and chiefly afflict the older age group. On the basis of probability alone, it is much more likely that an osteolytic lesion in a person over forty years of age is metastatic rather than primary. A complete examination of the skeletal remains should be undertaken, both grossly and radiographically to ascertain if other lesions are present. The vertebrae in particular should be examined as they are the major site of metastatic involvement. Metastatic carcincoma is predominantly multifocal with several lesions in one bone or several bones while the primary bone tumors are usually solitary lesions.

Posttraumatic conditions such as myositis ossificans and excessive callus formation may suggest osteosarcoma except there is no break in the continuity of the cortex and no cortical destruction. Finally, an acute osteomyelitis in the shaft of a long bone of an adult may resemble an osteolytic osteosarcoma.

Paleopathological Cases

Reports of osteosarcoma in archaeological material emphasize the difficulty in diagnosing bone tumors. In what is perhaps the first description of a pathological fossil bone, a cave bear femur with a large growth was diagnosed as an osteosarcoma (Esper, 1774). The lesion was later found to be excessive callus following a fracture. The Kanam mandibular fragment from East Africa exhibits an asymmetrical swelling on the lingual and labial surfaces of the mandible near the symphysis. The lesion on this human jaw fragment from the Lower or Middle Pleistocene was diagnosed as an osteosarcoma (Lawrence, 1935). More recent opinion suggests that the swelling is instead due to callus following a fracture.

A 5th Dynasty femur from Egypt was originally described as an osteosarcoma (Smith and Dawson, 1924). The large irregular bone mass is situated very near the fused distal femoral epiphysis on the medial side. The lack of periosteal reaction around the base of the tumor and the absence of spiculation make a diagnosis of osteosarcoma highly unlikely (see Fig. 125). The growth is probably a benign osteochondroma (Rowling, 1961). A large tumor on the skull of a pre-Columbian Peruvian was originally diagnosed as an osteosarcoma (MacCurdy, 1923). The lesion is more likely due to a slowly invading meningioma as discussed in the previous section on meningiomas (see Fig. 140).

A very probable case of osteosarcoma has been reported in an Iron Age skeleton from Switzerland (Hug cited by Brothwell, 1967). The larger portion of the tumor involves the left humeral head and extends down about a third of the shaft. Spiculation at right angles to the shaft is readily apparent in this specimen. A very large osteosarcoma has been described in a Saxon male aged twenty to thirty years (Brothwell, 1967). The spiculated tumor is located at the distal end of the left femur and measures approximately 25 by 28 centimeters. A young adult male from a Medieval churchyard exhibits extensive destruction and some new bone formation in the right ilium perhaps caused by an osteosarcoma (Gejvall, 1960).

Ewing's Sarcoma

Incidence and Age

Ewing's sarcoma is a primary malignant tumor of bone apparently arising from immature reticular cells of the medullary cavity. It is surpassed in frequency among other primary malignant bone tumors by multiple myeloma, osteosarcoma, and chondrosarcoma. Ewing's sarcoma comprised 7 percent of the total malignant tumors in the Mayo Clinic series and 5 percent of the total series (Dahlin, 1967). The Netherlands Committee lists a higher frequency of 11.2 percent for the malignant bone tumor series and 5.7 percent for the total series.

Most cases of Ewing's sarcoma occur between the ages of ten to twenty years. Indeed, the tumor is rare after the age of thirty. Similar lesions in individuals past the age of thirty are more likely due to metastatic carcinoma.

Location

The skeletal location of Ewing's sarcoma is highly variable. The tumor arises in the hematopoietic tissue of the medullary cavity and thus may affect any portion of a long bone or flat bone. The epiphyseal region is rarely involved. The long bones, particularly the femur, tibia and humerus, are very common sites in persons under twenty years. After this age, the major sites of red marrow (hematopoietic tissue) are the flat bones. Therefore, it is not surprising that Ewing's sarcoma is most commonly found in the pelvic bones, ribs, and vertebrae in persons over twenty years (Aegerter and Kirkpatrick, 1968). Ewing's sarcoma affecting the midshaft is more likely to be centrally located in the medullary cavity, whereas the tumor in the metaphysis is usually eccentric (Ridings, 1964).

Unlike other primary bone tumors, Ewing's sarcoma commonly metastasizes to other bone sites producing punched-out areas of bone lysis, particularly in the flat bones. Such lesions may in fact be produced by a multifocal origin of the tumor rather than metastatic spread from the primary site.

Appearance

Ewing's sarcoma may often involve the entire shaft of a long bone. Even in cases where the roentgenogram reveals only partial involvement, the tumor tissue occupies a large part of the bone and has not as yet produced bone lysis of the cortex and trabeculae. The radiographic and gross appearances of Ewing's sarcoma are highly variable. Bone destruction is the most common finding with mottled rarefaction of both the medullary canal and cortex. Much of the destruction is due to necrosis following ischemia and the tumor tissue rapidly replaces the destroyed bone.

As the tumor extends through the cortex, the periosteum is elevated and is often stimulated to produce a thin layer of nonneoplastic bone. Successive elevation of the periosteum

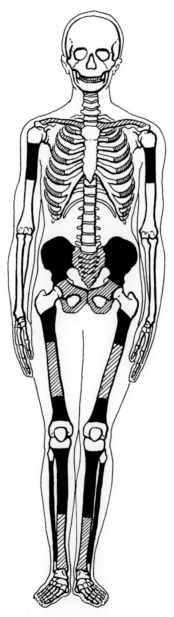

Figure 149. Skeletal distribution of Ewing's sarcoma. Black areas indicate the most frequent sites and diagonal lines mark occasional sites.

and reactive bone formation creates an "onion peel" appearance of the thickened cortex. This appearance was evident in all Ewing's sarcomas of long bones classified as central diaphyseal (Sherman and Soong, 1956). However, these periosteal changes are less common in Ewing's sarcoma of other bone sites.

As mentioned earlier, the periosteal response in Ewing's sarcoma occasionally produces a "sun burst" appearance very similar to osteosarcoma. In other instances the bone necrosis and subperiosteal bone apposition may produce a lesion resembling acute osteomyelitis or eosinophilic granuloma. Involvement of several bones with destructive lesions of Ewing's sarcoma may suggest the more common metastatic carcinoma. The large difference in the age groupings for the two processes helps separate them.

Multiple Myeloma (Plasma Cell Myeloma, Myelomatosis)

Incidence and Age

Multiple myeloma is a malignant neoplastic proliferation of the plasma cells in the bone marrow. Other common names for this condition are plasma cell myeloma and myelomatosis. Multiple myeloma is the most common primary malignant bone tumor. It comprised 43 percent of the malignant bone tumors in the Mayo Clinic series (Dahlin, 1967), although other series may have a smaller proportion of cases (Jacobson, 1971).

Multiple myeloma is a tumor of the older age group with almost 90 percent of the cases over the age of fifty years (*End Results in Cancer* #3, 1968). This tumor is uncommon below the age of forty and is indeed rare in adolescents and children (Geschickter and Copeland, 1949; Porter, 1963).

Location

The bones that contain hematopoietic marrow in adults are the major sites of involvement. The bones most frequently involved are vertebrae, rib, skull, and pelvis. The proximal portions of the femur and humerus are often involved in advanced cases when their yellow marrow is converted to red marrow in an effort to alleviate the anemia. Multiple myeloma lesions are

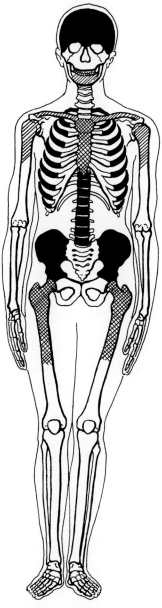

Figure 150. Skeletal distribution of multiple myeloma. Black areas indicate the most frequent sites; checked areas mark common sites; and diagonal lines indicate occasional sites.

rare in the bones distal to the elbows and knees (Carson et al., 1955; Aegerter and Kirkpatrick, 1968). In the shoulder, the scapula and clavicle may be perforated by lesions while the humerus is untouched. The scapular lesions are found almost exclusively along the axillary border and the acromion. The lesions of the clavicle are mainly on the lateral ends near the glenoid fossa of the scapula (Moseley, 1963).

Appearance

Uncontrolled proliferation of myeloma cells in the bone marrow results in various bone changes of which the most characteristic are the punched-out lesions of bone lysis. The variation in size of such lesions is considerable although most are relatively small compared to those of metastatic carcinoma. The sharply circumscribed myeloma lesions occur most frequently in the skull and long bones. Similar lesions may often involve the lateral portion of the clavicle and the acromion process of the scapula. Bone sclerosis around these lytic lesions is very uncommon.

Multiple myeloma lesions begin in the bone marrow and encroach upon the cortex often producing a scalloped appearance in the roentgenogram. There may be expansion of the cortex by this invasion, but more often there is complete perforation with sharply demarcated borders of destruction. In thin flat bones such as the bones of the skull vault, the tumor usually perforates both inner and outer tables to invade the contiguous soft tissues (see Figs. 152 & 153).

Diffuse lesions of multiple myeloma may occur, particularly in the vertebrae and ribs. The trabeculae become fine and sparse with thinning of the cortex. The generalized osteoporosis in these bones often produces pathologic fractures and collapse of the vertebral bodies. When the well-defined lesions in the sternum, clavicle, and long bones coalesce to form larger areas of destruction, pathologic fractures may occur here as well. Pathologic fractures have been found in 62 percent of ninety-seven patients (Snapper et al., 1953).

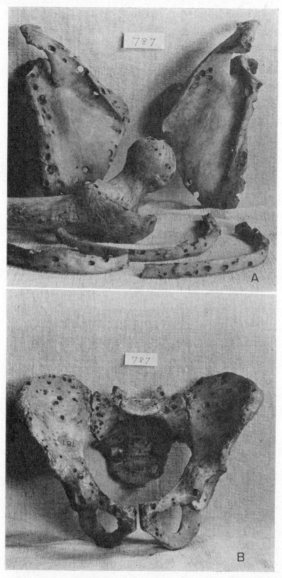

Figure 151. Multiple myeloma in a fifty-year-old male from the Terry Collection of autopsy skeletons. (A.) The scapulae have many punched-out lesions limited to the axillary and vertebral borders and the acromion spine. The proximal head of the femur is affected by the perforations and all parts of the ribs are involved. (B.) The pelvis is literally riddled by the sharply defined lesions which originate in the medullary space. (USNM-Terry Collection 787)

378 Paleopathological Diagnosis and Interpretation

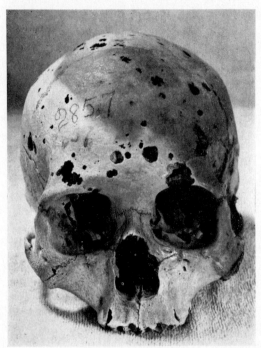

Figure 152. Multiple myeloma in a pre-Columbian Peruvian. This frontal view emphasizes the complete absence of bony reaction around the lesions. Many of the perforations have scalloped borders. (USNM 242,559)

Paleopathology

Eight possible cases of multiple myeloma have been reported in New World archaeological material. Two other previously unreported cases are described here. Table XV provides some basic information on each of these specimens. One of the individuals is only ten years old and is more likely a case of histiocytosis X. Three of the four cases described by Morse and his associates are individuals under the age of forty years, and one of these has a large area of erosion affecting the inner table and surrounded by a nodular periostitis quite atypical of the disease. The other cases listed in Table XV have extensive skeletal involvement fairly characteristic of multiple myeloma. Nearly identical lesions may be produced by carcinomatous metastases and in view of its high prevalence in populations

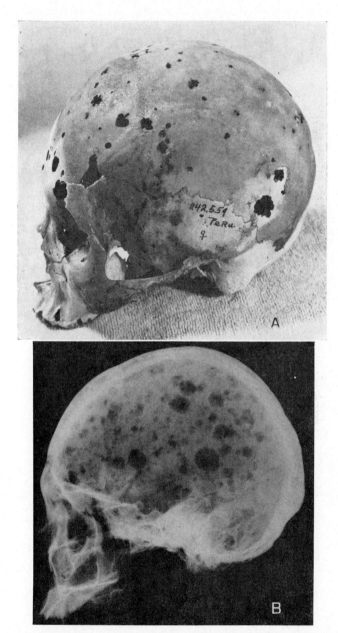

Figure 153. Multiple myeloma in the same individual as the previous figure. (A.) Lateral view of the skull. (B.) Roentgenogram reveals dozens of small lytic defects throughout the skull vault. There is no evidence of sclerosis. (USNM 242,559)

TABLE XV
REPORTED PALEOPATHOLOGICAL CASES OF MULTIPLE MYELOMA

Date	Sex	Age	Locality	Preservation	Report
~3300 B.C.	F	46-50	Indian Knoll, Ky.	complete	USNM 290,064
200-900 A.D.	F	25	Calico Hills, Fla.	incomplete	Morse et al., 1974
200-900 A.D.	F	<25	Calico Hills, Fla.	calvarium	Morse et al., 1974
500-1200 A.D.	M	45	Sowell Mound, Fla.	calvarium	Morse et al., 1974
300-1450 A.D.	F	45	Santa Cruz Is., Calif.	nearly complete	Morse, 1969
800 A.D.	M	>50	Binghamton, N.Y.	nearly complete	Ritchie and Warren, 1932
1200 A.D.	?	10	Rochester, N.Y.	nearly complete	Williams et al., 1941
1200 A.D.	F	>40	Kane Mound, Mo.	nearly complete	Brooks and Melbye, 1967
1300 A.D.	F	35	Mangum Mound, Miss.	nearly complete	Morse et al., 1974
1500 A.D.	F	>50	Peru	skull	USNM 242,559
2000-1500 B.C.	M	50	Pyrenees	skull	Fusté, 1955
900-1200 A.D.	M	>50	Kérpuszta, Hungary	incomplete	Nemeskéri and Harsányi, 1959
1100-1600 A.D.	?	adult	Ipswich, Britain	calvarium	Wells, 1964
1300-1600 A.D.	?	adult	Sandwich, Britain	calvarium	Wells, 1964

today, it seems wise to regard metastatic carcinoma as a possibility in all of these cases (see Brothwell, 1967).

Two previously unreported specimens of multiple myeloma are at the Smithsonian Institution. USNM 242,559 is the isolated skull of an adult Peruvian female, probably of pre-Hispanic date. Literally dozens of small lytic lesions perforate one or both tables of the skull vault (see Fig. 152). The bones of the base of the skull are also involved. The roentgenogram reveals many other small areas of bone lysis originating in the diploic space. There is no sign of bone sclerosis around the lesions.

USNM 290,064 is the complete skeleton of an adult female aged about fifty years. This individual is from Indian Knoll, Kentucky, a Middle Archaic site around 3300 B.C. The skull exhibits several small punched-out lesions in the frontal, parietal, and occipital bones measuring 2 to 12 mm. The roentgenogram reveals dozens of lytic lesions originating in the diploic tissue between the inner and outer tables (see Fig. 154). The mandible has large lesions near the left condyle and behind the right third molar.

The postcranial skeleton exhibits destructive lesions in many of the bones. Both clavicles have lesions at the medial and lateral ends. Both scapulae present extensive destructive lesions along the axillary border and smaller lesions near the glenoid cavity. A roentgenogram reveals the beginnings of bone lysis in the acromion of the right scapula. Both humeri have small lesions at the proximal end just inferior to the articular surface. The ulnae and radii are not involved.

Only six vertebrae remain in this individual: the 1st, 2nd, 4th, and 6th cervicals, the 1st thoracic and 1st lumbar. Though from different regions of the vertebral column, all exhibit extensive destruction, and these lesions probably riddled the entire spine.

Both femora have destructive lesions at the proximal end as shown in the roentgenogram. A close examination reveals several small foci of bone lysis in the diaphysis of the left femur. They originate in the medullary cavity. The bones of the lower legs and feet have no lesions.

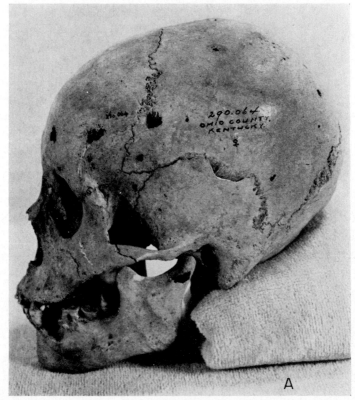

Figure 154. Multiple myeloma in an adult female from Indian Knoll, Kentucky. (A.) Lateral view of skull showing several perforations ranging in size from 1 to 15 mm. The mandible is also involved. (USNM 290,064)

The pelvis and ribs, both important sites of involvement in multiple myeloma, are not present. Fortunately, the collector of this skeleton, Clarence B. Moore, includes a photograph of this individual *in situ* in his report (Moore, 1916, burial 132). He makes no mention of the pathological lesions but uses this individual to illustrate the flexed position of the Indian Knoll burials. The pelvis and ribs are both visible in the photograph, and both exhibit destructive lesions, particularly in the pelvis near the anterior inferior iliac spine.

The size, location, and appearance of the destructive lesions

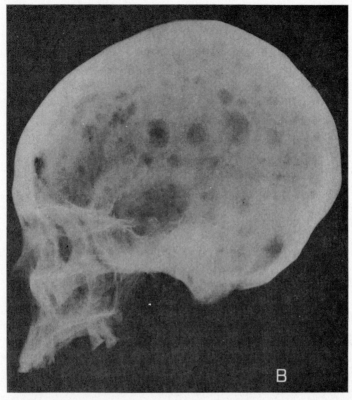

Figure 154B. Roentgenogram reveals many small areas of lysis without areas of bone sclerosis. (USNM 290,064)

in this Archaic Indian are highly indicative of multiple myeloma rather than metastatic carcinoma. There is no bone sclerosis surrounding any of the lesions. The vertebrae, clavicles, and ribs are involved in 90 percent of all myeloma cases and are evident in this person as well (Geschickter and Copeland, 1949). Lesions in the proximal ends of the femur and humerus with no bone lysis below the elbow or knee are highly characteristic of multiple myeloma. Thus, the diagnosis reached here is multiple myeloma in this fifty-year-old female with an alternative diagnosis of metastatic carcinoma.

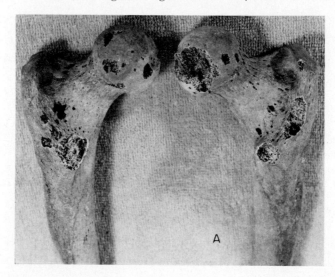

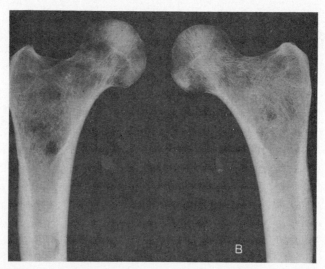

Figure 155. Multiple myeloma. (A.) Posterior view of both femora with many perforations of the cortex. (B.) Roentgenogram reveals several more areas of lysis located distally in the left femur. (USNM 290,064)

Metastatic Carcinoma

Incidence

Metastatic carcinoma of bone is much more common than the primary malignant bone tumors. Metastatic carcinoma lesions in bone are produced by detached fragments of a soft tissue cancer which are disseminated in the body by the bloodstream and set up secondary foci of neoplastic growth. As mentioned earlier, the skeleton is one of the most common sites of cancer metastasis. Indeed, one noted author states that approximately two-thirds of all malignant neoplasms metastasize to the skeleton. The figure may reach as high as 85 percent for the most common carcinomas: breast, lung, prostate, kidney, and thyroid (Jaffe, 1958). Many of these lesions are quite small and require sectioning of the gross specimen to be seen. Archaeological material will not reveal the small neoplastic foci present in the vertebrae even when sectioned. Using radiography and gross examination of the exterior rather than sectioning, one would expect a lower frequency of 25 to 30 percent for the incidence of neoplastic metastasis to bone. A similar frequency of 27 percent for metastasis to bone was found in 1000 consecutive autopsies of patients who died with carcinoma (Abrams et al., 1950).

Age and Location

The major carcinomas metastasizing to the skeleton affect the older age group, particularly after the age of fifty years. Metastatic lesions are found most often in the vertebral column, particularly in the thoracic region followed by the lumbar region and sacrum (Jaffe, 1958). Other major sites of metastasis are the ribs, pelvis, proximal end of the femur, sternum, and skull. Carcinomas rarely metastasize to bones distal to the elbow and knee.

The route of metastasis is very interesting and worth mentioning here. Very often the skeleton will have a multitude of metastatic lesions while the soft tissues are untouched. In fact, bone lesions are often the first clinical sign of carcinoma. Rather than first travelling to the lungs, cancer cell emboli may reach the bone directly through the bloodstream. This occurs via the

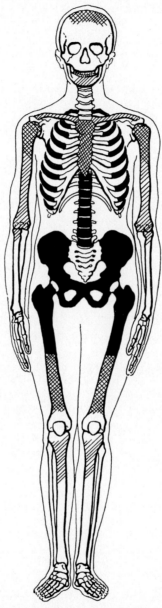

Figure 156. Skeletal distribution of metastatic carcinoma. Black areas indicate the most frequent sites; checked areas mark common sites; and diagonal lines indicate occasional sites.

"vertebral vein system" (Batson, 1942) also called the paravertebral venous plexus. This plexus indirectly links the inferior and superior vena cavae thus avoiding the liver and lungs. The paravertebral venous plexus is a network of valveless veins around the vertebrae with connections to the cranium and extremities. This low pressure, sluggish system may be temporarily static or even reverse flow allowing tumor emboli to set up secondary foci of proliferation. Through these interconnections, tumor emboli may pass from the systemic circulation to the paravertebral plexus and thus reach the vertebral column, pelvis, and cranium. The thin-walled veins allow tumor penetration more easily than the thicker arteries. Arterial spread of metastases may occur, particularly to the highly vascularized bone marrow.

Appearance

Metastatic lesions may be either osteolytic, osteoplastic, or a mixture of the two processes. The majority of metastatic bone lesions are predominantly destructive. Most osteolytic lesions are produced by carcinomas of the kidney, thyroid, lung, and gastrointestinal tract. Breast carcinoma produces mixed lesions while carcinoma of the prostate almost always produces osteoplastic lesions (Coley, 1960; Spjut et al., 1971). The bone production in osteoplastic lesions represents a reactive response by the normal bone-forming tissue to the tumor and is not caused by the malignant cells themselves. The bone destruction in osteolytic lesions is caused by growth pressure from the proliferating tumor tissue.

The radiographic and gross appearances of metastatic bone lesions do not permit specific differentiation of the different carcinomas except in a general way as outlined above. Furthermore, the multiple lesions often closely resemble the punched-out lesions of multiple myeloma. The tumor emboli are "seeded" in the bone marrow, and the proliferating tissue may create sharply demarcated round or oval areas of lysis or a generalized osteoporosis (particularly in the vertebrae) not recognized by the inexperienced observer. Pathological fractures and vertebral collapse are common.

Differential Diagnosis

A differential diagnosis between metastatic carcinoma and multiple myeloma is difficult. Both processes affect the same age group and involve essentially the same area of the skeleton. Multiple myeloma lesions are very rarely sclerotic while this often occurs in the slower growing carcinomas. The vertebral pedicles are often involved in metastatic carcinoma compared to little involvement in multiple myeloma. Lesions of the pedicles were found in 94 percent of seventy-four cases of metastatic carcinoma compared to only 15 percent in fifty-four cases of multiple myeloma (Jacobson et al., 1958). In general, the lesions of multiple myeloma appear more constant in size and more numerous than those of metastatic carcinoma, although considerable variation may be found.

Paleopathology

Although much more common than primary malignant bone tumors, very few cases of possible metastatic carcinoma have been recorded in the literature. Table XVI lists some basic information on these cases and an additional four specimens previously unreported. All are characterized by sharply defined lytic lesions ranging from a few millimeters to several centimeters in diameter and often surrounded by varying degrees of nodular periostitis or sclerosis.

In his classic study of the Pecos Pueblo Indians, Earnest Hooton briefly lists two individuals with metastatic disease. Both are at the Peabody Museum at Harvard and have been re-examined. PM 59802 is an adult female with metastatic lesions in the left ulna and radius and several vertebrae. The bones of the left forearm are much lighter than those of the right. The distal epiphyses of both bones have been destroyed and the remaining portions are very thin. This could easily be due to postmortem erosion. The proximal ends show slight periosteal proliferation of bone, particularly around the articular surface of the ulna. Radiography reveals no medullary areas of lysis. The cervical and upper thoracic vertebrae appear normal while six lower thoracics and two lumbars have many lytic areas in the vertebral bodies and pedicles. Most of the cortex has been

TABLE XVI
REPORTED PALEOPATHOLOGICAL CASES OF METASTATIC CARCINOMA

Date	Sex	Age	Preservation	Bones Involved	Locality	Report
~3000 B.C.	M	30-35	skull	skull	Egypt	Wells, 1963a
425-850 A.D.	?	?	skull and partial skeleton	skull, pelvis, and several vertebrae	Britain	Brothwell, 1967
1200-1500 A.D.	F	30	skull	skull	Denmark	Møller and Møller-Christensen, 1952
1100-1400 A.D.	M	40-50	skull and partial skeleton	skull, right ilium, right humerus, and left femur	Sweden	Gejvall, 1960
500-900 A.D.	F	40-50	skull and partial skeleton	left ulna and radius, eight vertebrae	Pecos Pueblo, New Mexico	PM 59802*
1000-1300 A.D.	F	30-35	skull and partial skeleton	right humerus and ulna, nine thoracic vertebrae	Pecos Pueblo, New Mexico	PM 59834*
before 1500	F	40-50	skull	skull	Llactashica, Peru	PM 8074
before 1500	M	40-50	skull and partial skeleton	skull, pelvis, and many vertebrae	Huacho, Peru	USNM 379,293
500-1500 A.D.	F	30-40	skull and partial skeleton	skull, ribs, clavicle, and eleven thoracic vertebrae	Hooper Bay, Alaska	USNM 339,122
500-1500 A.D.	M	40-50	skull	skull	Saint Lawrence Island, Alaska	USNM 280,691

* Originally reported by Hooton, 1930 and re-examined.

destroyed in the lumbar vertebrae revealing bone condensation of the spongy tissue around the many areas of bone destruction. The sacrum and remainder of the skeleton including the skull appear normal when examined radiographically.

PM 59834 is an adult Pueblo Indian female with extensive bone destruction affecting the right elbow and all twelve thoracic vertebrae (see Figs. 157 & 158). The vertebral destruction is almost entirely confined to the bodies and four of the vertebrae are fused by a pronounced bony reaction. There is marked sclerosis of the trabeculae in the upper thoracic vertebrae surrounding the many localized areas of destruction (Fig. 158). Hooton incorrectly refers to the right ulna as being a left ulna. He also mentions possible metastatic lesions on the frontal and left parietal which in the opinion of the present author are not

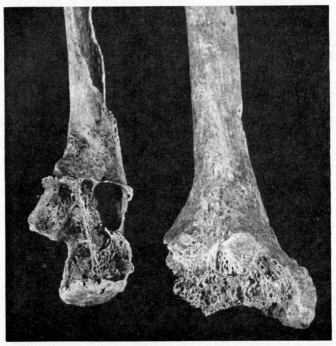

Figure 157. Anterior view of right ulna and humerus from an adult Pueblo Indian woman with metastatic carcinoma. The epiphyses are almost completely destroyed by the lytic metastases. The remaining trabeculae are thickened. (PM 59834)

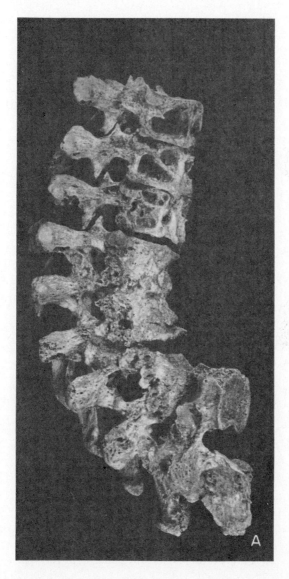

Figure 158. Thoracic portion of the vertebral column showing destruction produced by metastatic carcinoma. (A.) External view.

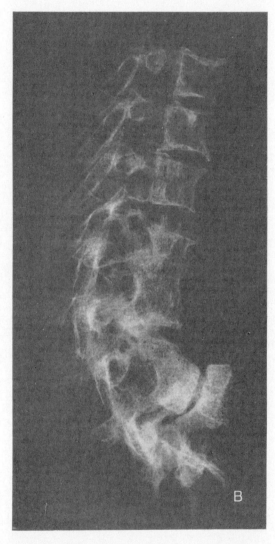

Figure 158B. Roentgenogram. (PM 59834)

related to the malignant process. Trauma could well account for the small, healed depression on the frontal bone and postmortem erosion has affected the left parietal in this very fragmentary skull. Radiography reveals no lytic defects in the bones of the skull and remainder of the skeleton.

Two ancient Peruvians present good evidence of metastatic carcinoma. A solitary skull from Llactashica, Peru has a large, irregular area of bone destruction mainly involving the left parietal and apparently originating in the outer table (Fig. 159). The surrounding bone is severely pitted by the destructive process with little evidence of bone regeneration. A skull from Huacho, Peru shows even more extensive destruction of bone involving most of the frontal bone (Fig. 160). The same pitting process of advancing destruction is seen around the lesion. The vertebrae and pelvis from this individual exhibit similar lesions.

Two Alaskan Eskimos from the large Smithsonian Institution series have bone metastases from carcinoma. The solitary skull of a male Eskimo from St. Lawrence Island has a large circumscribed lesion in the left parietal and several smaller lesions in other bones of the cranial vault (Fig. 161). The lesions are bevelled with the larger opening in the inner table. Several lesions may be seen endocranially which do not perforate the outer table.

A female Eskimo from Hooper Bay has seven perforations of the outer table and the roentgenogram reveals several more small zones of destruction (Fig. 162). Many of the lesions are surrounded by a nodular periostitis and sclerosis of the diploic tissue. Similar lesions are evident in the ribs, clavicle, and thoracic vertebrae.

Paleopathology and Tumorous Conditions

There is a paucity of reports on any tumorous conditions in ancient populations. Perhaps a major factor is the youthful age composition of many prehistoric and early historic human populations. The average life spans of individuals in these populations were much shorter than in most modern societies. Few people lived past the age of fifty years which is the predominant age group for carcinoma and multiple myeloma. However, indi-

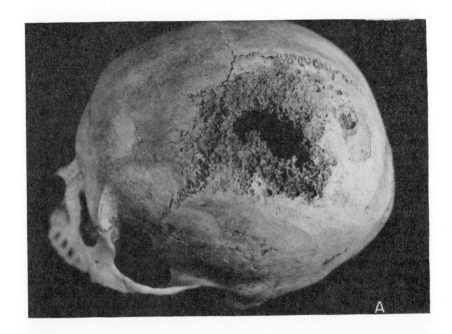

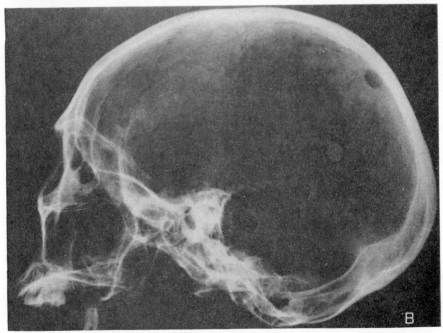

Figure 159. Metastatic carcinoma in an adult female from Llactashica, Peru. (A.) Large lytic defect in the left parietal surrounded by irregular pitting. A smaller lesion is evident posteriorly.
Figure 159B. Roentgenogram reveals very little bone reaction to the destruction. (PM N/8074)

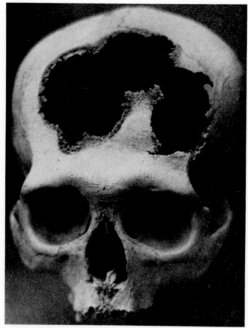

Figure 160. Metastatic carcinoma in an adult from Huacho, Peru. The skull has a large defect eroding much of the frontal bone and surrounded by a zone of pitting destruction. (USNM 379,293)

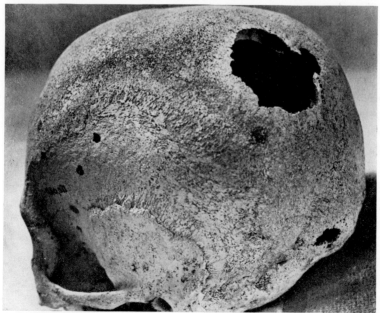

Figure 161. Metastatic carcinoma in a male Eskimo from St. Lawrence Island, Alaska. The solitary skull reveals a large defect in the left parietal and several smaller lesions in other bones of the vault. (USNM 280,091)

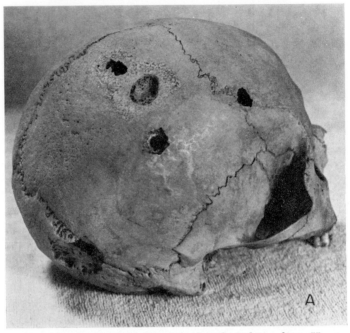

Figure 162. Metastatic carcinoma in a female Eskimo from Hooper Bay, Alaska. (A.) Lateral view showing lesions in several bones of the vault. Some lesions are surrounded by zones of nodular periostitis and sclerosis of the diploic tissue.

viduals in the younger age groups are particularly susceptible to certain tumorous growths such as osteochondroma, fibrous dysplasia, giant-cell tumor, histiocytosis X, Ewing's tumor, and osteosarcoma. Such tumors and tumor-like processes may be evident in early skeletal series, although the low frequency of bone tumors mentioned in the beginning of this chapter should be reemphasized.

With the exception of several classic studies of Egyptian skeletal remains (Ruffer, 1921; Smith and Dawson, 1924) and the Pueblo Indians (Hooton, 1930), few skeletal series have been examined thoroughly for tumorous processes. Even more unfortunately, at least three-fourths of all archaeological skeletal material consists of skulls without the postcranial skeleton. This imbalance was created by the 19th century anthropologist's bias for the skull and disinterest in the postcranial remains. Fortun-

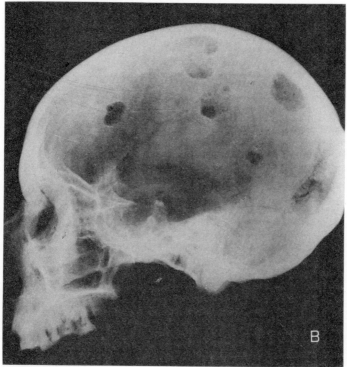

Figure 162B. Roentgenogram revealing several small areas of bone lysis surrounded by bony sclerosis. (USNM 339,122)

ately, modern archaeologists and physical anthropologists are more concerned about recovering the entire skeleton and indeed, preserving individuals of all ages rather than keeping only the adult skeletons. Thus, more intense studies of skeletal series now available are needed to determine the frequency of certain tumorous conditions in early human populations.

Such population studies certainly apply to other disease processes as well. Reports of isolated pathological specimens have little meaning unless placed in their archaeological and epidemiological context. Studies of such conditions as osteoarthritis, and dental disease are practically worthless unless an epidemiologic approach is employed. Hopefully, this book has provided a foundation in gross bone pathology for those interested in the study of bone diseases in ancient human populations, particularly within the framework of a population study.

BIBLIOGRAPHY
BONE TUMORS AND TUMOR-LIKE PROCESSES

Abbott, K. H. and C. B. Courville: "Historical notes on the meningiomas," *Bull. Los Angeles Neurol. Soc.*, 4:101-113, 1939.

———: "Notes on the pathology of cranial tumors. I. Osteomas of the skull with incidental mention of their occurence in the ancient Incas," *Bull. Los Angeles Neurol. Soc.*, 10:19-34, 1945.

Abrams, H. L., R. Spiro, and N. Goldstein: "Metastases in carcinoma. Analysis of 1000 autopsied cases," *Cancer*, 3:74-85, 1950.

Acsádi, G. and J. Nemeskéri: *History of Human Life-span and Mortality*, Akademiai Kiado, Budapest, 1970.

Aegerter, E. and J. Kirkpatrick: *Orthopedic Diseases*, W. B. Saunders, Philadelphia, 2nd ed., 1968.

Batson, O. V.: "The vertebral vein system as a mechanism for the spread of metastases," *Amer. J. Roentgenol.*, 48:715-718, 1942.

Béraud, C., P. Morel, and A. R. Boyer: "Ostéome géant fronto-ethmoïdal découvert sur un crâne médiéval du Var," *J. Radiol. Electrol.*, 42:45-47, 1961.

Berry, A. C. and R. J. Berry: "Epigenetic variation in the human cranium," *J. Anat.*, 101:361-379, 1967.

Brooks, S. T. and J. Melbye: "Skeletal lesions suggestive of pre-Columbian multiple myeloma," in W .D. Wade (ed.): *Miscellaneous Papers in Paleopathology; Mus. North. Ariz. Tech. Series*, 7, 1967.

Brothwell, D. R.: "The palaeopathology of early British man," *J. Roy. Anthrop. Inst.*, 91:318-344, 1961.

———: "Evidence for neoplasms," in D. R. Brothwell and A. T. Sandison (ed.): *Diseases in Antiquity*, Thomas, Springfield, 1967.

Bullough, P. G.: "Ivory exostosis of the skull," *Postgrad. Med. J.*, 41:277-281, 1965.

Caffey, J.: "On fibrous defects in cortical walls of growing tubular bones," *Advances Pediat.*, 7:13-51, 1955.

Carson, C. N., L. V. Ackerman, and J. D. Maltby: "Plasma cell myeloma. A clinical, pathologic and roentgenologic review of 90 cases," *Amer. J. Clin. Path.*, 25:849-888, 1955.

Coley, B. L.: *Neoplasms of Bone*, P. B. Hoeber, New York, 1960.

Cushing, H.: "The cranial hyperostoses produced by meningeal endotheliomas," *Arch. Neurol. Psych.*, 8:139-152, 1922.

Dahlin, D. C.: *Bone Tumors*, Thomas, Springfield, 2nd. ed., 1967.

Denninger, H. S.: "Osteitis fibrosa in a skeleton of a prehistoric American Indian," *Arch. Path.*, 11:939-944, 1931. Also in D. R. Brothwell and A. T. Sandison (ed.): *Diseases in Antiquity*, Thomas, Springfield, 1967.

Derry, D. E.: "Anatomical report," *Arch. Survey of Nubia Bull.*, 6:42, 1909.

End Results in Cancer (Report #3), U.S. Dept. H.E.W., Washington, D.C., 1968.
End Results in Cancer (Report #4), U.S. Dept. H.E.W., Washington, D.C., 1972.
Esper, E. J. C.: *Ausführliche Nachrichten von neuenteckten Zoolithen umbekannter vierfüssiger Thiere*, Nurenberg, 1774.
Fusté, M.: "Anthropología de las poblaciones pirenáicas durante el periodo neo-eneolitico," *Trab. Inst. 'Bernardino de Sohagún' Antrop. y Etnologia*, *14*:109, 1955.
Gejvall, N-G.: *Westerhus: Medieval Population and Church in the Light of Skeletal Remains*, Boktryckeri, Lund, 1960.
Geschickter, C. F. and M. M. Copeland: *Tumors of Bone*, Lippincott, Philadelphia, 1949.
Gregg, J. B. and W. M. Bass: "Exostoses in the external auditory canals," *Ann. Otol., Rhinol, Laryngol.*, 7:834-839, 1970.
Hooton, E. A.: *The Indians of Pecos Pueblo*, Yale Univ. Press, New Haven, 1930.
Hrdlička, A.: "Ear exostoses," *Smithson. Misc. Coll.*, 93, 1935.
Hug, E.: *Die Anthropologische Sammlung in Naturhistorischen Museum Bern*, Natural History Museum, Bern, 1956.
Jacobson, H. G., M. H. Poppel, J. H. Shapiro, and S. Grossberger: "The vertebral pedicle sign: A roentgen finding to differentiate metastatic carcinoma from multiple myeloma," *Amer. J. Roentgenol.*, 80:217, 1958.
Jacobson, S. A.: *The Comparative Pathology of the Tumors of Bone*, Thomas, Springfield, 1971.
Jaffe, H. L.: "Giant-cell reparative granuloma, traumatic bone cyst, and fibrous (fibro-osseous) dysplasia of the jaw bones," *Oral Surg.*, 6:159-175, 1953.
——————: *Tumors and Tumorous Conditions of the Bones and Joints*, Lea and Febiger, Philadelphia, 1958.
Johnson, L. C.: "A general theory of bone tumors," *Bull. N.Y. Acad. Med.*, 29:164-171, 1953.
Krogman, W. M.: "The skeletal and dental pathology of an early Iranian site," *Bull. Hist. Med.*, 8:28-48, 1940.
Lawrence, J. W. P.: "Appendix A," in L. S. B. Leakey: *Stone Age Races of Kenya*, Oxford Univ. Press, London, 1935.
Lichtenstein, L.: "Histiocytosis X. Integration of eosinophilic granuloma of bone, Letterer-Siwe disease, and Schüller-Christian disease as related manifestations of a single nosologic entity," *Arch. Path.*, 56:84-102, 1953.
——————: *Bone Tumors*, C. V. Mosby, St. Louis, 2nd ed., 1970.
Lieberman, P. H., C. R. Jones, H. W. Dargeon, and C. F. Begg: "A reappraisal of eosinophilic granuloma of bone, Hand-Schüller-Christian

syndrome, and Letter-Siwe syndrome," *Medicine*, 48:375-400, 1969.

MacCurdy, G. G.: "Human skeletal remains from the highlands of Peru," *Amer. J. Phys. Anthrop.*, 6:217, 1923.

McKenna, R. J., C. P. Schwinn, K. Y. Soong, and N. L. Higinbotham: "Sarcomata of the osteogenic series (osteosarcoma, fibrosarcoma, chondrosarcoma, parosteal osteogenic sarcoma, and sarcomata, arising in abnormal bone," *J. Bone Joint Surg.*, 48A:1-26, 1966.

Møller, P. and V. Møller-Christensen: "A medieval female skull showing evidence of metastases from a malignant growth," *Acta Path. Microbiol. Scand.*, 30:336-342, 1952.

Moodie, R. L.: "Tumors of the head among the pre-Columbian Peruvians," *Ann. Med. Hist.*, 8:394-412, 1926.

Moore, C. B.: "Some aboriginal sites on Green River, Kentucky," *J. Phila. Acad. Nat. Sci.*, 16:431-511, 1916.

Morse, D.: *Ancient Disease in the Midwest*, Illinois State Mus. Rep. of Invest., #15, 1969.

Morse, D., R. C. Dailey, and J. Bunn: "Prehistoric multiple myeloma," *Bull. N.Y. Acad. Med.*, 54:447-458, 1974.

Moseley, J. E.: *Bone Changes in Hematologic Disorders*, Grune and Stratton, New York, 1963.

Nemeskéri, J. and L. Harsányi: "Die Bedeutung paläopathologischer untersuchungen für die historische Anthropologie," *Homo*, 10:203-226, 1959.

Netherlands Committee on Bone Tumors. Radiological Atlas of Bone Tumors, Williams and Wilkins, Baltimore, 1966.

Porter, F. S.: "Multiple myeloma in a child," *J. Pediat.*, 62:602-604, 1963.

Price, C. H. G.: "The incidence of osteogenic sarcoma in southwest England and its relationship to Paget's disease," *J. Bone Joint Surg.*, 44B:366-376, 1962.

Putschar, W. G. J.: "Problems in the pathology and paleopathology of bone," in S. Jarcho (ed.): *Human Palaeopathology*, Yale Univ. Press, New Haven, 1966.

Ridings, G. R.: "Ewing's tumor," *Radiol. Clin. N. Amer.*, 2:315-325, 1964.

Ritchie, W. A. and S. L. Warren: "The occurrence of multiple bony lesions suggesting myeloma in the skeleton of a pre-Columbian Indian," *Amer. J. Roentgenol.*, 28:622-628, 1932.

Roche, A. F.: "Aural exostoses in Australian aboriginal skulls," *Ann. Otol., Rhinol., Laryngol.*, 73:82-91, 1964.

Rogers, L.: "Meningiomas in Pharaoh's people: hyperostosis in ancient Egyptian skulls," *Brit. J. Surg.*, 36:423-424, 1949.

Rowling, J. T.: "Pathological changes in mummies," *Proc. Roy. Soc. Med.*, 54:409-415, 1961.

Rubin, P.: *Dynamic Classification of Bone Dysplasias,* Year Book Med. Publ., Chicago, 1964.
Ruffer, M. A.: *Studies in the Palaeopathology of Egypt,* R. L. Moodie (ed.), Univ. of Chicago Press, Chicago, 1921.
Selby, S.: "Metaphyseal cortical defects in the tubular bones of growing children," *J. Bone Joint Surg., 43A*:395-400, 1961.
Sherman, R. S. and K. Y. Soong: "Ewing's sarcoma: its roentgen classification and diagnosis," *Radiology, 66*:529-539, 1956.
Sherman, R. S. and D. Wilner: "The roentgen diagnosis of hemangioma of bone," *Amer. J. Roentgenol., 86*:1146-1159, 1961.
Smith, E. G. and W. R. Dawson: *Egyptian Mummies,* Allen and Unwin, London, 1924.
Smith, E. G. and F .W. Jones: *The Archaeological Survey of Nubia Report for 1907-1908,* Vol. II, *Report on the Human Remains,* Cairo, 1910.
Snapper, I., L. B. Turner, and H. L. Moscovitz: *Multiple Myeloma,* Grune and Stratton, New York, 1953.
Spjut, H. J., H. D. Dorfman, R. E. Fechner, and L. V. Ackerman: *Tumors of Bone and Cartilage,* Fascicle 5 of *Atlas of Tumor Pathology,* Armed Forces Institute of Pathology, Washington, D.C., 2nd series, 1971.
Stradford, H. T.: "Bone tumors," in P. Rubin (ed.): *Clinical Oncology for Medical Students and Physicians,* Amer. Cancer Soc., Rochester, 1971.
Urteaga, O. and J. E. Moseley: "Craniometaphyseal dysplasia (Pyle's disease) in an ancient skeleton from the Mochica culture of Peru," *Amer. J. Roentgenol., 99*:712-716, 1967.
Vyhnánek, L.: "Osteoma osteoidum. Eine Kasuistik aus dem frühmittelalterichen Skelettmaterial," *Z. Orthop., 109*:922-923, 1971.
Weinfield, M. S. and H. R. Dudley: "Osteogenic sarcoma. A follow-up study of the 94 cases observed at the Massachusetts General Hospital from 1920 to 1960," *J. Bone Joint Surg., 44A*:269-276, 1962.
Wells, C.: "Ancient Egyptian pathology," *J. Laryngol. Otol., 77*:261-265, 1963a.
―――――: "Polyostotic fibrous dysplasia in a 7th century Anglo-Saxon," *Brit. J. Radiol., 36*:925-926, 1963b.
―――――: "Two medieval cases of malignant disease," *Brit. Med. J., 1*: 1611-1612, 1964.
―――――: "A pathological Anglo-Saxon femur," *Brit. J. Radiol., 38*:393-394, 1965.
Williams, G. D., W. A. Ritchie, and P. F. Titterington: "Multiple bony lesions suggesting multiple myeloma in a pre-Columbian Indian aged ten years," *Amer. J. Roentgenol., 46*:351-355, 1941.
Willis, R. A.: *Pathology of Tumours,* Butterworth, London, 3rd ed., 1960.
Zariquiey, M. O.: "Magicians and meningiomas," *Med. Radiog. Photog., 34* (3):70-72, 1958.

AUTHOR INDEX

A

Abbott, K. H., 328, 353, 355, 397
Abel, O., 82, 83
Abrams, H. L., 385, 397
Acheson, R. M., 47, 55
Ackerknecht, E. H., 16, 301, 312
Acsádi, G., 110, 160, 397
Aegerter, E., 15, 18, 55, 273, 321, 327, 350, 362, 363, 372, 376, 397
Akrawi, F., 160
Aksoy, M., 221, 231, 232, 238, 248
Allison, A. C., 234, 248
Allison, M. J., 50, 55, 173, 188, 237, 248
Andersen, J. G., 97, 160, 190, 193, 197, 198, 199, 201, 207, 209
Anderson, J. E., 23, 55, 95, 160, 235, 248
Angel, J. L., 42, 55, 214, 234, 235, 242, 245, 248
Armelagos, G. J., 16, 235, 242, 246, 247, 248, 249
Asada, T., 46, 47, 55
Ash, J. E., 145, 160
Ashley Montagu, M. F., 42, 55
Ashmead, A. S., 95, 160, 193, 209
Auderbach, O., 188

B

Ball, G. V., 309, 312
Barber, C. G., 36, 37, 55
Barlow, T., 101, 160, 254, 274
Barnetson, J., 209
Bartels, P., 172, 188
Barton, R. P., 203, 209
Basedow, H., 254, 274
Bass, W. M., 15, 16, 23, 27, 57, 82, 83, 94, 163, 332, 398

Batson D. V., 179, 188, 387, 397
Baumhoff, M. A., 50, 56
Beekman, F., 78, 83
Benedict, E. B., 74, 83
Bennett, G. A., 312
Béraud, C., 329, 397
Berry, A. C., 332, 397
Bittner, L. H., 145, 160
Blast, T. H., 18, 56
Bloom, W., 15
Blumberg, B. S., 302, 312
Blumberg, J. M., 243, 244, 248
Bohrer, S. P., 235, 248
Boland, E. W., 296, 312
Bosworth, 188
Bourke, J. B., 188, 302, 304, 312
Bowers, W. F., 159, 160
Boyd, M. F., 248
Bradlow, D. R., 107, 160
Brailsford, J. F., 75, 83
Breed, R. S., 170, 188
Brooks, S. T., 380, 397
Brothwell, D. R., xii, 16, 35, 37, 38, 42, 56, 82, 83, 160, 172 188, 194, 209, 322, 336, 353, 355, 371, 381, 389, 397
Brown, J. S., 151, 160
Browne, S. G., 210
Bruce-Chwatt, L. J., 236, 249
Brues, A. M., 94, 96, 160, 265, 274
Brühl, G., 95, 160
Bruusgaard, E., 109, 160
Buchman, J., 160
Bullen, A. K., 96, 161
Bullough, P. G., 328, 397
Buret, F., 87, 88, 161
Burton, R., 25, 56
Bushnell, G. H. S., 175, 188
Butler, C. S., 90, 161
Bywaters, E. G. L., 306, 312

403

C

Caffey, J., 40, 56, 99, 161, 220, 223, 249, 343, 397
Campbell, W. C., 161
Cannefax, G. R., 92, 161
Carlson, D. S., 235, 242, 246, 247, 249
Carroll, D. S., 222, 226, 229, 249
Carson, C. P., 376, 397
Castiglioni, A., 89, 161
Cavalli-Sforza, L. L., 220, 225, 235, 249
Cave, A. J. E., 171, 188
Chalmers, J., 260, 274
Chamberlain, W. E., 198, 207, 210
Chapman, F. H., 302, 312
Chatterjee, S., 243, 249
Chaussinand, R., 198, 210
Cheyne, W. W., 175, 177, 188
Church, F. H., 145, 161
Clark, G., 171, 188
Cochrane, R. G., 192, 193, 197, 210
Cockburn, T. A., 16, 86, 161, 170, 174, 188
Cofield, R. B., 182, 188
Cole, H. N., 95, 98, 161
Coley, B. L., 319, 344, 363, 387, 397
Collins, D. H., 18, 56, 287, 299
Cooley T. B., 249
Cooney, J. P., 205, 210
Copeman, W. S. C., 312
Cornet, L., 153, 161
Coulson, W. J., 129, 161
Crosby, A. W., 87, 161
Csonka, G. W., 138, 139, 141, 161
Cule, J., 235, 249
Currarino, G., 225, 249
Cushing, H., 353, 397

D

Dahlin, D. C., 319, 321, 325, 348, 351, 352, 354, 363, 369, 371, 374, 397
Davey, T. F. 196, 203, 210
Davidson, S., 230, 238, 249, 254, 274
Dawson, W. R., 302, 310, 314, 324, 371, 395, 400

De Blecourt, J. J., 296, 312
Delahaye, R. P., 143, 161
Dellinger, S. C., 95, 167
Denninger, H. S., 95, 161, 338, 397
De Palma, A. F., 20, 56
Derry, D. E., 171, 188, 397
Derums, V. Y., 97, 161, 172, 188
Doull, J. A., 196, 210
Dubos, R., 174, 189
Dunn, F. L., 236, 249

E

Eaton, G. F., 95, 161
Eckert, W. G., 16
Eichenholtz, S. N., 136, 162
Eng, L. L., 231, 249
Enlow, D. H., 15
Epstein, B. S., 260, 274
Esguerra-Gómez, G., 198, 210
Esper, E. J. C., 370, 398
Evans, W. A., 99, 162

F

Faget, G. H., 198, 210
Fasal, P., 197, 210
Fawcett, D. W., xii, 15
Feldman, F., 208, 210
Ferguson, R. G., 174, 189
Fisher, A. G. T., 312
Foote, J. A., 177, 264, 275
Fournier, A., 98, 106, 162
Fox, H., 300, 312
Friederici, G., 24, 26, 56
Friedman, S. A., 209, 210
Frost, H. M., 15, 261, 275
Fusté, M., 380, 398

G

Gaidner, W., 309, 312
Gann, T., 95, 162, 238, 249
Garcia-Frias, J. E., 173, 189
Garn, S. M., xii, 43, 47, 49, 50, 56

Author Index

Garner, M. F., 145, 162
Garrod, A. B., 292, 309, 313
Gejvall, N-G., 95, 97, 159, 162, 172, 189, 235, 249, 264, 275, 371, 389, 398
Geschickter, C. F., 321, 348, 353, 354, 374, 383, 398
Gibson, A., 83
Gilfillan, S. C., 313
Gill, E., 56
Gilmour, W. N., 65, 78, 83
Gindhart, P. S., 47, 49, 56
Girdlestone, G. R., 174, 189
Goff, C. W., 95, 160, 162
Gofton, J. P., 304, 313
Goldman, C. H., 143, 145, 153, 162
Goldstein, M. S., 16, 23, 56, 82, 83, 96, 162
Gondos, B., 206, 209, 210
Gramberg, K. P. C. A., 88, 162
Gray, P. H. K., 37, 56, 235, 249, 262, 275
Green, M., 65, 83
Green, W. T., 61, 62, 83
Greenfield, G. B., 15
Gregg, J. B., 332, 398
Grin, E. I., 93, 111, 123, 138, 162
Guthe, T., 141, 142, 162

H

Hackett, C. J., xii, 16, 138, 142, 143, 145, 158, 162, 163
Hall, M. C., 15
Hallock, H., 179, 189
Haltom, W. L., 95, 163
Hamerman, D., 291, 313
Hamlin, H., 90, 163
Hamperl, H., 26, 27, 56, 214, 249
Hancox, N. M., 15
Hare, R., 171, 189
Harley, G. W., 142, 163
Harris, H. A., 46, 47, 57
Harrison, L. W., 87, 163
Harrison, M. H. M., 282, 313
Hazen, H. H., 110, 115, 163
Heberling, J. A., 78, 83

Heine, J., 279, 313
Helfet, A. J., 143, 153, 163
Hengen, O. P., 239, 242, 246, 247, 249
Henschen, F., 16, 95, 159, 162, 213, 244, 245, 254, 275
Hess, A. F., 254, 264, 275
Hilpert, 65, 83
Hoadley, M. F., 97, 165
Hobo, T., 62, 83
Hodgson, A. R., 175, 179, 182, 189
Hogan, M. J., 243, 250
Holcomb, R. C., 87, 88, 89, 90, 96, 163
Hollander J. L., 313
Honeij, J. A., 210
Hooton, E. A., 23, 57, 82, 83, 95, 163, 173, 189, 216, 236, 237, 246, 250, 256 265, 275, 302, 313, 328, 388, 395, 398
Hopkins, R., 210
Howe, P., 246, 256
Hoyme, L. E., 23, 27, 57, 82, 83, 94, 163
Hrdlička, A., 38, 57, 96, 163, 173, 175, 189, 213, 216, 242, 250, 264, 275, 302, 313, 332, 399
Hudson, E. H., xii, 87, 88, 90, 91, 94, 110, 111, 138, 139, 163, 164
Huebler, 65, 83
Hug, E., 371, 399
Hulse, E. V., 193, 210
Hurtado, E. D., 38, 57
Hussein, M. K., 97, 164
Hutchinson, J., 107, 164
Hyde, J. N., 95, 164

I

Iler, D. H., 20, 59
Inkster, R. G., 142, 165, 195, 211
Isager, K., 172, 189

J

Jacobs, M. L., 86
Jacobson, H. G., 399
Jacobson, S. A., 327, 351, 362, 388, 399

Jacono, G., 175, 189
Jaffe, H. L., 15, 61, 83, 101, 104, 164, 256, 260, 261, 266, 275, 287, 313, 319, 325, 327, 344, 348, 350, 351, 352, 363, 399
Janssens, P. A., 24, 57
Jarcho, S., 16, 89, 110, 164, 193, 210, 250
Jeans, P. C., 101, 102, 106, 110, 164
Job, C. K., 201, 204, 210
Johansson, S., 176, 189
Johnson, L. C., 242, 250, 280, 313, 318, 399
Joklik, W. K., 164, 170, 189
Jones, F. W., 38, 57, 97, 166, 171, 194, 212, 216, 251, 276 340, 401
Jones, J., 95, 164
Jones, L. G. G., 140, 164
Jones, P. R. M., 47, 57
Jopling, W. H., 97, 165, 210

K

Kail, F., 141, 164
Kampmeier, R. H., 113, 164
Karnosh, L. J., 107, 164
Karsh, R. S., 303, 313
Kato, K., 254, 275
Kerley, E. R., 16, 242, 243, 244
Keefer, C. S., 279, 313
Kellgren, J. H., 292, 313
Keyes, E. L., 109, 112, 115, 164
Khanalkar, V. R., 196, 210
Kidd, K. E., 309, 313
Kidder, A., 175, 189
Knaggs, R. L., 15, 61, 83, 174, 189, 270, 275, 313
Kobert, R., 310, 313
Koganei, Y., 246, 250
Krogman, W. M., 399
Krumbhaar, E. B., 95, 164

L

La Fond, E. M., 175, 177, 189
Lamb, D. S., 95, 165

Landells, J. W., 284, 313
Lanzkowsky, P., 233, 238, 250
Larsen, R., 66, 83
Laughlin, W. S., 26, 27, 56
Lawrence, J. S., 278, 314
Lawrence, J. W. P., 370, 399
Lawson, J. B., 238, 250
Layrisse, M., 237, 238, 250
Lechat, M. F., xii, 201, 204, 210
Lee, T., 264, 275
Lichtenstein, L., 333, 356, 257, 400
Lichtor, J., 173, 189
Lieberman, P. H., 357, 400
Lisowski, F. P., 30, 35, 57
Litzmann, C., 272, 275
Livingstone, F., 236, 237, 250
Lodge, T., 260, 275
Loiacono, P. L., 226, 250
Luck, J. V., 15, 79, 83, 266, 275
Lunn, H. F., 143, 165
Lurie, M. B., 211

M

MacCurdy, C. G., 36, 55, 95, 165, 355, 371
Macfarlane, W. V., 131, 165
Magnuson, H. J., 123, 165
Mankin, H. J., 263, 275, 280, 314
Manouvrier, L., 36, 57
Margetts, E. L., 30, 31, 57
Marshall, W. A., 57
Mattingly, D. J., 315, 400
Maul, H. G., 143, 145, 153, 165
May, J. M., 247, 250
May, W. P., 303, 314
McDonagh, J. E. R., 165
McHenry, H., 44, 50, 57
McKay, C. V., 141, 158, 165
McKenna, R. J., 400
McLaren, D. S., 250
McLean, F. C., 15, 253, 256, 275
McLean, S., 98, 99, 101, 165
Means, H. J., 95, 165
Meinecke, B., 171, 189
Merriweather, A. M., 139, 165

Author Index

Milgram, J. E., 74, 84
Møller, P., 389, 400
Møller-Christensen, V., xii, 37, 38, 56, 57, 97, 142, 165, 172, 190, 193, 195, 198, 199, 201, 207, 211, 238, 241, 242, 243, 250, 264, 275
Moodie, R. L., 16, 36, 57, 82, 173, 190, 302, 314, 329, 354, 355, 400
Mooney, J., 25, 27
Moore, C. B., vii, 382, 400
Moore, J. G., 237, 250
Moore, S., 216, 217, 250
Morant, G. M., 97, 165
Morgan, E. L., 95, 165
Morgan, J. D., 62, 84
Morse, D., xii, 16, 20, 23, 24, 28, 38, 42, 57, 58, 82, 84, 96, 165, 166, 171, 173, 190, 277, 304, 305, 314, 353, 360, 380, 400
Morton, R. S., 93, 138, 166
Moseley, J. E., 218, 219, 220, 222, 231, 237, 242, 246, 250, 251, 340, 355, 376, 400
Motulsky, A. G., 251
Murdock, J. R., 198, 203, 207, 211
Murray, J. F., 138, 139, 140, 166
Murrill, R. L., 159, 166

N

Nadeau, G., 25, 58
Nathan, H., 239, 241, 242, 244, 251
Nathanson, L., 190
Nemeskéri, J., 110, 160, 380, 400
Neumann, G. K., 27, 58

O

Oakley, K. P., 35, 58
O'Bannon, L. G., 173, 190
Ong, H. A., 106, 166
Oosthuizen, S. F., 143, 153, 166
Orton, S. T., 95, 166

P

Pales, L., 16, 42, 58, 82, 97, 166
Park, E. A., 44, 46, 49, 58
Parmalee, A. H., 98, 166

Parramore, T. C., 166
Parrot, J., 100, 101, 166, 216, 251
Paterson, D. E., 198, 204, 211
Pendergrass, E. P., 102, 106, 166
Pesce, H., 193, 211
Phemister, D. B., 62, 79, 84, 185, 190
Pietrusewsky, M., 159, 166
Piggott, S., 30, 58
Pizzi, T., 237, 251
Platt, B. S., 47, 58
Plessis, J. .D, 209, 211
Pomeranz, M. M., 185, 190
Pommer, G., 284, 314
Poppel, M. H., 190
Porter, F. S., 374, 400
Post, P. W., 195, 212
Price, C. H. G., 363, 400
Pusey, W. A., 87, 166
Putkonen, T., 196, 107, 108, 166
Putschar, W. G. J., 15, 62, 84, 329, 400
Pyrah, L. N., 65, 78, 84

R

Rabkin, S., 96, 199
Ragir, S., 82, 84
Rat, J. H., 166
Reese, H. H., 27, 58
Requena, A., 173, 190
Reynolds, J., 227, 251
Rich, A. R., 170, 190
Richards, P., 193, 212
Richter, C. P., 46, 58
Ridings, G. R., 372, 400
Riordan, D. C., 212
Riseborough, A. W., 212
Ritchie, W. A., 173, 190, 380, 400
Ritvo, M., 15
Robinson, W. D., 292, 314
Roche, A. F., 332, 401
Roche, M. B., 302, 314
Rodger, F. C., 246, 251
Rodnan, G. P., 309, 314
Rogers, L., 354, 401
Rokhlin, D. G., 97, 166, 172, 190
Roney, J. G., 82, 84, 96, 166, 167, 173, 190
Rosahn, P. D., 109, 167

Rosen, R. S., 209, 212
Rosencrantz, E., 177, 190
Rost, G. S., 139, 167
Rowling, J. T., 171, 190, 324, 371, 401
Rubin, P., 340, 401
Ruffer, M. A., ix, 16, 82, 84, 171, 190, 275, 302, 303, 314, 395, 401

S

Sager, P., 172, 190
Samuels, R., 237, 251
Sanchis-Olmos, V., 176, 177, 191
Sandison, A. T., 239, 242, 243, 250, 276
Saul, F. P., 38, 58, 95, 167, 237, 243, 251, 254, 276, 305, 314
Scrimshaw, N. S., 237, 238, 251
Selby, S., 342, 401
Seltzer, C. C., 279, 314
Senn, N., 191
Shandling, B., 65, 84
Shands, A. R., 42, 58, 95, 163
Shanks, S. C., 15, 76, 84, 106, 185, 191
Shattuck, G. C., 110, 166, 237, 238, 241
Sherman, R. S., 351, 374, 401
Short, C. L., xii, 303, 304, 314
Sigerist, H. E., 16, 81, 84, 264, 276
Singer, C., 89, 166
Sissons, H. A., 253, 276
Sjövall, E., 172, 191
Smith, G. E., 97, 166, 171, 191, 194, 212, 216, 260, 264, 276, 302, 310, 314, 324, 340, 371, 395, 400
Smith, R. W., 258, 276
Snapper, I., 376, 401
Snorrason, E. S., 305, 314
Snow, C. E., 23, 27, 38, 42, 58, 95, 96, 159, 166, 236, 252, 264, 276, 314
Sokoloff, L., 278, 314
Sontag, L. W., 47, 58
Speed, J. S., 61, 84, 166
Spoehr, A., 153, 166
Spjut, H. J., 317, 336, 344, 238, 363, 365, 369, 387, 401
Squier, E. G., 29

Squire, T., 86
Starr, C. L., 66, 84
Stead, W. W., 174, 191
Stecher, R. M., 278, 314
Steinbach, H. L., 84
Steindler, A., 166
Stewart, D. M., 166
Stewart, T. D., vii, viii, xi, 20, 30, 32, 35, 38, 58, 59, 95, 96, 97, 153, 166, 174, 191, 302, 315
Still, G. F., 246, 252
Stokes, J. H., 102, 106, 110, 113, 122, 127, 129, 131, 166
Stradford, H. T., 319, 401
Straus, W. L., 301, 315
Sudhoff, K., 36, 59, 87, 167
Symmers, D., 109, 167

T

Taneja, B. L., 142, 145, 167
Taylor, R. W., 101, 167
Tello, J. C., 95, 167
Thoma, K. H., 107, 167
Tobin, W. J., 315
Todd, T. W., 20, 37, 59, 215
Toldt, C., 239, 252
Tordeman, B., 172, 191
Traut, E. F., 315
Trendel, 61, 64, 84
Trowell, H. C., 247, 252
Trueta, J., 62, 79, 84, 85
Turner, C., 305
Turner, T. B., 91, 167

U

Urteaga, 340, 401

V

Vallois, H. V., 301, 315
Van Gerven, D. P., 235, 242, 246, 247, 249

Van Nitsen, R., 142, 167
Vinson, H. A., 195, 209, 212
Virchow, R., 82, 85, 113, 129, 167, 216
Viteri, F. E., 238, 252
Vogelsang, T. M., 212
Von Zeissl, H., 127, 167
Vyhnánek, L., 38, 59, 336, 401

W

Wakefield, E. G., 95, 167, 236, 252
Waldron, H. A., 310, 315
Waldvogel, F. A., 61, 66, 85
Watson-Jones, R., 20, 59
Weinfield, M. S., 401
Weinmann, J. P., 15, 261, 266, 276
Weiss, D. L., 172, 191, 198, 212
Welcker, H., 241, 252
Wells, C., xii, 16, 42, 50, 55, 59, 82, 85, 172, 191, 194, 212, 262, 265, 276, 302, 310, 315, 335, 338, 380, 389, 401
West, H. F., 296, 315
Whitman, R., 175, 191
Whitney, J. L., 109, 110, 126, 167
Whitney, W. F., 82, 85, 95, 167, 173, 191
Wile, U. J., 109, 135, 167

Wilensky, W. F., 61, 64, 65, 73, 76, 85
Willcox, R. R., 93, 138, 141, 167
Willey, G. R., 94, 167
Williams, G. D., 361, 380, 402
Williams, H. U., 87, 95, 97, 113, 129, 142, 145, 167, 215, 216, 236, 252
Willis, R. A., 402
Wilson, D. C., 264, 276
Wilson, J., 143, 169
Wilson, P. W., 143, 169
Wilson, T., 24, 59
Winters, J. L., 61, 62, 65, 85
Wolbach, J. B., 47, 59
Wolff, E., 239, 252
Wolff, J., 261, 276
Wright, D. J. M., 169

Z

Zaborowski, 36, 59
Zadek, I., 61, 85
Zaino, D. E., 217, 235, 236, 242, 252
Zariquiey, M. O., 355, 402
Zilva, S. S., 246, 252
Zimmermann, E. L., 88, 169
Zinsser, H., 87, 169
Zorab, P. A., 305, 315

SUBJECT INDEX

A

Acroosteolysis, 204
Actinomycosis, 179, 208
Africa, 30, 36, 90, 92, 97, 138, 142, 152, 193, 194, 197, 201, 235
Alabama, 27, 96, 236
Alaska, 96, 130, 174, 254, 302
Albright's syndrome, 336
Aleuts, 96
Alexander the Great, 192
Altar de Sacrificios, 38, 237, 247, 305
American Indians, 23, 27, 28, 31, 37, 38, 42, 50, 95, 96, 101, 130, 135, 173, 184, 213, 214, 216, 217, 236-328, 254, 258, 259, 305, 329, 332, 338, 353, 355, 260, 380, 381-384
Amphiarthrosis, 277
Amputation, 36-39
  bone chanes in, 36-37
  causes of, 36
  paleopathological evidence of, 38-39
  pseudarthrosis confused with, 39-40
Ancylostoma duodenale, 246
Androgen, 256
Anemia, 213-252
  bibliography of, 248-252
  cribra orbitalia and, 239-248
  iron deficiency, 230-234, 236-239, 246-248
  malaria and, 234-236
  paleopathological evidence of, 235-236, 241-248
  sickle-cell, 225-230, *see also* Sickle-cell anemia
  spongy hyperostosis, 213-219
  thalassemia, 220-225, *see also* Thalassemia
Angioma, *see* Hemangioma

Ankylosing spondylitis, 294-298
  age incidence of, 296
  bibliography of, 312-315
  bone changes in, 296
  differential diagnosis of vertebral osteophytosis and, 289, 296-298
  joints affected in, 294-295
  paleopathological evidence of, 304-309
  pathogenesis of, 296
  prevalence of, 296
  sex incidence of, 296
Annulus fibrosus, 287, 296
Archaic period, vii, 23, 42, 50, 96, 264, 381-384
Aretaios, 192
Aribasios, 192
Arikara Indians, 27
Arizona, 96
Arkansas, 28, 134, 332
Armed Forces Institute of Pathology, xi, 61
Army Medical Museum, vii, 61, 113
Arrow wounds, 24
Arthritis, 227-315, *see also* Ankylosing spondylitis, Gout, Osteoarthritis, Osteophytosis, and Rheumatoid arthritis
  bibliography of, 312-315
  classification of, 277
  infectious, 78-79, 299
  paleopathological evidence of, 300-312
  traumatic, 289
Arthritis deformans, *see* Osteoarthritis
Arthrosis, *see* Osteoarthritis
Ascaris lumbricoides, 246
Ascorbic acid deficiency, *see* Scurvy
Aseptic necrosis, 13, 62, 73, 229
Asia, 30, 36, 97, 138, 142, 192, 194, 294

411

Atrophy of bone, 260-261
  amputation and, 37
  bone changes in, 261
  dislocations and, 39
  mechanical stress and, 12
  Wolff's law and, 261
Auditory tori, 329-333
Australian aborigines, 138, 141, 142, 145, 149, 150, 254
Aztecs, 96

B

Balantidium coli, 246
Bamboo spine, see Ankylosing spondylitis
Basketmaker culture, 96
Bedouins, 138
Bejel, 86, 138-142, see also Syphilis, nonvenereal
Benign bone cyst, see Solitary bone cyst
Bible, 171, 192
Biparietal thinning, 260
Bone
  atrophy of, see Atrophy
  blood supply of, 12-13, 62-64
  formation of, 8-9
  functions of, 3
  gross anatomy of, 4-7
  inflammation of, 13
  resorption of, 9-10
  types of, 5-7
Boomerang leg, 145
Bosnia, 93, 111, 138
Britain, 30, 38, 42, 50, 82, 97, 172, 193, 194, 235, 264 302 305, 309, 310, 335, 338, 371, 380
British Columbia, 24
Brittle bones, see Osteogenesis imperfecta
Brodie's abscess, 74-76
  appearance of, 75-76
  differential diagnosis of,
    benign bone cyst and, 74, 346
    osteoid-osteoma and, 333

tuberculosis and, 184, 186
  size of, 75
Bronze age, 42, 50, 172, 235, 305
Bubas, see Yaws
Button scurvy, 91, 93, 138, see also Syphilis, nonvenereal

C

Calcium, 3, 12, 13
California, 50, 82, 96, 173, 332, 355, 380
Callus, 18, 19, 20
Cancellous bone, 7
Catacombs of Paris, 97
Celsus, 192
Charcot joint, 135, 208, 299
Chichen Itza, 216, 236, 247
Chile, 195
China, 30, 192, 194, 264
Chondroma, 235
  appearance of, 235
  incidence of, 235
Cloaca, 66, 69
Clutton joint, 135, 299
Coccidioidomycosis, 179, 208
Collagen, 253
Columbia, 198
Columbian theory of syphilis, 87, 88
Columbus, Christopher, 87, 89
Comminuted fracture, 17-18
Comox Indians, 195
Compact bone, 7
Compound fracture, 18
Congenital hip dislocation, see Dislocation
Congenital syphilis, see Syphilis, congenital
Cooley's anemia, see Thalassemia major
Cortex, 5
Coxa vara, 270
Craniometaphyseal dysplasia, 340
Craniotabes, 266
Cribra cranii, 213, see also Cribra orbitalia

Cribra orbitalia, 213, 239-248
  bibliography of, 248-252
  causes of, 243-248
  classification of, 239-241
  iron deficiency anemia and, 242, 246-248
  paleopathological evidence of, 241-242, 243-248
  pathogenesis of, 239
  primates and, 244, 248
  spongy hyperostosis and, 213, 239-243, 247
  studies of, 241-242
Cromwell, 93
Crusades, 88, 90, 193
Crushing injuries, 24
Czechoslovakia, 38, 336

D

Degenerative joint disease, *see* Osteoarthritis
Denmark, 30, 97, 172, 193, 195, 198, 264, 305
Dentinogenesis imperfecta, 262
Diabetes mellitus, 209
Diaphyseal aclasis, 321
Diaphysis, 5
Diarthrosis, 278, 279
Dichuchwa, 91, 93, *see also* Syphilis, nonvenereal
Dimetrodon, 82
Diplodocus longus, 302
Diploë, 214
Dislocation, 39-43
  atrophy in, 39
  bone changes in, 39
  congenital hip, 39-40
  paleopathological evidence of, 42-43
  skeletal distribution of, 39

E

Ear exostoses, 329-333
  appearance of, 330-332
  etiology of, 332

Easter Island, 159
Eburnation, 282-283, 294
Ecuador, 101
Egypt, 37, 38, 97, 171, 194, 235, 240, 260, 264, 301, 304, 310, 322, 324, 340, 353, 354
Elephantiasis, 192
Elephas, 192
Enchondroma, *see* Chondroma
Endemic syphilis, *see* Syphilis, nonvenereal
Endochondral ossification, 9
Endosteum, 5
England, *see* Britain
Entamoeba histolytica, 246
Eosinophilic granuloma, 355-362
Epilepsy, 31, 36
Epiphyseal plate, 9, 43, 46, 62
Epiphysis, 4
Erythroblastic anemia, *see* Thalassemia major
Eskimos, 96, 130, 133, 254, 302
Espundia, 151
Estrogen, 256
Ewing's sarcoma, 371-374
  appearance of, 372-374
  differential diagnosis of,
    eosinophilic granuloma and, 374
    osteomyelitis and, 68, 81, 374
    osteosarcoma and, 370, 374
  incidence of, 374
  skeletal distribution of, 374-376
Exostosis, 321, 325, 329

F

Facies leprosa, 194, 199, 201-203
False joint, *see* Pseudarthrosis
Fibroma, nonossifying, *see* Nonossifying fibroma
Fibrous bone, 5
Fibrous cortical defect, *see* Nonossifying fibroma
Fibrous dysplasia, 336-341
  Albright's syndrome and, 336
  appearance of, 338
  incidence of, 336

paleopathological evidence of, 338-340
skeletal distribution of, 336-338
Fish vertebra radiographic sign, 227
Florida, 96, 215, 236, 332, 380
Folic acid deficiency, 237
Fournier's tooth, 108
Fracture, 17-23
   age estimation of, 21-22
   callus formation in, 18
   classification of, 17-18
   healing of, 18, 20
   incidence in American Indian populations of, 22-23
   nonunion of, 20
   pseudarthrosis in, 21
   rate of repair in, 20-21
Framboesia, see Yaws
France, 30, 36, 42, 89, 305, 329
Frenga, 93, 138, see also Syphilis, nonvenereal
Frostbite, 195, 209

G

Gangosa, 140, 145, 151
Germany, 30, 172, 247
Giant-cell tumor, 346-350
   appearance of, 348-350
   classification of, 346-348
   differential diagnosis of, 350
      aneurysmal bone cyst and, 350
      fibrosarcoma and, 350
      hyperparathyroidism, 350
      nonossifying fibroma and, 350
      solitary bone cyst and, 350
   incidence of, 348
   skeletal distribution of, 348, 349
Giardia intestinalia, 246
Goundou, 152, 153
Gout, 300, 309-312
   bibliography of, 312-315
   causes of, 300, 309, 312
   historical figures and, 309
   joints affected in, 300
   lead poisoning and, 309-312
   paleopathological evidence of, 309-312
   pathogenesis of, 300
Granulation tissue, 13, 99, 182, 292
Greece, 42, 138, 192, 235, 244
Greek literature, 87, 171, 192
Greenstick fracture, 17
Ground glass radiographic appearance, 256, 293
Growth arrest lines, see Transverse lines
Guatemala, 38, 237, 305
Guiac wood and history of syphilis, 87
Gumma, syphilitic, 123

H

Haida Indians, 304
Hair-on-end radiographic pattern, 217-219
Harris's lines, see Transverse lines
Harvard Medical School, xi, xii, 61, 113
Haversian canals, 5, 7, 66
Hawaii, 38, 42, 159, 198, 242
Heberden's nodes, 282
Helminthic infestation, 237
Hemangioma, 350-353
   appearance of, 351-353
   classification of, 350-351
   differential diagnosis of, 352-354
      meningioma and, 352, 354
      osteosarcoma and, 352
   incidence of, 351
   paleopathological evidence of, 352-353
   skeletal distribution of, 351
Hematogenous osteomyelitis, see Osteomyelitis
Hematologic disorders, see Anemia
Hemenway expedition, 305
Hereditary spherocytosis, 218, 219
Hippocrates, 192, 309
Histiocytosis X, 355-362
   appearance of, 358, 360
   classification of, 355, 357
   paleopathological evidence of, 360-362

Subject Index

skeletal distribution of, 358, 359
Homo erectus, 322
Hong Kong, 175, 198
Hookworm infection, 237, 246
Hopewell culture, 37, 94, 95, 96, 305
Hormones, 13, 14
Hungary, 380
Huron ossuary, 309
Hutchinson's teeth, 107-108
Hydroxyproline, 253
Hyperemia, 12
Hyperostosis spongiosa orbitae, *see* Cribra orbitalia
Hyperparathyroidism, 14, 336, 350

I

Illinois, 23, 27, 37, 42, 96, 135, 173, 184, 304, 305, 306, 338, 353, 360
Incas, 94, 96
India, 142, 171, 192, 197, 264
Indian Knoll, vii, viii, 23, 96, 332, 333, 381-384
Indians, *see* American Indians
Industrial revolution, 82, 263-264
  osteomyelitis and, 82
  rickets and, 263-264
Infectious arthritis, *see* Arthritis
Infectious disease in natural selection, 86, 234-236
Inflammation, 13
Intramembranous ossification, 8
Involucrum, 66, 68
Iraq, 88, 138, 140, 141
Ireland, 93, 138, 254
Iroquois, 23
Iron age, 42, 264, 302, 371
Iron deficiency anemia, 230-234, 236-238, 246-248
  American Indians and, 236-238
  bibliography of, 248-252
  causes of, 231, 237-238, 246-248
  cranial changes in, 231, 246-248
  cribra orbitalia and, 242, 246-248
  differential diagnosis of,
    rickets and, 233-234
    sickle-cell anemia and, 232

thalassemia and, 222, 232
paleopathological evidence of, 235-236, 241-242, 243-248
parasitic infestation and, 237, 246
postcranial changes in, 231-234
rickets associated with, 232-234
spongy hyperostosis and, 218-219, 231, 246-248
Ischemia, 12
Italy, 30, 89, 90, 235
Ivory exostosis, *see* Osteoma

J

Japan, 30
John of Damascus, 164, 193

K

Kanam mandible, 370
Kentucky, vii, 23, 51, 95, 96, 264, 332, 333, 381-384
King Charles VIII, 89
Koch's postulates, 196

L

La Chapell-aux-Saints, 301
Lacrimal gland inflammation, 243
Lamellar bone, 6
Leishmaniasis, 151, 193
Lepra, 192
Leprosaria, 88, 172, 193, 198, 199, 243
Leprosy, 192-212
  bibliography of, 209-212
  bone changes in, 201-207
  classification of, 197
  cribra orbitalia and, 243
  differential diagnosis of, 208-209
    diabetes mellitus and, 209
    frostbite and, 209
    fungal infections and, 208-209
    neoplastic conditions and, 209
    pyogenic osteomyelitis and, 199, 208
    syphilis and, 137, 208
    yaws and, 208

facies leprosa in, 194, 199, 201-203
foot changes in, 203-206
frequency of skeletal involvement in, 198-199
geographical distribution of, 194
hand changes in, 203-204
history of, 192-195
lepromatous form of, 197
osteomyelitis complicating, 199
paleopathological evidence of, 193-195
  Britain, 194
  Denmark, 195
  Egypt, 194
  New World, 195
periostitis in, 207
prevalence of, 197
skeletal distribution of, 199-201
syphilis confused with, 88
transmission of, 196
tuberculoid form of, 197
tuberculosis and decline of, 197-198
Letterer-Siwe disease, 355-352, *see also* Histiocytosis X
Lithuania, 97, 172
Louisiana, 173, 195, 198, 332

M

Malaria, 234-236
Mariana Islands, 150, 153, 158
Marie-Strümpell disease, *see* Ankylosing spondylitis
Maryland, 82
Mayan Indians, 38, 95, 216, 217, 236-238, 247, 254, 258, 259, 305
Mechanical stress affecting bone, 12
Mediterranean disease, *see* Thalassemia major
Medullary cavity, 5
Meningioma, 353-355
  appearance of, 353-355
  differential diagnosis of, 353-355
    hemangioma and, 352, 354
    osteosarcoma and, 354
  paleopathological evidence of, 354-355, 356

Mercury and the history of syphilis, 88
Metabolic bone disease, 253-276, *see also under* Atrophy, Osteogenesis imperfecta, Osteomalacia, Rickets, and Scurvy
Metaphysis, 5
Metastatic carcinoma affecting bone, 385-396
  appearance of, 387
  differential diagnosis of multiple myeloma and, 388
  incidence of, 385
  paleopathological evidence of, 388-396
  skeletal distribution of, 385, 386
  spread of, 386, 387
Mexico, 38, 91, 95, 96, 159, 236, 237
Middle ages, 30, 42, 50, 97, 172, 193, 195, 196, 235, 264, 329, 371
Mississippian culture, 23, 27, 28, 96, 135, 184, 304, 305, 338, 360
Missouri, 380
Moon's molar, 108
Morbus gallicus, 89
Mulberry tooth, 108
Multiple myeloma, 374-384
  appearance of, 376
  differential diagnosis of metastatic carcinoma and, 388
  incidence of, 374
  paleopathological evidence of, 378-384
  skeletal distribution of, 374, 375
Mummies, 171, 173, 195, 235, 303, 310
Mumps and cribra orbitalia, 243
Mycobacterium leprae, 195-196
Mycobacterium tuberculosis, 170
Myelomatosis, *see* Multiple myeloma
Myositis ossificans, 322-323
  etiology of, 322
  in *Homo erectus*, 322
  skeletal distribution of, 323

N

Nazca culture, 173
Neanderthal man, 300-301

Nebraska, 27
Necator americanus, 246
Negroes, 225, 236
Neolithic age, 30, 82, 171, 172, 173, 235, 264, 301, 305
Neoplasms, *see* Tumors, bone
New Guinea, 142
New Mexico, 23, 82, 95, 173, 216, 265, 305, 327
New York, 173, 360, 380
Nonossifying fibroma, 340-344
   appearance of, 343-344
   classification of, 340, 342
   incidence of, 342
   skeletal distribution of, 342
Nonunion of fractures, 20
Nonvenereal syphilis, *see* Syphilis, nonvenereal
Norway, 93, 138, 196, 264
Nucleus pulposus, 287, 296

O

Oceania, 30, 142, 159
Ohio, 96, 305
Oklahoma, 96
Ontario, 23, 309
Order of St. Lazarus and history of leprosy, 89
Ossification,
   endochondral, 9
   intramembranous, 8-9
Ossification centers, 9
Osteitis, *see* Osteomyelitis
Osteoarthritis, 278-286
   age incidence of, 278
   bibliography of, 312-315
   bone changes in, 279-286
   cause of, 278
   definition of, 278
   differential diagnosis of rheumatoid arthritis and, 279, 293-294
   joints involved in, 279
   paleopathological evidence of, 300-302
   pathogenesis of, 279
Osteoblast, 5, 9

Osteochondroma, 319-324
   appearance of, 321-322
   hereditary form of, 321
   incidence of, 319
   paleopathological evidence of, 322-324
   skeletal distribution of, 319-321
Osteoclast, 10
Osteoclastoma, *see* Giant-cell tumor
Osteogenesis imperfecta, 261-262
   bone changes in, 261-262
   genetics of, 261
   paleopathological evidence of, 262
   rickets and, 262
   Wormian bones in, 262
Osteogenic sarcoma, *see* Osteosarcoma
Osteoid, 9, 253
Osteoid-osteoma, 333-336
   appearance of, 333-335
   incidence of, 333
   paleopathological evidence of, 335-336
   resembling nonsuppurative osteomyelitis, 333
Osteoma, 325-333
   appearance of, 328-330
   "button" form of, 327-328
   classification of, 325-327
   incidence of, 327-328
   paleopathological evidence of, 327, 329-333
   skeletal distribution of, 326, 327
Osteomalacia, 272-274
   bone changes in, 272-273
   causes of, 272
   osteoarthritis resulting from, 274
   pregnancies and, 272
   rickets and, 272
Osteomyelitis, 60-85
   acute hematogenous form of, 61-72
      bone changes in, 66, 68, 70
      pathogenesis of, 64, 66
      skeletal distribution of, 62-65
   bacterial causes of, 60-61
   bibliography of, 83-85
   Brodie's abscess, 75-76
   classification of, 60
   chronic form of, 74-78

bone changes in, 74
causes of, 74
epithelioid carcinoma and, 74
definition of, 60
differential diagnosis of, 79-81
  Ewing's sarcoma and, 68, 81
  leprosy and, 208
  osteoid-osteoma and, 333
  osteosarcoma and, 81, 370
  Paget's disease and, 81
  syphilis and, 77, 81, 137
  subperiosteal ossifying hematoma and, 81
  tuberculosis and, 80, 81, 176, 184-186
direct infection form of, 72-74
  gross morphology of, 73-74
  pathogenesis of, 73-74
  sequestrum formation in, 73
non-suppurative (sclerosing osteomyelitis of Garré), 76-77
paleopathological evidence of, 81-82
pyogenic arthritis resulting from, 64-65, 78-79, 299
sickle-cell anemia and, 229
Osteopenia, 253-262
congenital, 261-262
dietary, 253-256
endocrine, 256-260
stress deficiency, 260-261
Osteophytosis, vertebral, 287-289
age incidence of, 287
bibliography of, 312-315
bone changes in, 287-289
differential diagnosis of ankylosing spondylitis and, 289, 296-298
grading of severity in, 303
location of, 287
paleopathological evidence of, 302-303
pathogenesis of, 287-289
racial variation of, 302
Osteoporosis,
definition of, 12, 256
post-menopausal, 256-260
senile, 256-260
Osteosarcoma, 362-371
appearance of, 365
classification of, 362
differential diagnosis of, 365-370
  chondrosarcoma and, 369
  Ewing's sarcoma and, 370
  fibrosarcoma and, 369
  fibrous dysplasia and, 369
  hemangioma and, 352
  meningioma and, 354
  metastatic carcinoma and, 370
  myositis ossificans and, 370
  osteomyelitis and, 81, 370
  syphilis and, 117, 122, 137
incidence of, 362-363
Paget's disease and development of, 363
paleopathological evidence of, 370-371
size of, 365
skeletal distribution of, 363, 364

P

Paget's disease
osteomyelitis and, 81
osteosarcoma and, 363, 369
syphilis and, 137
tuberculosis and, 179
Paleolithic age, 30, 235, 301
Paleopathology
bibliographies of, 16
definition of, ix
Pannus, 292
Papyri, Egyptian, 171
Parathyroid hormone, 262
Paravertebral venous plexus, 179, 387
Pathologic fracture, 17
Paulos of Aegina, 192, 312
Peabody Museum, xi, 31, 61, 113, 305
Pecos Pueblo, 23, 216
Periosteum, 5
Periostitis, see Osteomyelitis
Peru, 29, 31, 35, 36, 50, 95, 101, 173, 213, 214, 233, 236, 302, 329, 332, 340, 355, 356, 371, 380
Phytic acid and iron deficiency anemia, 238, 246
Pian, see Yaws

Pinta, 86, 90, 91, 92
Plasma cell myeloma, *see* Multiple myeloma
Plinius, 192, 310
Poker back spine in ankylosing spondylitis, 296
Porotic hyperostosis, *see* Spongy hyperostosis
Portugal, 90
Post-menopausal osteoporosis, *see* Senile osteoporosis
Pott's disease, 176-182, *see also* Tuberculosis
Pre-Columbian theory of syphilis, 87-90
Primary stratum of transverse line, 46
Proline and ascorbic acid, 253
Protein deficiency anemia, 237, 245
Pseudarthrosis, 21, 38
Pueblo Indians, 23, 94, 96, 216, 236, 305, 327, 332
Puerto Rico, 95, 159
Pyogenic arthritis, *see* Arthritis
Pyogenic osteomyelitis, *see* Osteomyelitis

R

Rachitic rosary, 270
Radesyge, 93, 138, *see also* Syphilis, nonvenereal
Raynaud's disease, 209
Renaissance, 193, 197
Rheumatoid arthritis, 289-294
  bibliography of, 312-315
  bone changes in, 293
  differential diagnosis of osteoarthritis and, 279, 293-294
  etiology of, 289-292
  incidence of, 292
  paleopathological evidence of, 303-304
  pathogenesis of, 292-293
  radiographic appearance of, 293
  rheumatoid factor in, 289-292
  skeletal distribution of, 292

Rheumatoid spondylitis, *see* Ankylosing spondylitis
Rhinopharyngitis mutilans in yaws, 145, 151
Rickets, 262-273
  adult form of, *see* Osteomalacia
  age incidence of, 265
  bibliography of, 274-276
  causes of, 263
  cranial changes in, 266
  dental changes in, 270
  dwarfism in, 272
  history of, 263-265
  industrial revolution and, 263-264
  paleopathological evidence of, 263-265
    Egypt, 264
    New World, 264-265
    Old World, 264
  pathogenesis of, 265
  pseudopathology and, 265
  rib changes in, 270
  social customs and prevalence of, 264
  vitamin D and, 263
Roman literature, 87, 171, 192, 310
Russia, 30, 97, 138, 172

S

Saber-shin tibia, 102-104, 139, 145
Sacred Cenote at Chichen Itza, 216, 236, 247
Salmonella, 229
Saltfluss, 93, *see also* Syphilis, nonvenereal
Saracen ointment and history of syphilis, 88
Saxons, 50, 55, 172, 194, 262, 302, 305, 335, 338, 371
Scalping, 24-29
  bone changes in, 26-27
  geographical distribution of, 24-25
  paleopathological evidence of, 27-28
  techniques of, 25-26
Schüller-Christian syndrome, 355-362, *see also* Histiocytosis X

Scleroderma, 209
Scorbutism, see scurvy
Scotland, 93, 138, 242
Scurvy, 253-256
  adult form of, 253
  Australian aborigines and, 254
  bibliography of, 274-276
  cranial changes in, 256
  cribra orbitalia and, 246, 256
  disuse atrophy complicating, 256
  fractures in, 255-256
  ground glass radiographic appearance in, 256
  history of, 254
  infantile form of, 254-256
  ocean voyages and, 254
  paleopathological evidence of, 254
  rib changes in, 256
  rickets confused with, 254, 256
  subperiosteal hemorrhages in, 254
  vitamin C and, 253
Senile osteoporosis, 256-260
  age incidence of, 257-258
  bone changes in, 260
  parietal thinning in, 260
  physical activity and, 258, 260
Sequestrum, 66, 117, 123, 131, 137, 185
Sibbens, 91, 93, 138, see also Syphilis, nonvenereal
Sickle-cell anemia, 225-230
  bibliography of, 248-252
  cranial changes in, 226
  differential diagnosis of,
    iron deficiency anemia and, 232
    multiple myeloma and, 226
    thalassemia major and, 221, 222, 226, 229
  genetics of, 225
  malaria and, 234-236 235-236
  paleopathological evidence of, 235
  pathogenesis of, 225-226
  postcranial changes in, 226-229
  pyogenic osteomyelitis associated with, 229
  spongy hyperostosis and, 217, 219
  thrombosis in, 226, 229, 235
  vertebral changes in, 227
Sincipital T, 35-36
  description of, 35
  geographical distribution of, 36
  ritualistic significance of, 36
Smithsonian Institution, xi, 61, 113, 305, 381, 393
Solitary bone cyst, 344-246
  appearance of, 344, 346
  differential diagnosis of, 346
    aneurysmal bone cyst and, 346
    Brodie's abscess and, 346
    enchondroma and, 356
    eosinophilic granuloma and, 346
    giant-cell tumor and, 346, 350
  incidence of, 344
  skeletal distribution of, 344, 345
Somatotropic hormone (STH), 14
South America, 29, 31, 35, 36, 38, 50, 91, 95, 96, 101, 159, 173, 195, 198, 213, 233, 237, 302, 305, 329, 332, 340, 355, 356, 371, 380
South Dakota, 27, 332
Spain, 90
Spina ventosa in tuberculosis, 176
Spirocolon, 138, see also Syphilis, nonvenereal
Spondylitis, see Ankylosing spondylitis
Spondylosis deformans, see Osteophytosis, vertebral
Spongy bone, 6
Spongy hyperostosis, 213-219, 239-248
  American Indians and, 236-238
  bibliography of, 248-252
  cribra orbitalia and, 213, 239-243, 247
  etiology of, 215-219
  iron deficiency anemia and, 218-219, 231, 246-248
  paleopathological evidence of, 234-236
  pathogenesis of, 213-214
  sickle-cell anemia and, 217-219
  thalassemia and, 216-219
Staphylococcus aureus, 60, 61, 62, 82

Subject Index

Streptococcus, 61
Stress deficiency, 260-261, *see also* Atrophy
Strongyloides, 246
Subluxation, 293
Sweden, 93, 97, 138, 235, 245, 264
Switzerland, 371
Symmetrical osteoporosis, 213, *see also* Spongy hyperostosis
Symphysis, 277
Synarthrosis, 277
Synchondrosis, 277
Synovial fluid, 278
Syphilis, acquired, 86-97, 108-137
  bibliography of, 160-169
  bone changes in, 113-136
  cranial changes in, 127
  differential diagnosis of, 136-137
    hereditary anemias and, 137
    leprosy and, 137, 208
    meningioma and, 137
    metastatic carcinoma and, 137
    multiple myeloma and, 137
    osteosarcoma and, 117, 122, 137
    pyogenic osteomyelitis and, 77, 81, 137
    tuberculosis and, 137, 187-188
    yaws and, 145, 153
  epidemic of 1500, 88-90
  frequency of skeletal involvement in, 123-126
  gummatous osteomyelitis in, 123-126
  joint changes in, 131, 135-136, 299
  origin of, 86-94
    Columbian theory and, 87-88
    pre-Columbian theory and, 88-90
    Unitarian theory and, 90-94
  osteitis in, 115-122
  paleopathological evidence of, 94-97
    New World, 94-97
    Old World, 97
  periostitis in, 115
  prevalence in early human populations of, 110-112
  skeletal distribution of, 112-114
  treponematosis and, 86, 90-94
  vertebral changes in, 126
Syphilis, congenital, 98-108
  bibliography of, 160-169
  cranial changes in, 99-100
  dental stigmata of, 106-108
  early form of, 98-100
  frequency of skeletal involvement in, 98
  gummatous osteomyelitis in, 106
  hyperplastic osteoperiostitis in, 102-106
  late form of, 101-108
  nonvenereal syphilis and, 139
  osteochondritis in, 98-99
  osteomyelitis in, 99
  periostitis in, 98-99
  rickets and, 100-101
Syphilis, nonvenereal, 86, 90-93, 138-142
  bibliography of, 160-169
  bone changes in, 139-140
  congenital syphilis and, 139
  frequency of skeletal involvement in, 111, 138-139
  geographical distribution of, 138
  nasal-palatal destruction in, 139-140
  paleopathological evidence of, 94, 140-142
  prevalence of, 111
  skeletal distribution of, 139
  treponematosis and, 86, 90-94
  yaws and, 139
Syria, 140
Syringomyelia, 209

T

Tello Collection of trephined Peruvian skulls, 31
Tennessee, 23, 95, 173
Texas, 23, 82, 96, 332
Thalassemia major, 220-225
  bibliography of, 248-252
  cranial changes in, 220-221
  differential diagnosis of,
    iron deficiency anemia and, 222, 232

sickle-cell anemia, 221, 222, 226, 229
  early bone changes in, 222
  genetics of, 220
  late bone changes in, 222-225
  paleopathological evidence of, 234-236
  postcranial changes in, 222-225
  spongy hyperostosis and, 216-219
Thalassemia minor, 220-225
  genetics of, 220
  iron deficiency anemia confused with, 225
  thalassemia major and, 225
Tophi in gout, 300, 310
Torulosis, 209
Trabeculae, 7
Trachoma, 243-244
Transverse lines, 43-55
  adults with, 49
  appearance of, 43-45
  etiology of, 46-47
  infants with, 47, 49
  mechanism of formation of, 46
  morbidity and, 50
  paleopathological studies of, 49-50
  skeletal distribution of, 43
Trauma, 17-59
  amputation, 36-39, *see also* Amputation
  arrow and spear wounds, 24
  bibliography of, 55-59
  crushing injuries, 24
  dislocations, 39-43
  fractures, 17-23, *see also* Fracture
  scalping, 24-29, *see also* Scalping
  Sincipital T, 35-36
  transverse lines, 43-55, *see also* Transverse lines
  trephination, 29-35, *see also* Trephination
Trepanation, *see* Trephination
Trephination, 29-35
  differential diagnosis of, 35
  geographical and chronological distribution of, 30-31
  reasons for, 31
  reviews of, 29-30
  success of, 31
  techniques of, 31
Treponarid, *see* Syphilis, nonvenereal
Treponema pallidum, 90, 91, 92
Treponema pertenue, 142
Treponematosis, 90, *see also* Syphilis and Yaws
Trichuris trichuris, 246
Tropical ulcer, 151
Tsara'ath, 192
Tuberculosis, 170-191
  age incidence of, 175
  American Indians and, 174-175
  bibliography of, 188-191
  bovine form of, 170, 174
  differential diagnosis of, 176, 177, 186-188
    Brodie's abscess and, 184, 186
    congenital syphilis and, 98
    leprosy and, 208
    pyogenic osteomyelitis and, 80, 81, 176, 184-186
    syphilis and, 127, 187-188
  frequency of bone involvement in, 175
  joint changes in, 182-186, 299
    pathogenesis of, 182-183
    periostitis and, 185
    sequestrum formation and, 185
  leprosy and prevalence of, 197-198
  multiple bone lesions in, 176
  paleopathological evidence of, 171-174
    Egypt, 171
    Europe, 172-173, 198
    New World, 173-175
  Pott's disease in, 176-182
  skeletal distribution of lesions in, 176, 177, 178
  vertebral changes in, 176-182
    gross appearance of, 179-182
    location of, 179
    pathogenesis of, 180-182
    vertebral fusion in, 182
Tumors, bone, 316-401, *see also under specific headings*
  bibliography of, 397-401
  classification of, 316-317

Subject Index

incidence of, 318-319
paleopathology and, 394-396
prevalence of, 318-319

U

Unitarian theory of treponematosis, 90-94
United States, *see under specific states*
Usura orbitae, 213, 239, *see also* Cribra orbitalia
Uta, 151
Utah, 265

V

Vedic hymns, 171
"Venereal leprosy," 88
Venereal syphilis, *see* Syphilis, acquired and congenital
Venezuela, 173
Vertebral osteophytosis, *see* Osteophytosis
Virginia, 23, 27, 332
Vitamin B_{12} deficiency, 237
Vitamin C,
   deficiency of, *see* Scurvy
   function of, 253
   sources of, 253-254
Vitamin D, 263
Volkmann canal, 66

W

Warren Anatomical Museum, xi, 61, 113
Weanling diarrhea, 247
Weeden Island Complex, 96
Wimberger's sign in congenital syphilis, 99
Wolff's law, 261
Woodland period, 23, 353
Wormian bones, 262
Woven bone, 5

Y

Yaws, 86, 91, 92, 142-160
   bibliography of, 160-169
   bone changes in, 143-153
   cranial changes in, 145, 150
   dactylitis in, 143-145, 146-147
   frequency of skeletal involvement in, 142-143
   gangosa in, 145, 151
   geographical distribution of, 142
   goundou in, 152, 153
   gummatous osteomyelitis in, 153
   paleopathological evidence of, 153, 158-160
   prevalence of, 142
   saber-shin tibia in, 145, 148
   skeletal distribution of, 143, 144
   syphilis and, 145, 153
   treponematosis and, 86, 90-94